An Integrated Approach to Artificial Intelligence

An Integrated Approach to Artificial Intelligence

Edited by Mick Benson

CLANRYE
INTERNATIONAL
www.clanryeinternational.com

Clanrye International,
750 Third Avenue, 9ᵗʰ Floor,
New York, NY 10017, USA

ISBN: 978-1-63240-646-0

Cataloging-in-Publication Data

An integrated approach to artificial intelligence / edited by Mick Benson.
        p. cm.
Includes bibliographical references and index.
ISBN 978-1-63240-646-0
1. Artificial intelligence. I. Benson, Mick.
TA347.A78 I58 2018
006.3--dc23

For information on all Clanrye International publications
visit our website at www.clanryeinternational.com

Contents

Preface..IX

Chapter 1   **Association Rule Based Flexible Machine Learning Module for Embedded System Platforms like Android**..1
Amiraj Dhawan, Shruti Bhave, Amrita Aurora, Vishwanathan Iyer

Chapter 2   **Rescue System with Health Condition Monitoring Together with Location and Attitude Monitoring as well as the other Data Acquired with Mobile Devices**...........8
Kohei Arai, Taka Eguchi

Chapter 3   **Blocking Black Area Method for Speech Segmentation**...15
Dr. Md. Mijanur Rahman, Fatema Khatun, Dr. Md. Al-Amin Bhuiyan

Chapter 4   **For a Better Coordination Between Students Learning Styles and Instructors Teaching Styles**..21
Sylvia Encheva

Chapter 5   **Wavelet Compressed PCA Models for Real-Time Image Registration in Augmented Reality Applications**...25
Christopher Cooper, Kent Wise, John Cooper, Makarand Deo

Chapter 6   **Sensitivity Analysis for Aerosol Refractive Index and Size Distribution Estimation Methods based on Polarized Atmospheric Irradiance Measurements**.....................35
Kohei Arai

Chapter 7   **Evaluation of Cirrus Cloud Detection Accuracy of GOSAT/CAI and Landsat-8 with Laser Radar: Lidar and Confirmation with Calipso Data**................................43
Kohei Arai, Masanori Sakashita

Chapter 8   **Fuzzy Soft Sets Supporting Multi-Criteria Decision Processes**..............................51
Sylvia Encheva

Chapter 9   **An Empirical Comparison of Tree-based Learning Algorithms: An Egyptian Rice Diseases Classification Case Study**...54
Mohammed E. El-Telbany, Mahmoud Warda

Chapter 10   **Innovative Processes in Computer Assisted Language Learning**.............................59
Khaled M. Alhawiti

Chapter 11   **Speed and Vibration Performance as well as Obstacle Avoidance Performance of Electric Wheel Chair Controlled by Human Eyes Only**...66
Kohei Arai, Ronny Mardiyanto

Chapter 12   **Bidirectional Extraction of Phrases for Expanding Queries in Academic Paper Retrieval**...74
Yuzana Win, Tomonari Masada

Chapter 13   **Comparative Analysis of Improved Cuckoo Search(ICS) Algorithm and Artificial Bee Colony (ABC) Algorithm on Continuous Optimization Problems**......................81
Shariba Islam Tusiy, Nasif Shawkat, Md. Arman Ahmed, Biswajit Panday, Nazmus Sakib

Chapter 14   **Vital Sign and Location/Attitude Monitoring with Sensor Networks for the Proposed Rescue System for Disabled and Elderly Persons who Need a Help in Evacuation from Disaster Areas**......................87
Kohei Arai

Chapter 15   **Analysis and Prediction of Crimes by Clustering and Classification**......................97
Rasoul Kiani, Siamak Mahdavi, Amin Keshavarzi

Chapter 16   **Zernike Moment Feature Extraction for Handwritten Devanagari (Marathi) Compound Character Recognition**......................104
Karbhari V. Kale, Prapti D. Deshmukh, Shriniwas V. Chavan, Majharoddin M. Kazi, Yogesh S. Rode

Chapter 17   **Method and System for Human Action Detections with Acceleration Sensors for the Proposed Rescue System for Disabled and Elderly Persons who Need a Help in Evacuation from Disaster Area**......................113
Kohei Arai

Chapter 18   **Application of Distributed Lighting Control Architecture in Dementia-Friendly Smart Homes**......................120
Atousa Zaeim, Samia Nefti-Meziani, Adham Atyabi

Chapter 19   **Speech Emotion Recognition in Emotional Feedback for Human-Robot Interaction**......................128
Javier G. Rázuri, David Sundgren, Rahim Rahmani, Aron Larsson, Antonio Moran Cardenas, Isis Bonet

Chapter 20   **Application of Machine Learning Approaches in Intrusion Detection System**......................136
Nutan Farah Haq, Musharrat Rafni, Abdur Rahman Onik, Faisal Muhammad Shah, Md. Avishek Khan Hridoy, Dewan Md. Farid

Chapter 21   **Appropriate Tealeaf Harvest Timing Determination Referring Fiber Content in Tealeaf Derived from Ground based Nir Camera Images**......................146
Kohei Arai, Yoshihiko Sasaki, Shihomi Kasuya, Hideto Matusura

Chapter 22   **Human Lips-Contour Recognition and Tracing**......................153
Md. Hasan Tareque, Ahmed Shoeb Al Hasan

Chapter 23   **Some More Results on Fuzzy k-competition Graphs**......................158
Sovan Samanta, Madhumangal Pal, Anita Pal

Chapter 24   **Locality of Chlorophyll-A Distribution in the Intensive Study Area of the Ariake Sea, Japan in Winter Seasons based on Remote Sensing Satellite Data**......................166
Kohei Arai

Chapter 25   **Attribute Reduction for Generalized Decision Systems**......................174
Bi-Jun REN, Yan-Ling FU, Ke-Yun QIN

Chapter 26   **A Trust-based Mechanism for Avoiding Liars in Referring of Reputation in Multiagent System**..179
Manh Hung Nguyen, Dinh Que Tran

Chapter 27   **A New Trust Evaluation for Trust-based RS**...188
Sajjawat Charoenrien, Saranya Maneeroj

Chapter 28   **Comparative Study of Optimization Methods for Estimation of Sea Surface Temperature and Ocean Wind with Microwave Radiometer Data**..193
Kohei Arai

Chapter 29   **Digital Library of Expert System based at Indonesia Technology University**........................199
Dewa Gede Hendra Divayana, I Putu Wisna Ariawan, I Made Sugiarta,
I Wayan Artanayasa

Chapter 30   **New Concepts of Fuzzy Planar Graphs**..207
Sovan Samanta, Anita Pal, Madhumangal Pal

Chapter 31   **Driver's Awareness and Lane Changing Maneuver in Traffic Flow based on Cellular Automaton Model**..215
Kohei Arai, Steven Ray Sentinuwo

Permissions

List of Contributors

Index

Preface

This book aims to highlight the current researches and provides a platform to further the scope of innovations in this area. This book is a product of the combined efforts of many researchers and scientists, after going through thorough studies and analysis from different parts of the world. The objective of this book is to provide the readers with the latest information of the field.

One of the definitive leaps in the progress of science and technology in the past few decades has been the development of machines which exhibit intelligence. This intelligence displayed by machines is known as artificial intelligence. The capabilities classified as artificial intelligence are the competence of understanding human speech, understanding complex data, etc. Artificial intelligence in the 21st century has become an integral part of the industrial sector as it helps in solving a number of issues. The various advancements in artificial intelligence are glanced at in this book, and their applications as well as ramifications are booked at in detail. With state-of-the-art inputs by acclaimed experts of this field, this book targets students and professionals.

I would like to express my sincere thanks to the authors for their dedicated efforts in the completion of this book. I acknowledge the efforts of the publisher for providing constant support. Lastly, I would like to thank my family for their support in all academic endeavors.

Editor

Association Rule Based Flexible Machine Learning Module for Embedded System Platforms like Android

Amiraj Dhawan[1] Shruti Bhave[2]   Amrita Aurora[3] Vishwanathan Iyer[4]

Abstract—The past few years have seen a tremendous growth in the popularity of smartphones. As newer features continue to be added to smartphones to increase their utility, their significance will only increase in future. Combining machine learning with mobile computing can enable smartphones to become 'intelligent' devices, a feature which is hitherto unseen in them. Also, the combination of machine learning and context aware computing can enable smartphones to gauge users' requirements proactively, depending upon their environment and context. Accordingly, necessary services can be provided to users.

In this paper, we have explored the methods and applications of integrating machine learning and context aware computing on the Android platform, to provide higher utility to the users. To achieve this, we define a Machine Learning (ML) module which is incorporated in the basic Android architecture. Firstly, we have outlined two major functionalities that the ML module should provide. Then, we have presented three architectures, each of which incorporates the ML module at a different level in the Android architecture. The advantages and shortcomings of each of these architectures have been evaluated. Lastly, we have explained a few applications in which our proposed system can be incorporated such that their functionality is improved.

Keywords—*machine learning; association rules; machine learning in embedded systems; android, ID3; Apriori; Max-Miner*

I. INTRODUCTION

Smartphones today are equipped with a number of features which have made it possible for users to obtain information at their fingertips. Incorporation of context aware computing in smartphones can give rise to innumerable new applications in mobile computing. A context aware system uses context to provide information and/or necessary services to users. The information and services provided depend on the current tasks the user is performing on the smartphone [1].

To utilize context awareness to its fullest potential, the system should have knowledge about how the device is used by the user and in what context. This can be achieved through machine learning. In machine learning, the system learns to make associations between the various tasks performed by the user and the corresponding context, which can be the inputs given to the device or other environmental factors. Android being a widely used mobile platform, in this paper, we have proposed methods to incorporate machine learning in the Android architecture. This is achieved through a machine learning (ML) module.

We have identified two major functionalities that the ML module should provide to achieve context awareness in the

system and accordingly have described two modes of operation of the ML module. Next, we have proposed three variations in the Android architecture, which will enable machine learning to be included in the system. In each of these architectures, the ML module is placed at different levels in the Android architecture, which determines how the module will interact with the system as a whole. The placement of the ML module in the Android architecture determines which mode of operation it will be best suited for, as well as the applications that the ML module can be used for.

Finally, we have explained a few applications that can use context aware computing for a more user friendly smartphone experience. These applications make use of the ML module to learn about the device usage patterns and the context of the user. Accordingly, it forms certain associations and rules, using which these applications are prompted to the user proactively.

II. LITERATURE SURVEY

A. Android Architecture

Figure. 1 shows the principal components and levels in the Android architecture.

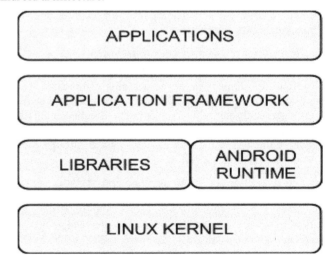

Fig. 1.   The Android Architecture

The basic working and functionalities of each component are described as follows [2]:

- Linux Kernel:

Linux kernel forms the bottom layer of the architecture. It provides basic system functionality like process management, device management and memory management. Also, the kernel handles functions like networking and a huge array of device drivers, to which peripheral hardware is interfaced.

- Libraries:

There is a set of libraries above the linux kernel, which includes WebKit- an open source web browser engine, libc, SQLite database-a useful repository for sharing and storage of application data, libraries to record and play audio as well as video, SSL libraries to monitor internet security, etc.

- Android Runtime:

This is the third section and belongs to the second layer from the bottom. It provides an important component named Dalvik Virtual Machine which is like Java Virtual Machine, designed and optimized specially for Android.

The Dalvik Virtual Machine uses core features of Linux like memory management and multi-threading. The Dalvik Virtual Machine enables every Android application to run in its own process, with its own instance of the Dalvik Virtual Machine. Additionally, the Android runtime provides a set of core libraries. Using these libraries, Android application developers can write Android applications using the standard Java programming language.

- Application Framework:

The Application Framework layer provides a number of higher level services to applications as Java classes. Application developers can use these services to write their applications.

- Applications:

All applications are strictly installed at the top layer. Home, Messages, Contact Books and Games are a few examples of these applications.

B. ID3 (Iterative Dichotomiser 3) Algorithm

ID3, developed by Ross Quinlan, is a decision tree learning algorithm used in the domains of machine learning and language processing. The ID3 algorithm employs a top-down search through the given datasets to test each attribute at every tree node. In this way, a decision tree is constructed. Every tuple in the database is then passed through the tree, which results in its classification.

To determine the input attribute that provides maximum information about the output, a metric called entropy is used. In information theory, entropy is used to measure the order or certainty in a given database set. A higher value of entropy indicates poor classification. A metric called information gain decides which input attribute is to be split. It is calculated as the difference between the entropy of the original dataset and the weighted sum of the entropies of the subdivided datasets. Information gain for each attribute is determined, and the attribute with the highest gain is split [3].

C. Apriori Algorithm

Apriori is a classical algorithm used for association rule learning. This algorithm tries to successively create larger and larger item sets that appear sufficiently in the database.

The algorithm starts with simple association rules of input to output mapping and tries to create more complex rules from these simple rules by increasing the size of the input and output item sets if they appear in the training data sufficient number of times. Every apriori algorithm requires two parameters, first is the support threshold and second is a confidence threshold. Apriori algorithm takes a bottom up approach since it starts with individual items in the item sets and successively adds multiple items in the item sets if they satisfy the support threshold.

This algorithm is one of the most popular association rule generation algorithms. But, the only disadvantage of this algorithm is in its bottom up approach which requires a lot of processing since it enumerates over all the combinations of the item sets [4].

D. Max-Miner Algorithm

Max-Miner algorithm is used to generate association rules with a complexity which is linear to the number of patterns or rules present in the training data.

The complexity is independent of the maximum length of these rules. The algorithm makes use of set-enumeration tree. It heuristically orders the items and dynamically reorders them on a pre-node basis, which leads to an improvement in performance. Hence, this algorithm is preferred [5].

E. Papers

In [6], how context aware systems enhance human computer interaction has been explained. Since context aware computing relies not only on the explicit current input given to a device, but also on the history of actions performed by the device and is capable of modifying the output with changing situations, the unnecessary interaction with the user is reduced.

In [7], the working of 'SenSay' has been described. SenSay is a mobile phone that is capable of manipulating its profile settings in accordance with the user's environment and physiological state. Various sensors, placed on the body give inputs about the users' state to a sensor box mounted on the waist. Based on the sensor information, a decision module computes the resultant action to be taken. However, SenSay does not incorporate machine learning, but relies solely on a set of predefined rules to determine the output. Also, SenSay gets inputs from sensors mounted on the body to obtain information about the user's context. Since our method does not require the user to carry any additional device on their person, it is more convenient to handle.

[8] explores the concept of using machine learning to train a system to choose songs to be played depending on the current activity and physiological state of the user. This device stores all the information relevant to a song, including the time of the day at which it is played, the corresponding activity of the user as well as the user's rating of the song. This is the training period of the system.

Once the training is complete, the device can select and play songs depending on the user's context using machine learning. However, its scope is limited only to song selection and music playlists.

III. PROPOSED SYSTEM

A. Functionalities Required

Broadly, the proposed Machine Learning Module is supposed to provide two major functionalities as described below:

1) In the first case, any third party application should be able to use the module by providing the following information:

a) Set of inputs and/or parameters that are critical for the application.

b) Set of valid outputs/actions that the application understands.

c) Training data in the form of rules such as:

Inputs and/or Parameters =>Output/Action, which are used to learn and automatically generate rules with their confidence level.

2) In the second case, the machine learning module works on its own in a global scope and tries to learn how and in what context the user uses the interface and the android system. Depending on the learning algorithm it should be able to automatically generate rules. Consider an example where the user frequently uses an application near a location (like a train station). Here, the system tries to learn this association. Accordingly, it generates a rule that if the user is near the same location, then the used application is invoked automatically.

B. Modes of operation of ML module

In order to provide the above mentioned functionalities, the proposed machine learning module has two modes of operation as follows:

- Application Level Learning:

The module is used in this mode whenever other applications are to use the learning module for purposes internal to the application. This mode fulfills the first required functionality.

- System Level Learning:

The module constantly monitors the system and checks if the current context i.e., the current inputs and/or parameters match an already learned generated rule. If a match is found, then the required output/action is performed. Also, if an event occurs due to the users' intervention, like invocation of an application, the module adds the current inputs and/or parameters and the action i.e., the invocation of application as an entry in the training data. In this manner, the system learns new rules on the runtime.

C. Steps to be followed by the third party application in order to use the ML Module

To effectively use the ML module, any third party application needs to follow a set of steps to ensure proper configuration of the module and explain the kind of input and the expected list of outputs to the module. Since multiple applications should be able to use the ML module, a method is required through which the ML module can uniquely identify each application.

Accordingly, the module will load the proper context and use the inputs provided by the application. The process is as follows:

- Register_App:

In this step the third party application is required to call the Register_App API of the ML module in order to register itself with the ML module. This procedure generates a random alphanumeric string and stores this alphanumeric string along with the application name passed. This random alphanumeric string (identification key) is passed to the third party application which is required to remember this string. In any further communication with the ML module, this string is to be passed. The string helps the ML module to uniquely identify applications when the requests are coming through a common API.

In all the further requests, the string helps the ML module to load the proper context of the application for proper usage. This step needs to be done only once for initialization. After this step, the third party application is always required to send the identification key with the requests.

- Set_Input_Output:

In this step the third party application is required to send an array of parameters with their data types that are supposed to be used as inputs for the ML algorithm. It should also send an array of data types of the outputs which the ML module should produce, along with the identification key with which the applications are registered. This key helps the ML module to know which application is trying to access the API. This request is supposed to be sent only once, before calling Generate_Rules.

- Load_Training_Data:

In this step the third party can send any training data if available, to the ML module. This data should be in the format specified in the Set_Input_Output call. This step can be skipped and the data for generating the rules can be set as and when it is available as individual rows/tuples.

- Set_Training_Data_Row:

This step can be done repeatedly at any time or even skipped. This request is used to insert a new row in the training data for the applications, determined by the accompanying identification key. The format of the row should be as per the structure defined using Set_Input_Output.

Either Load_Training_Data or Set_Training_Data_Row should be performed (single or multiple times) to ensure that the ML module has some training data to generate rules. If the application does not have a training data set available then it can insert rows of training data whenever an event occurs. Hence this step is essential.

- Generate_Rules:

This step asks the ML module to generate the rules as per the support threshold and confidence threshold provided by the third party application. This request returns 'False' if the training data set is empty. If the training data for the application has at least one row, the returned data is the set of rules generated according to the support and confidence thresholds passed along with the request. This request also stores the generated rules along with the identification key of the application for use. The design decision of allowing the application to provide the required support threshold and confidence threshold is to allow various applications to generate rules with a confidence level which is required by the application. This ensures that the generated rules are flexible according to the need of the application.

Consider a case in which an application does repeated calls to Generate_Rules with varying support and confidence threshold to smartly select which rule to follow and give some response to the event. This request is essential to be called at least once with success. This request can be configured in two modes:

1) *Automated:In this mode, the request is generated internally by the ML module whenever Set_Training_Data or Set_Training_Data_Row is called. The third party applications do not need to call this function directly*

2) *Manual: In this mode, the application is required to call this request explicitly whenever required.*

- Get_Current_Output:

The application after generating the rules from the training data can then use the rules internally to get the output for a given set of inputs. Else, it can ignore the returned rules and call this request Get_Current_Output and pass an array of the inputs to get the inferred output from the ML module. The module on receiving this request first loads the context of the application with the help of the identification key.

Next it loads the generated rules for this application and parses the rules to check if it has a rule with the inputs passed along with the request. If such a rule is found then the output of the rule is sent back to the application else'Null'is passed. The application, depending on the expected output sent by the ML module can then take actions for the event(s) (set of inputs passed to this request).

- Send_Feedback_Last_GCO_Request:

This stands for send feedback of last Get_Current_Ouput request. This request is used to send a feedback to the ML module for the last call to the request Get_Current_Output. If the output expected by the ML module was correct then a positive feedback is sent to the module else a negative feedback is sent. For a positive feedback, the module increases the confidence of the rule by a predefined amount. For a negative response the confidence of the rule is dropped by some predefined amount. This request is optional.

Some more non critical requests are as follows:

- Delete_Training_Data:

  This request is used to delete all the training data.

- Delete_Training_Data_Row:

This request is used to delete a row from the training data. The set of inputs needs to be provided to find the row required

for deleting. Either the first matched row is deleted or all the rows with the same values of the set of inputs are deleted.

- Change_Inputs_Outputs:

In case the application requires to change the structure of the training data then this request can be used to specify the new structure of the training data required. The missing inputs from the new structure are deleted from the training data and any new input is kept as null for the previous training data. Any changes in the output are also handled the same way as inputs.

D. System Level Learning: A special case of Application Level Learning

System level learning is a special case of application level learning in which the entire system is one application and the requests are auto generated by listeners on the sensors (inputs/parameters) and events (outputs like vibrator or speakers, invocation of an application). If there is an event like invocation of an application by the user, the state of the sensors with some other parameters like time etc, are used as the input. The current inputs are monitored constantly. At any point if the inputs match a rule, the output as per the rule can be used to automatically generate an event without the users' intervention.

E. Architectures

Placing the Machine Learning Module in the android architecture is a critical decision to be made. The placement affects the way other applications would interact with the ML module and also how the module can work independently in the System Level Learning mode.

- Application Level Architecture

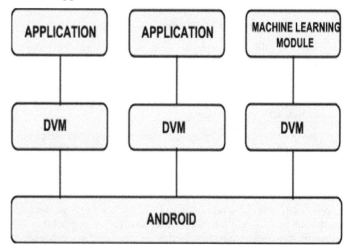

Fig. 2. Application level architecture

In this proposed architecture, the machine learning module is used as an application which executes over its own Dalvik Virtual Machine (DVM). The third party applications would be required to communicate with the Machine Learning Module Application (MLMA). This can be achieved using the Intent class of the Android Framework for Inter Application Communication.

The following snippet shows how inter application communication can be achieved.

```
<activity android:name=".SecondActivity">
<intent-filter>
<action
android:name="com.machineLearning.action.REGISTER_AP
P_ML" />
<action
android:name="com.machineLearning.action.SET_INPUTS_
OUTPUTS" />
<action
android:name="com.machineLearning.action.GET_CURREN
T_OUTPUT" />
/* and so on for all the requests mentioned above */
</intent-filter>
</activity>
```

This method requires the third party applications to specify an action for each of the functions provided by the MLMA. This approach is not preferable since it requires a lot of effort from the third party application. Also in this approach since the ML module runs as an independent application, it cannot access all the various sensors and input data sources without proper permissions. Thus in this case the ML module would require access to all the sensors and in return permissions to all the sensors which is difficult to manage.

In this case since the ML module works as a separate application, using the module for system wide learning is difficult. For system level learning the operating system would be required to use the functionality of the module as if it is another third party application. Thus system level learning cannot work independently without the need for the operating system to use it.

- System Level Architecture

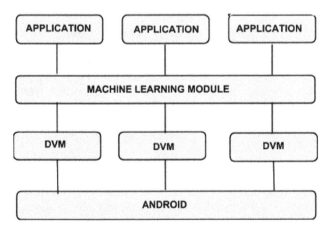

Fig. 3. System Level Architecture

In this approach, the ML module works as a layer between the DVM and the applications. This gives easy access to the machine learning module from the applications.

In this case since the ML module does not act as a standard application, it can be bundled with the operating system itself and surpass the permissions required as in the previous approach.

The module being part of the operating system can easily access any sensor. Since in this approach the ML module is between the DVM and the third party applications, it can easily monitor the usage of the applications and can work well for system level learning. However since the ML module is shared across the third party applications, providing access to the module from the applications for application level learning is challenging. Separating the applications context from each other would be difficult since all the applications share the same ML module layer.

This approach would require a lot of change in how the applications interact with the DVM instances. The module would be required to monitor all the interactions between the applications and the corresponding DVM machine. This would require changes in the core android operating system.

- Hybrid Architecture

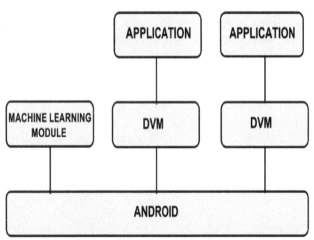

Fig. 4. Hybrid Architecture

In this approach, the module works as an independent system level library which is available as an API for applications for usage. Since the module is an independent library, the operating system too can use the functionality and perform system level learning. This approach can be used effectively for both system level as well as application level learning. Since the ML module is a library, it does not require permissions to access the sensors or other input sources. The applications can communicate with the module using all the requests explained earlier.

The operating system can register itself and send training data to learn usage patterns. Thus, it can deduce which application is highly likely to be used at the current instance. This output or deduced information can be used to automatically start applications as per the current input and sensor state. This architecture not only provides a good and clean interface for application level learning but also supports system level learning.

F. Generation of Rules

Once the application (or operating system in case of system level learning) creates a training data set for the ML module using either Load_Training_Data or Set_Training_Data_Row, it can instruct the machine learning module to generate association rules. These can be used to predict the current usage output as per the current input state. This procedure may take considerable time depending on the size of the training data. The module at this point starts parsing the training data for the particular application and tries to generate association rules that cross the support and confidence threshold values. The ML module generates these association rules by using algorithms like Apriori, ID3 or Max-Miner, which are popular association rule generation algorithms. For this application Max-Miner algorithm would be preferable since it uses less resources to come up with good association rules as compared to Apriori algorithm. Since embedded systems have lesser resources, performance should be the most important criteria while selecting the association rule generation algorithm.

Once the rules are generated for an application, it can then request for the current output possible according to the rules generated by providing the input list. If a rule exists with similar input state as provided by the application, the corresponding output is returned by the machine learning module along with its confidence level. The application can then decide what is supposed to be done with the output.

In case of the system level learning, the operating system performs all the tasks just like an ordinary application can work.

The actual end action is decided by the applications depending on the output of the machine learning module. This output can also be ignored by the application thus proving flexible usage of the module.

IV. APPLICATIONS

The following examples provide an overview of the varied use of the machine learning module in mobile applications. Depending on the users everyday usage of the device, the machine learning module trains itself and provides an output that saves time and is comfortable to access by the user.

A. Music player application

A user has an everyday schedule of listening to particular songs at particular instants of time, for example, while travelling to work at 8:00 AM or returning home at 7:00 PM. The machine learning module stores all the information relevant to opening of the music player and playing the song.

This includes the time at which it is played and the activity of the user at that time. Inputs of inserting the headphones in the device and opening the player at that particular time are provided to the module. It then trains itself according to this users schedule and generates an application ID which denotes that the music player application is running currently. If the same task is performed by the user every day at the same time after the training period, the module automatically provides a widget to the user at that time of the day to open the player thus saving his time.

B. Automatic Profile Settings

Automatic profile settings can be extremely beneficial to students in schools and colleges as well as employees in companies. The predefined time of a daily lecture or meeting is known to the user. Inputs of these lecture/meeting timings as well as the location of the school/office is provided as input to the machine learning module. Combining the users' environmental location and the timing of the users usage on the mobile device, the module trains itself and depending on the application ID provided, it displays a widget suggesting to the user to change the phone settings to silent mode. This prevents the unnecessary ringing of mobiles in important meetings and lectures making it immensely user friendly.

C. Task List

The user experience can be drastically improved by displaying relevant information on the device's screen rather than spending time searching for that information. Machine learning module takes input of a task list and the time for execution of this task from the user. Following the users usage pattern, it trains itself to display a widget of this task list needed at the time of its implementation. For example, if the user purchases groceries from the market every Friday at 7:00 PM, then the mobile will automatically display the list of groceries to be purchased by the user at this time every week. This saves time and makes it user friendly for the user.

D. Messaging

The user habitually sends the same alert message at a particular instant every day at the same time for a week. For example, a user before leaving for school/work informs his parents/spouse that he has left home. The ML module saves a draft of this message. The time, day and message are provided as the input to the ML module. Depending on the time span when the user leaves for school/work, the module displays an output i.e., the drafted message on the users screen with the recipients entered. The user can then just press the send option to send the message thus saving his time rather than retyping the same message every day and then searching for the recipients.

E. Location Based Profile Settings

If the user traverses the same path every day eg: from home to work or vice versa, the ML module keeps a record of his path taken regularly which is provided as input. With the help of Google plus, if there is a huge amount of traffic on that path on any particular day, it informs the user to traverse another path with lesser traffic. Thus the users' energy and time are saved.

F. Alarm

The user sets the alarm every day in a specific span of time. Thus the time and date of entering the event is provided to the ML module as input. The ML module then learns and trains itself according to the input of these events. The module provides an application ID which indicates the application being that of the alarm. The output is provided to the user in form of a widget that suggests the needed occurrence of the alarm. Thus the module trains itself to display a widget of the alarm every day at certain time interval before the person sets the alarm.

V. CONCLUSION

Incorporation of context awareness in mobile computing has a wide scope in a number of smartphone applications. The ability to learn about users' preferences and usage patterns and suggest services accordingly will facilitate a more user-friendly smartphone experience.

Although this paper describes context awareness and machine learning specific to the android framework, it can also be extended to include other embedded systems like set-top box or a set-top unit. The machine learning module can learn about the channel preferences of the viewer, on the basis of the previous viewing history, time as well as day of the week. Accordingly, it can suggest which channel is to be played at what time. As most viewers have fixed television schedules as well as fixed preferences in terms of channels and television shows, the channels can be tailored according to each user's needs and demands. Thus, the machine learning module can be incorporated in various embedded systems for a more personalized and simplified usage experience.

VI. FUTURE WORK

The paper in its current scope focuses on the importance of a machine learning module in an embedded system platform like android. The main focus is on how to design the module to be flexible so that it can be used by other applications as well as by the system. It also deals with the interface of the module to allow maximum flexibility and allow efficient use of the module.

Taking this proposed system as the base, future efforts include generalizing the module for generic embedded system platforms, analyze and compare the performance of the module using various association rule generation algorithms like ID3, Apriori and Max-Miner from a qualitative perspective. The future work on this system also includes working on the privacy issues of this machine learning module since it handles user data. Another research area could be to design and implement the machine learning module in a hardware integrated circuit to offload the heavy processing required by the module from the main processing unit. The

hardware IC can be designed to support processors based on the ARM architecture which is widely used on embedded systems. This could lead to better support for machine learning on such systems and provide unparalleled support to understand the user/environment better in order to take smart decisions by the embedded platforms.

Robotic platforms can leverage the machine learning capabilities to be slightly closer to achieving artificial intelligence. This is because these platforms can then accurately predict future events based on past events. If the module is implemented as a hardware IC, then the interface between the robotic platforms and this module can be simplified.

REFERENCES

[1] G. D. Abowd, A. K. Dey, P. J. Brown, N. Davies, M. Smith, and P. Steggles, "Towards a Better Understanding of Context and Context-Awareness," in Proceedings of the 1st international symposium on Handheld and Ubiquitous Computing Karlsruhe, Germany: Springer-Verlag, 1999.

[2] "Android Architecture," http://www.tutorialspoint.com/android/android_architecture.htm

[3] KalpeshAdhatrao, Aditya Gaykar, AmirajDhawan, RohitJha and VipulHonrao, "Predicting Students' Performance Using ID3 and C4.5 Classification Algorithms," in International Journal of Data Mining & Knowledge Management Process (IJDKP) Vol.3, No.5, September 2013.

[4] "Apriori Algorithm," http://en.wikipedia.org/wiki/Apriori_algorithm

[5] Roberto J, and BayardoJr, "Efficiently Mining Long Patterns from Databases," IBM Almaden Research Center.

[6] H. Lieberman, T. Selker, "Out of Context: Computer Systems that Adapt to and Learn from, Context," in IBM Systems Journal, Volume 39, Nos. 3 and 4, 2000.

[7] DanielSiewiorek, AsimSmailagic, Junichi Furukawa, NeemaMoraveji, Kathryn Reiger, and Jeremy Shaffer, "Sen-Say: A Context-Aware Mobile Phone," Human Computer Interaction Institute and Institute for Complex Engineered Systems ,Carnegie Mellon University.

[8] SandorDornbush, Jesse English, Tim Oates, ZarySegall and Anumpam Joshi, "XPod: A Human Activity Aware Learning Mobile Music Player," University of Maryland, Baltimore County.

2

Rescue System with Health Condition Monitoring Together with Location and Attitude Monitoring as Well as the Other Data Acquired with Mobile Devices

Kohei Arai[1]
1 Graduate School of Science and Engineering
Saga University
Saga City, Japan

Taka Eguchi[1]
1 Graduate School of Science and Engineering
Saga University
Saga City, Japan

Abstract—Rescue system with health condition monitoring together with location and attitude monitoring as well as the other data acquired with mobile devices is proposed. Backup system for location estimation is also proposed. On behalf of GPS receivers and WiFi beacon receivers, ZigBee is used as a backup system. Attitude can be monitored with acceleration-meters equipped in the commercially available smart phones and i-phones. Also, the number of steps and calorie consumptions can be monitored with the commercially available smart phones and i-phones. By using these body attached sensors, health condition of the persons who need a help for rescue when the emergency situations can be monitored and used for rescue planning and triage. Overall system configuration is proposed together with the detailed system descriptions with some of the experimental data.

Keywords—Rescue system; Location estimation; Attitude estimation; Health monitoring; Mobile applications; Triage; Rescue planning

I. INTRODUCTION

There are previously proposed methods and systems which allow physical health monitoring [1]-[5]. Most of previous methods and systems are not wearable and do not allow psychological status monitoring. The proposed physical and psychological health monitoring system is intended to monitor these five major vital signs. Instead of direct blood pressure measurement, indirect blood pressure measurement is proposed by using a created regressive equation with the measured body temperature, heart rate and the number of steps because it is hard to measure the blood pressure directly. Also, consciousness can be monitored by using acquired eye images and its surroundings on behalf of using EEG sensors, because EEG signals are used to be suffered from noises.

There are previously proposed evacuation and rescue methods and systems [6]-[8]. It may be possible to find that multi agent-based simulation makes it possible to simulate the human activities in rescue and evacuation process [9],[10]. A multi agent-based model is composed of individual units, situated in an explicit space, and provided with their own attributes and rules [11]. This model is particularly suitable for modeling human behaviors, as human characteristics can be presented as agent behaviors. Therefore, the multi agent-based model is widely used for rescue and evacuation simulation [9]-[13].

In this study, GIS map is used to model objects such as road, building, human, fire with various properties to describe the objects condition. With the help of GIS data, it enables the disaster space to be closer to a real situation [13]-[16].

A rescue model for people with disabilities in large scale environment is proposed. The proposed rescue model provides some specific functions to help disabled people effectively when emergency situation occurs. Important components of an evacuation plan are the ability to receive critical information about an emergency, how to respond to an emergency, and where to go to receive assistance. Triage is a key for rescue procedure. Triage can be done with the gathered physical and psychological data which are measured with a sensor network for vital sign monitoring. Through a comparison between with and without consideration of triage, it may be possible to find that the time required for evacuation from disaster areas with consideration triage is less than that without triage. The following section describes the proposed rescue system with triage followed by examples of the monitored data of health conditions together with the location of attitude monitoring. Then alternative location determination with ZigBee receiver and transmitter is described with some experimental data. Finally, conclusion is described together with some discussions.

II. PROPOSED RESCUE SYSTEM

A. Basic Idea

Fig.1 shows the concept of the proposed rescue system. There are three major components, persons who need a help for evacuation, Information Collection Center: ICC for health, traffic, and the other conditions together with the location and attitude information of the persons who need a help and the rescue peoples. Body attached sensors allow measurements of health conditions and the location and attitude of the persons who need a help. The measured data can be transmitted to the ICC through smart-phone, or i-phone, or tablet terminals of which the persons who need a help are carrying. By using the collected health condition and the location/attitude as well as traffic condition information, most appropriate rescue peoples are determined by the person by the person. It is better to consider a triage in the emergency rescue stages. Therefore, health condition monitor is necessary. Fig.2 shows the proposed health condition monitoring system together with the acquired data transmission system.

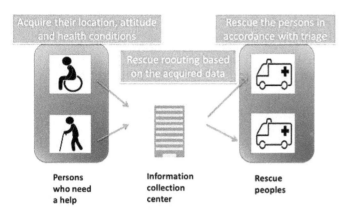

Fig. 1.   Concept of the proposed rescue system

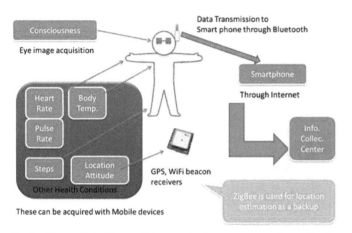

Fig. 2.   Proposed health condition monitoring system together with the acquired data transmission system

B. Sensors

There are four major items of the vital signs for triage. Those are Body temperature, Blood pressure, Pulse rate, Number of blesses. Therefore, these four items are mandatory for triage. Other than these, Number of steps, calorie consumptions would be better to monitor together with the location and attitude as well as types of movement (walking, standing, sitting, laying, running, jumping, etc.). These data can be transmitted to the ICC through WiFi networks (Bluetooth for transmission from sensors to smart-phone and WiFi network for transmission from the smart-phone to ICC). Location measurement can be done with GPS receiver in the smart-phone and with WiFi beacon receiver is the same smart-phone. The GPS receiver does not work indoor situation. Also, both GPS receiver and WiFi receiver based location determination accuracy is not good enough. Therefore, some of alternative method would be better to add to the rescue system. In this paper, ZigBee of transmitter and receiver is used for location determination as a backup as shown in Fig.2.

Outlooks of body attached sensors are shown n Fig.3. (a)Pulse Rate   (b)Heart Rate   (c)Body   Temperature (d)Blood Pressure (e)Step, Calorie Consumption, Location and Attitude can be measured with these sensors.

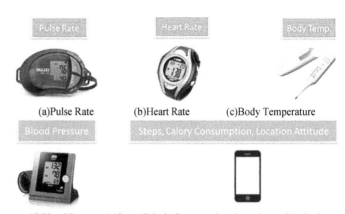

(a)Pulse Rate   (b)Heart Rate   (c)Body Temperature

(d)Blood Pressure (e)Step, Calorie Consumption, Location and Attitude

Fig. 3.   Used body attached sensors for health monitoring

C. Triage

In the triage stage, the types of disabilities which are shown in Table 1 are taken into account. Through a consideration of these types of disabilities, 10 grades of disabilities are taken into account in the triage.

TABLE I.   TYPES OF DISABILITY

Types of Disability
Cognitive Disorder
Neuropathy
Movement Disorder
Elderly Condition
Hearing Loss
Language, Visual Impairment

D. Examples of Measured Data

Other than smart-phone based step monitoring, there are some body attached step monitoring sensors. Fig.4 shows an example of the step monitoring sensor. The sensor allows measurement not only the number of steps but also calorie consumption can be measured. Also, these measured data are archived and referred through Bluetooth communications. One of example of the archived steps and calorie consumption is shown in Fig.5.

Fig. 4.   Alternative step monitoring sensor

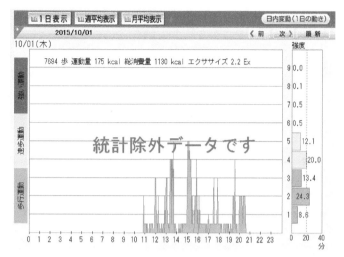

Fig. 5.   Example of the acquired step data together with calorie consumption

Moreover, the number of steps can be measured with i-phone application software tools. Fig.6 shows an example of the measures steps in a month with iOS8 of i-phone.

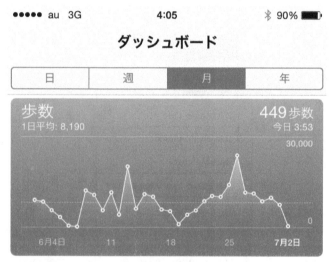

Fig. 6.   Step monitoring with iOS8 of i-phone

There is health condition monitoring application software tool so called HealthKit under the iOS8. The menu of the HealthKit is shown in Fig.7.

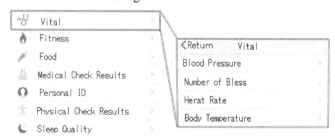

Fig. 7.   Health condition monitoring with HealthKit with iOS8

Other than the vital records, Fitness, Food, Medical check results, physical check results, sleep quality and personal ID

can be referred. Process flow of the proposed health monitoring system together with the acquired data transmission system is shown in Fig.8.

Fig. 8.   Process flow of the proposed health monitoring system together with the acquired data transmission system

On the other hand, Android OS of smart-phone which is shown in Fig.9 provides API which allows the step count and step detector as shown in Fig.10. That is Android4.4kit-kat.

Fig. 9.   Example of Android OS of smart-phone

Android4.4 kitkat

Steps can be get with Acceleration meter

· TYPE_STEP_COUNTER

Accumulated steps can be get

· TYPE_STEP_DETECTOR

Once step is acquired, send the data

It is available to send the steps in the time interval

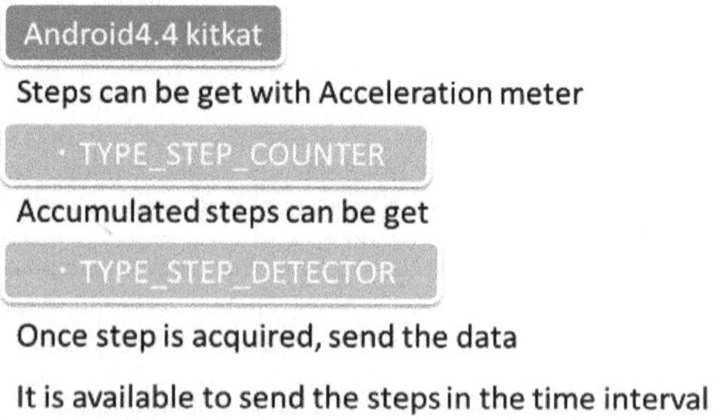

Fig. 10.   Additionally available APIs by using Android4.4kitkat

Using the API of Android4.4kit-kat together with acceleration meter, types of movements can be determined as shown in Fig.11. Also, identification of attitude type by using the difference between actual and reference power spectrum derived from acceleration meter is available as shown in Fig.12.

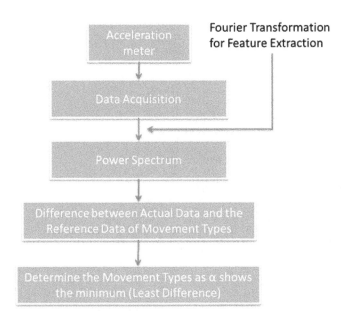

Fig. 11. Method for attitude detection

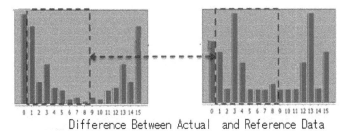

Fig. 12. Identification of attitude type by using the difference between actual and reference power spectrum derived from acceleration meter

Acquired acceleration meter data is compared to the previously acquired reference data of the designated several movement types in the frequency domain (Power spectrum). The frequency components between a and b are compared followed by the summation of the different between both, a and b calculation as shown in equation (1).

$$\alpha = |a2-b2| + |a3-b3| + \cdots + |a9-b9| \qquad (1)$$

Thus movement types are discriminated. Also, movement types can be discriminated with ZigBee. Fig.13 shows outlook of the ZigBee used for the experiments. Movement types, in this case, can be identified as "Stop", "Begin to move" and "Freely falling down" as shown in Fig.14. One of the examples of the receiving signal for the movement type of the "Freely falling down" is shown in Fig.15 while that of the "Stop" on the floor is shown in Fig.16, respectively. Thus the movement types can be identified with receiving signal strength of ZigBee. Meanwhile, one of examples of the measured pulse rate with the Pulse Coach which is shown in Fig.3 (a) is shown in Fig.17. Pulse per minute can be measured every one second.

Fig. 13. Outlook of ZigBee

Fig. 14. Movement type identification with ZigBee

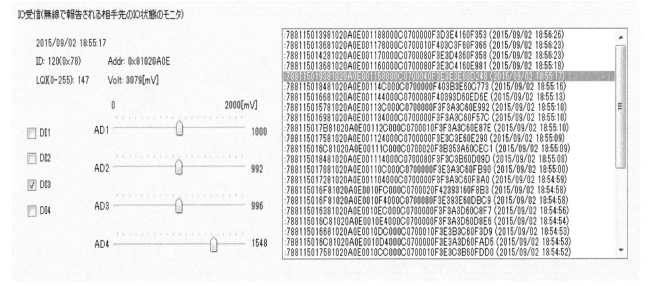

Fig. 15. Example of the receiving signal from the ZigBee when it is falling freely (DI3)

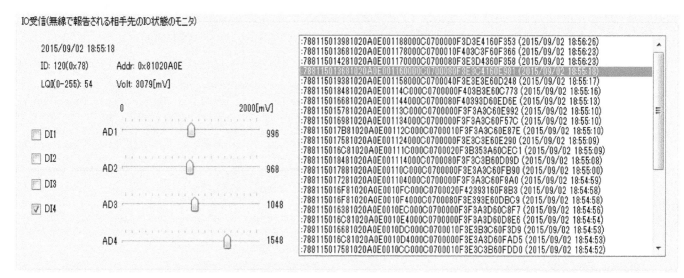

Fig. 16. Example of the receiving signal from the ZigBee when it is on the floor (DI4)

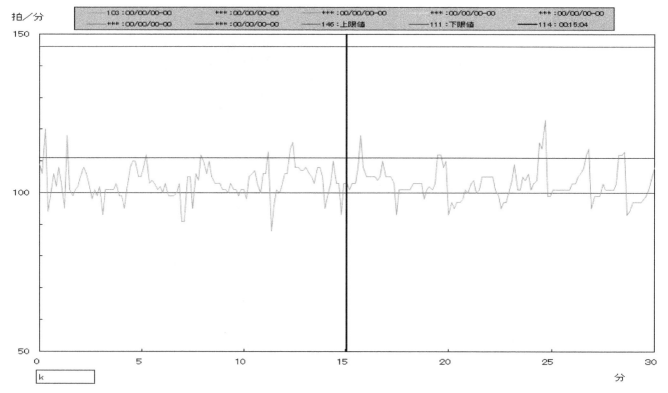

Fig. 17. Example of Pulse Rate measured data (1/minute)

III.   DISTANCE MEASUREMENTS WITH ZIGBEE

A. Basic Idea

ZigBee coverage is limited up to around 100 m. Therefore, the location of the ZigBee receiver can be identified from the surrounding three ZigBee transmitters as shown in Fig.18. Also, one pair of ZigBee transmitter and receiver makes distance measurements with signal strength. Also, ZigBee coverage can be expanded with through repeaters as shown in Fig.19.

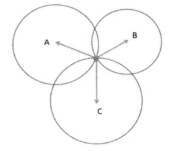

Fig. 18. Location determination concept with three ZigBee stations

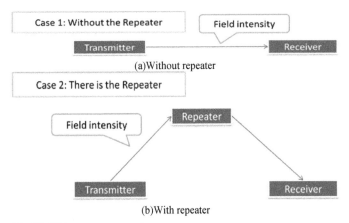

(a)Without repeater

(b)With repeater

Fig. 19. Distance measurements

B. Measurement Data

Fig.20 shows example of the received signal of the repeater. The data are aligned as shown n Fig.20 (b). The signal includes not only signal strength but also the repeater ID. Therefore, it can be identified that the data is received through which repeaters.

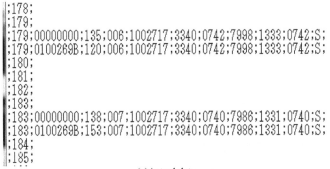

(a)Actual data

The order of the actual data from the top right to left bottom
Time stamp
ID of the repeater
Signal Strength: LQI
Continuous No.
ID of the receiver
Supply voltage of the receiver
AI3(mV)
AI1(three times of voltage)
AI1(mV)
AI3(mV)
Packet ID

(b)Alignment of the data

Fig. 20. Example of the receiving signal

Fig.21 (a) shows example of the distance measurement results for the case of without repeater (Outdoor) while Fig.21 (b) shows that of with repeater (Indoor), respectively. There is

much electro-magnetic interference from the wall, pillar, etc. for the case of "Indoor". Therefore, location estimation accuracy is not so good for the "Indoor" case.

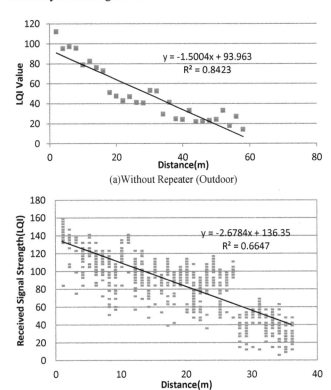

$y = -1.5004x + 93.963$
$R^2 = 0.8423$

(a)Without Repeater (Outdoor)

$y = -2.6784x + 136.35$
$R^2 = 0.6647$

(b)With Repeater (Indoor)

Fig. 21. Example of relation between receiving signal strength and the distance

IV. CONCLUSION

Rescue system with health condition monitoring together with location and attitude monitoring as well as the other data acquired with mobile devices is proposed. Backup system for location estimation is also proposed. On behalf of GPS receivers and WiFi beacon receivers, ZigBee is used as a backup system. Attitude can be monitored with acceleration-meters equipped in the commercially available smart phones and i-phones. Also, the number of steps and calorie consumptions can be monitored with the commercially available smart phones and i-phones. By using these body attached sensors, health condition of the persons who need a help for rescue when the emergency situations can be monitored and used for rescue planning and triage. Overall system configuration is proposed together with the detailed system descriptions with some of the experimental data.

Also, it is found that the distance measurements can be done with ZigBee. Moreover, it is found that the coverage of the ZigBee location identification can be expanded with ZigBee transmitter and receiver (Repeater).

ACKNOWLEDGMENT

The author would like to thank Dr. Trang Xuang Sang of Vinh University in Vietnam for his effort to conduct simulation studies.

REFERENCES

[1] K.Arai,WeIarable healthy monitoring sensor network and its application to evacuation and rescue information server system for disabled and elderly person, International Jmynal of Research and Review on Computer Science, 3, 3, 1633-1639, 2012.

[2] Kohei Arai, Wearable computing system with input output devices based on eye-based Human Computer Interaction: HCI allowing location based Ib services, International Jmynal of Advanced Research in Artificial Intelligence, 2, 8, 34-39, 2013.

[3] Kohei Arai, Vital sign and location/attitude monitoring with sensor networks for the proposed rescue system for disabled and elderly persons who need a help in evacuation from disaster areas, International Jmynal of Advanced Research in Artificial Intelligence, 3, 1, 24-33, 2014.

[4] Kohei Arai, Method and system for human action detection with acceleration sensors for the proposed rescue system for disabled and elderly persons who need a help in evacuation from disaster areas, International Jmynal of Advanced Research in Artificial Intelligence, 3, 1, 34-40, 2014.

[5] Kohei Arai, Frequent physical health monitoring as vital sign with psychological status monitoring for search and rescue of handicapped, disabled and elderly persons, International Jmynal of Advanced Research in Artificial Intelligence, 2, 11, 25-31, 2013

[6] .J. Kaprzy Edt., Kohei Arai, Rescue System for Elderly and Disabled Persons Using Iarable Physical and Psychological Monitoring System, Studies in Computer Intelligence, 542, 45-64, Springer Publishing Co. Ltd., 2014.

[7] Obelbecker G., & Dornhege M., "Realistic cities in simulated environments - an Open Street Map to Robocup Rescue converter", Online-Proceedings of the Fmyth International Workshop on Synthetic Simulation and Robotics to Mitigate Earthquake Disaster, 2009.

[8] Sato, K., & Takahashi, T., "A study of map data influence on disaster and rescue simulation's results", Computational Intelligence Series, vol. 325. Springer Berlin / Heidelberg, 389–402, 2011.

[9] Ren C., Yang C., & Jin S., "Agent-Based Modeling and Simulation on emergency", Complex 2009, Part II, LNICST 5, 1451 – 1461, 2009.

[10] Zaharia M. H., Leon F., Pal C., & Pagu G., "Agent-Based Simulation of Crowd Evacuation Behavior", International Conference on Automatic Control, Modeling and Simulation, 529-533, 2011.

[11] Quang C. T., & Drogoul A., "Agent-based simulation: definition, applications and perspectives", Invited Talk for the biannual Conference of the Faculty of Computer Science, Mathematics and Mechanics, 2008.

[12] Cole J. W., Sabel C. E., Blumenthal E., Finnis K., Dantas A., Barnard S., & Johnston D. M., "GIS-based emergency and evacuation planning for volcanic hazards in New Zealand", Bulletin of the New Zealand society for earthquake engineering, vol. 38, no. 3, 2005.

[13] Batty M., "Agent-Based Technologies and GIS: simulating crowding, panic, and disaster management", Frontiers of geographic information technology, chapter 4, 81-101, 2005.

[14] Patrick T., & Drogoul A., "From GIS Data to GIS Agents Modeling with the GAMA simulation platform", TF SIM 2010.

[15] Quang C. T., Drogoul A., & Boucher A., "Interactive Learning of Independent Experts' Criteria for Rescue Simulations", Jmynal of Universal Computer Science, Vol. 15, No. 13, 2701-2725, 2009.

[16] Taillandier T., Vo D. A., Ammyoux E., & Drogoul A., "GAMA: a simulation platform that integrates geographical information data, agentbased modeling and multi-scale control", In Proceedings of Principles and practice of multi-agent systems, India, 2012.

3

Blocking Black Area Method for Speech Segmentation

Dr. Md. Mijanur Rahman

Dept. of Computer Science &
Engineering
Jatiya Kabi Kazi Nazrul Islam
University
Trishal, Mymensingh, Bangladesh

Fatema Khatun

Dept. of Electrical & Electronic
Engineering
Hamdard University Bangladesh
Sonargoan, Narayanganj,
Bangladesh

Dr. Md. Al-Amin Bhuiyan

Dept. of Computer Engineering
King Faisal University
Al Ahssa 31982, Saudi Arabia

Abstract—**Speech segmentation is an important sub problem of automatic speech recognition. This research is concerned with the development of a continuous speech segmentation system using Bangla Language. This paper presents a dynamic thresholding algorithm to segment the continuous Bngla speech sentences into words/sub-words. The research uses Otsu's method for dynamic thresholding and introduces a new approach, named blocking black area method to identify the voiced regions of the continuous speech in speech segmentation. The developed system has been justified with continuously spoken several Bangla sentences. To test the performance of the system, 100 Bangla sentences have been recorded from 5 (five) male speakers of different ages and 656 words have been presented in the 100 Bangla sentences. So, the speech database contains 500 Bangla sentences with 3280 words. All the algorithms and methods used in this research are implemented in MATLAB and the proposed system has been achieved the average segmentation accuracy of 90.58%.**

Keywords—Blocking Black Area; Boundary Detection; Dynamic Thresholding; Otsu's Algorithm; Speech Segmentation

I. INTRODUCTION

Automated Speech Recognition (ASR) is a popular and challenging area of research in developing human computer interactions. The main challenge of speech recognition lies in modeling the variations of the uttered speech, such as different geographical boundaries, social background, age, gender, occupation etc. Automated segmentation of speech signals has been under research for over 30 years [1]. It is a necessity for phonetic analysis of speech [2, 3], audio content classification [4] and many applications in the field of automatic speech recognition (ASR), including word recognition [5, 6]. Speech Recognition system requires segmentation of Speech waveform into fundamental acoustic units [7]. Segmentation is the very basic step in any voiced activated systems like speech recognition system and speech synthesis system. The set of fundamental acoustic units into which the speech waveform can be segmented are words, phonemes or syllables. Word is the preferred and natural unit of speech, because word units have well defined acoustic representation. So, this research chooses word as the basic unit for segmentation. Speech segmentation was done using wavelet [8], fuzzy methods [9], artificial neural networks [10] and Hidden Markov Model [11].

This paper will present the proposed dynamic thresholding algorithms for segmenting continuous Bangla speech sentences into words/sub-words. For speech segmentation, this research introduces a new approach, named *blocking black area method* to properly detect word boundaries in continuous speech segmentation. The paper is organized as follows: Section I describes the introduction of speech processing and the organization of this paper. In Section II, we will discuss about speech segmentation and types of segmentation. Section III will describe thresholding. In Section IV, Otsu's thresholding method will be discussed. Section V will present the blocking black area method. The implementation of the proposed system will be described in Section VI. Sections VII and VIII will describe the experimental results and conclusion, respectively.

II. SPEECH SEGMENTATION

Speech segmentation is the process of identifying the boundaries between words, syllables, or phonemes in spoken natural languages. The general idea of segmentation can be described as dividing something continuous into discrete, non-overlapping entities [12]. In speech segmentation, the basic idea of segmentation is to divide a continuous speech signal into smaller parts, where each of these segments has phonetical or acoustical properties that distinguishes it from neighboring segments. Segmentation can be performed, for example, at the *segment, phone, syllable, word, and sentence* or *dialog turn* level. In isolated word recognition systems, accurate detection of the endpoints of a spoken word is important for two reasons, namely: Reliable word recognition is critically dependent on accurate endpoint detection and the computation for processing the speech is less, when the endpoints are accurately located [13]. Automatic speech segmentation methods can be classified in many ways, but one very common classification is the division to blind [14] and aided segmentation algorithms [15]. A central difference between aided and blind methods is in how much the segmentation algorithm uses previously obtained data or external knowledge to process the expected speech.

III. DYNAMIC THRESHOLDING

In general, thresholding is the simplest method of image segmentation.

This research proposes thresholding techniques on speech segmentation. From a grayscale image, thresholding can be used to convert binary image [16]. In order to convert the image into a binary representation, the technique first converts the image into a grayscale representation and performs a particular threshold analysis process in order to determine which pixels are turned into black or which are white. This research proposes *dynamic thresholding* to convert 256 gray-levels images into monochrome ones. Two important thresholding techniques are fixed or static thresholding and dynamic thresholding. In fixed or static thresholding, the systems usually uses 127 (say) as default threshold value, but you could change this value and obtain darker or lighter images. In dynamic thresholding, the system uses a different threshold value for each pixel of the image. This value is selected automatically, analyzing the sub-image area around each pixel and finding the local contrast. If the contrast of this area is low, the pixel is binarized using a global pre-calculated threshold value, otherwise, when the contrast is high, the local threshold value is calculated and used. In thresholding technique, the output image replaces all pixels in the input image with luminance greater than a threshold with the value of 1 (white) or 0 (black). The problem is how to choose the desired threshold value. Different dynamic thresholding techniques have been used to compute the threshold value [17]. Hence, the research proposes Otsu's thresholding algorithm to compute the desired threshold.

IV. OTSU'S ALGORITHM

Otsu's method is a simple and effective automatic thresholding method, used in image segmentation [18], invented by Nobuyuki Otsu in 1979 [19], also known as binarization algorithm. It is used to automatically perform histogram shape-based image thresholding (i.e. the reduction of a grayscale image into a binary image). The algorithm assumes that the image is composed of two basic classes; such as foreground and background [19]. It then computes an optimal threshold value that minimizes the weighted within class variance; also maximizes the between class variance of these two classes. The algorithmic steps for calculating the threshold is given in Figure-1.

The mathematical formulation of the algorithm for computing the optimum threshold will explain in this section. Let $P(i)$ represents the image histogram of speech spectrogram. The two class probabilities $w_1(t)$ and $w_2(t)$ at level t are computed by:

$$w_1(t) = \sum_{i=1}^{t} P(i) \tag{1}$$

$$\text{and} \quad w_2(t) = \sum_{i=t+1}^{I} P(i) \tag{2}$$

The class means, $\mu_1(t)$ and $\mu_2(t)$ are:

$$\mu_1(t) = \sum_{i=1}^{t} \frac{iP(i)}{w_1(t)} \tag{3}$$

$$\text{and} \quad \mu_2(t) = \sum_{i=t+1}^{I} \frac{iP(i)}{w_2(t)} \tag{4}$$

Individual class variances:

$$\sigma_1^2(t) = \sum_{i=1}^{t} [i - \mu_1(t)]^2 \frac{P(i)}{w_1(t)} \tag{5}$$

$$\text{and} \quad \sigma_2^2(t) = \sum_{i=t+1}^{I} [i - \mu_2(t)]^2 \frac{P(i)}{w_2(t)} \tag{6}$$

The within class variance (σ_w) is defined as a weighted sum of variances of the two classes and given by:

$$\sigma_w^2(t) = w_1(t)\sigma_1^2(t) + w_2(t)\sigma_2^2(t) \tag{7}$$

Now we will calculate the *between class* variance. The between class variance (σ_b) is defined as a difference of total variance and within class variance and given by:

$$\sigma_b^2(t) = \sigma^2(t) - \sigma_w^2(t)$$

$$= w_1(t)[\mu_1(t) - \mu]^2 + w_2(t)[\mu_2(t) - \mu]^2$$

$$= w_1(t)w_2(t)[\mu_1(t) - \mu_2(t)]^2 \tag{8}$$

$$where \quad \mu = w_1(t)\mu_1(t) + w_2(t)\mu_2(t) \tag{9}$$

These two variances $\boldsymbol{\sigma_w}$ and $\boldsymbol{\sigma_b}$ are calculated for all possible thresholds, $t = 0 \dots I$ (max. intensity). Otsu finds the best threshold that *minimizes the weighted **within class** variance ($\boldsymbol{\sigma_w}$), also maximizes the weighted **between class** variance ($\boldsymbol{\sigma_b}$)*. Finally, the pixel luminance less than or equal to threshold is replaced by 0 (black) and greater than threshold is replaced by 1 (white) to obtain the binary or B/W image.

V. BLOCKING BLACK AREA METHOD

For speech segmentation, this research introduces a new approach, named *blocking black area method*. This method is used to block the voiced regions of the continuous speech, so that we can easily separate the voiced parts of the speech from silence or un-voiced parts in the continuous speech. The edges of the block are used as word boundaries in the continuous speech. The main task of speech segmentation is to detect the boundaries of speech units (i.e., start and end points detection). The algorithm is applied in the thresholded spectrogram image that produces rectangular black boxes in the voiced regions of the speech sentence, as shown in Figure-2. Each black box represents a speech unit (i.e., word or sub-word) of a speech sentence. The method works as follows:

- Summing the column-wise intensity values of thresholded spectrogram image.

- Find the image columns with fewer white pixels based on summing value and replace all pixels on this column with luminance 0 (black).

- Find the image columns with fewer black pixels based on summing value and replace all pixels on this column with luminance 1 (white).

- Detect the boundaries of voiced block and separate the

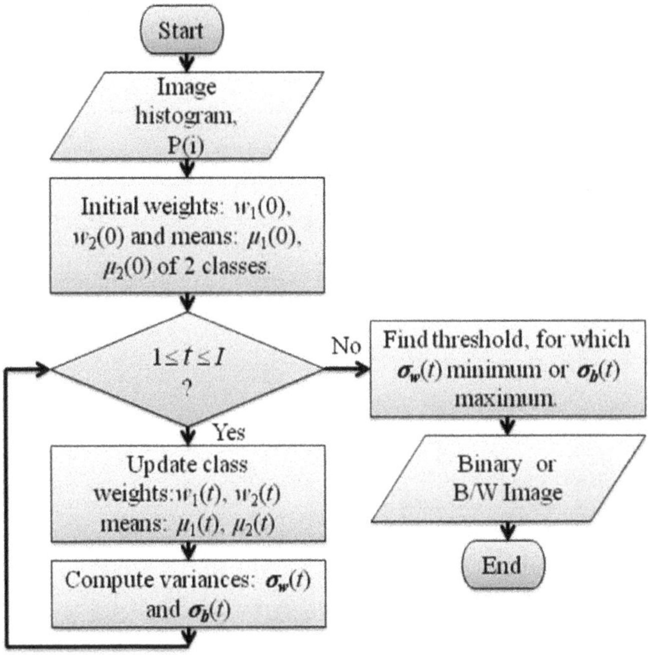

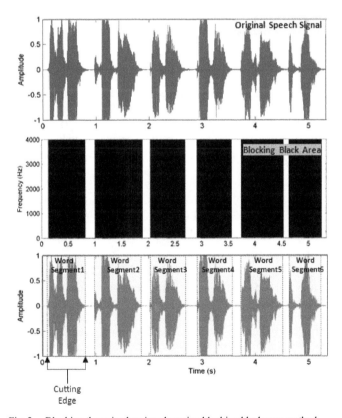

Fig. 1.   Otsu's Thresholding Algorithm

voiced block as speech units.

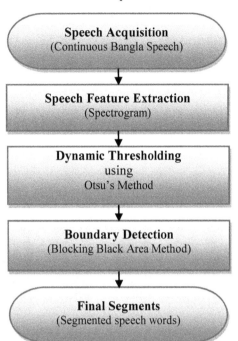

Fig. 3.   Proposed Speech Segmentation Procedure

VI.   IMPLEMENTATION

The proposed segmentation system, shown in   Figure-3, has the following major steps and will discuss in the following sub-sections.

A. *Speech Acquisition*

B. *Feature Generation and Thresholding*

C. *Word Boundary Detection*

D. *Speech Segment Separation*

A. *Speech Acquisition*

Speech acquisition is acquiring of continuous Bangla speech sentences through the microphone. Recording was done by 5 (five) native male speakers of Bengali. The sampling frequency is 16 KHz; sample size is 8 bits, and mono channels are used. The time-domain plot of a speech sentence ('আমাদের জাতীয় কবি কাজী নজরুল ইসলাম') is shown in Figure-4(a).

Fig. 2.   Blocking the voiced regions by using blocking black area method

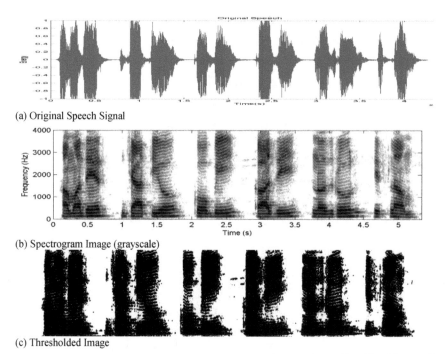

(a) Original Speech Signal

(b) Spectrogram Image (grayscale)

(c) Thresholded Image

Fig. 4.   Thresholded Spectrogram Images of the Speech Sentence 'আমাদের জাতীয় কবি কাজী নজরুল ইসলাম'

B. Speech Feature Generation and Thresholding

The feature extraction process generates spectrogram features from Bangla speech sentences. The grayscale spectrogram image of the speech sentence ('আমাদের জাতীয় কবি কাজী নজরুল ইসলাম') is shown in Figure-4(b). Spectrograms can be used to identify spoken words phonetically. For further processing of the spectrogram image, the labels of the image, such x-label, y-label and tile of the image, have been omitted, that's why label or title of the image is not shown in Figure-4(c). The thresholding algorithm is used to separate voiced regions from silence/un-voiced on continuous speech. The Matlab's 'graythresh' function is used to implement the Algorithm-3. This algorithm returns a level (i.e., threshold) value for which the intra-class variance of the black and white pixels is minimum. The output image replaces all pixels in the input image with luminance greater than or equal to the threshold with the value of 1 (fully white) and less than threshold with 0 (fully black) to get fully black/white image (i.e., thresholded image). The thresholded image of the above speech sentence is shown in Figure-4(c).

C. Word Boundary Detection

The newly introduced *blocking black area method* and *shape identification* techniques to properly detect word boundaries in continuous speech and label the entire speech sentence into a sequence of words/sub-words. The *block black area* method is applied in the thresholded spectrogram image that produces rectangular black boxes in the voiced regions of the speech sentence, as shown in Figure-5. Each rectangular black box represents a speech word or sub-word.

The method uses Matlab's 'regionprops' function to identify each rectangular object in the binary image that represents speech words/sub-words. The function 'regionprops' measures the properties of each connected object in the binary image. Different shape measurements properties, such as 'Area', 'BoundingBox', 'Centroid' are used to identify each rectangular object in the binary image. The 'Extrema' measurement, which is a vector of [top-left top-right right-top right-bottom bottom-right bottom-left left-bottom left-top], is used to detect the start (bottom-left) and end (bottom-right) points of each rectangular object, as shown in Figure-6.

D. Word Segment Separation

Each rectangular black box represents a speech segment, such as a word or sub-word. After detecting the start and end points of each black box, the word boundaries in the original speech sentence are marked automatically by these two points and separated each speech segment from the speech sentence. Figure-7 shows that 6 (six) black boxes represent 6 (six) word segments in the speech sentence 'আমাদের জাতীয় কবি কাজী নজরুল ইসলাম'.

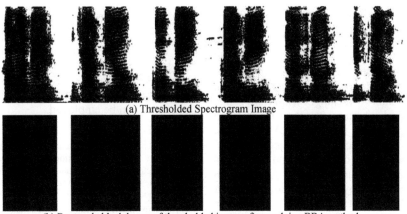

(a) Thresholded Spectrogram Image

(b) Rectangle black boxes of thresholded image after applying BBA method

Fig. 5. Effect of applying Blocking Black Area (BBA) Method – Producing rectangle black boxes in voiced regions. (a) Before applying Blocking Black Area Method and (b) After applying Blocking Black Area Method – Each black box represents a word/sub-word of the continuous speech

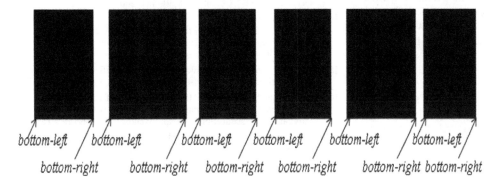

bottom-left / bottom-left / bottom-left / bottom-left / bottom-left / bottom-left /

bottom-right   bottom-right   bottom-right   bottom-right   bottom-right bottom-right

Fig. 6. Star and End point Detection of rectangular object

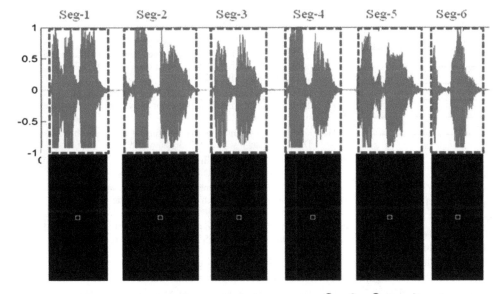

Fig. 7. Word Segments - 6 word segments in speech sentence 'আমাদের জাতীয় কবি কাজী নজরুল ইসলাম'

VII. EXPERIMENTS AND RESULTS

For speech segmentation, this research proposes the dynamic thresholding algorithm with newly introduced *blocking black area method* to segment the continuously spoken Bangla sentence into words or sub-words. All the programs related to the speech segmentation approaches have been implemented in Matlab. The '*myspectrogram.m*' program computes spectrogram image from the original speech signal.

This research uses MATLAB's '*graythresh*' function to implement modified Otsu's algorithm that returns the desired threshold. The output image replaces all pixels in the input image with luminance greater than or equal to the threshold with the value of 1 (fully white) and less than threshold with 0 (fully black). The '*Blocking Black Area*' method has been implemented in the program '*blockingBlackArea.m*' that produces rectangular black boxes in the thresholded spectrogram image. The research uses MATLAB function '*regionprops*' to identify each rectangular object and the function's '*Extrema*' is used to detect the start and end points of each black box. The word boundaries of the original speech sentence are marked automatically by these two points and cut the word segments from the speech sentence and finally, the speech segments are save as .wav file format.

The developed system has been justified with continuously spoken several Bangla sentences. To test the performance the system, 100 Bangla sentences have been recorded from 5 (five) male speakers of different ages and 656 words have been presented in the 100 Bangla sentences. So, the speech database contains 500 (100x5) Bangla sentences with 3280 (656x5) words. Each sentence has been recorded separately and saved as .wav file format to make the speech database. In segmentation this research expects only properly segmented words as segmentation output, but the program produced some sub-words. The developed system achieved the average segmentation accuracy of **90.58%**; the details result of segmentation is given in   Table-1.

TABLE I.   SPEECH SEGMENTATION RESULTS

Speaker ID	No. of Sentences	No. of Words Present	No. of Properly Segmented Words	Accuracy (%)
S1	100	656	517	78.81
S2	100	656	601	91.62
S3	100	656	612	93.29
S4	100	656	619	94.36
S5	100	656	622	94.82
Total	500	3280	2971	90.58

VIII. CONCLUSION

The main objective of this research is to develop an efficient system that can automatically segments words from the continuously spoken Bangla sentences. This research introduces some ideas to develop the system. This research proposes dynamic thresholding algorithm a new approach, named "Blocking Black Area" method to detect proper word/sub-word boundaries in speech segmentation. Some words are not properly segmented. No or very little gap between two successive words causes two or more words in a single segment. Also the gap within a word causes sub-word segmentation. This is due to some sources of variability is speech, such as, Phonetic identity (two samples might correspond to different phonetic segments), Pitch and Amplitude, Speaker (based age, sex, emotion, etc.), Microphone and Media, and Environment (including background noise, room acoustics, distance from microphone, etc).

For further improvements and expansions of the speech segmentation developed system, this research can be employed by using noise reduction algorithms in a noisy environment. Also a fuzzy logic based speech segmentation approach can be employed.

REFERENCES

[1]   Okko Rasanen, "Speech Segmentation and Clustering Methods for a New Speech Recognition Architecture", M.Sc Thesis, Department of Electrical and Communications Engineering, Laboratory of Acoustics and Audio Signal Processing, Helsinki University of Technology, Espoo, November 2007.

[2]   Mermelstein P, "Automatic segmentation of speech into syllabic units", Journal of Acoustical Society of America, Vol. 58, No. 4, pp. 880-883, Oct. 1975.

[3]   S L Mattys, P W Jusczyk, "Phonotactic cues for segmentation of fluent speech by infants", Cognition 78, 91–121, 2001.

[4]   Zhang T and Kuo C C J, "Hierarchical classification of audio data for archiving and retrieving", Proceedings of the Acoustics, Speech, and Signal Processing 1999 on 1999 IEEE International Conference, Vol. 6, pp. 3001-3004, 1999.

[5]   Antal M, "Speaker Independent Phoneme Classification in Continuous Speech", Studia Univ. Babes-Bolyal, Informatica, Vol. 49, No. 2, 2004.

[6]   D Dahan and M R Brent, "On the discovery of novel word like units from utterances: an artificial-language study with implications for native-language acquisition", J. Exp. Psychol. 128 (1999) 165–185.

[7]   Thangarajan R and Natarajan A M, "Syllable Based Continuous Speech Recognition for Tamil", South Asian Language Review VOL.XVIII, No.1, 2008.

[8]   Hioka Y and Namada N, "Voice activity detection with array signal processing in the wavelet domain", IEICE TRANSACTIONS on Fundamentals of Electronics, Communications and Computer Sciences, 86(11):2802-2811, 2003.

[9]   Beritelli F and Casale S, "Robust voiced/unvoiced classification using fuzzy rules", In 1997 IEEE workshop on speech coding for telecommunications proceeding, pages5-6, 1997.

[10] Qi Y and Hunt B, "Voiced-unvoiced-silence classification of speech using hybrid features and a network classifier", IEEE Transactions on Speech and Audio Processing, I(2):250-255, 1993.

[11] Basu S, "A linked-HMM model for robust voicing and speech detection", In IEEE international conference on acoustics, speech and signal processing (ICAASSP'03), 2003.

[12] Kvale K, "Segmentation and Labeling of Speech", PhD Dissertation, The Norwegian Institute of Technology, 1993.

[13] Lawrence Rabiner and Biing-Hwang Juang, "Fundamentals of speech Recognition", Prentice Hall, Englewood Cliffs, N.J., 1993.

[14] Sharma M and Mammone R, "Blind speech segmentation: Automatic segmentation of speech without linguistic knowledge", Spoken Language, 1996. ICSLP 96. Proceedings. Vol. 2, pp. 1237-1240, 1996.

[15] Schiel F, "Automatic Phonetic Transcription of Non-Prompted Speech", Proceedings of the ICPhS 1999. San Francisco, August 1999. pp. 607-610, 1999.

[16] Shapiro, Linda G. and Stockman, George C., "Computer Vision", Prentice Hall, ISBN 0-13-030796-3, 2002.

[17] Md. Mijanur Rahman and Md. Al-Amin Bhuiyan, "Dynamic Thresholding on Speech Segmentation", IJRET: International Journal of Research in Engineering and Technology, Volume: 02 Issue: 09, Sep-2013.

[18] Gonzalez, Rafael C. & Woods, Richard E, "Thresholding", In Digital Image Processing, pp. 595–611. Pearson Education, 2002.

[19] Nobuyuki Otsu, "A threshold selection method from gray-level histograms", IEEE Trans. Sys., Man., Cyber. 9 (1): 62–66, 1979.

4

For a Better Coordination Between Students Learning Styles and Instructors Teaching Styles

Sylvia Encheva
Stord/Haugesund University College
Bjørnsonsg. 45,
5528 Haugesund,
Norway

Abstract—**While learning has been in the main focus of a number of educators and researches, instructors' teaching styles have received considerably less attention. When it comes to dependencies between learning styles and teaching styles the available knowledge is even less. There is a definite need for a systematic approach while looking for such dependencies. We propose application of refinement orders and relational concept analysis for pursuing further investigations on the matter.**

Keywords—Refinement orders; Relational concept analysis; Learning

I. INTRODUCTION

Learning styles [7] in general refer to how people learn. In [18] they are described as - visual, aural, verbal, physical, logical, social and solitary. Students' learning styles in particular have been wildly discussed and structured in a number of models, [18]. According to Felder and Soloman's model [4] learners can be: active or reflective, depending on their tendencies to retain and understand information; sensing or intuitive, depending on whether they prefer learning facts or discover possibilities and relationships; visual or verbal, depending on their preferences to information been presented visually and verbally; sequential or global, depending on whether they are more comfortable gaining understanding in linear steps or in large jumps. Instructional methods for coping with different learning styles are also included.

The importance of addressing most common learning styles is emphasized in [5]. Students, whose learning styles are compatible with the teaching style of a course instructor tend to retain information longer, apply it more effectively, and have more positive post-course attitudes toward the subject than do their counterparts who experience learning/teaching style mismatches, [5]. It is pointed that students also differ in their preferences to the way presented information is organized: inductive - where facts and observations are given, and underlying principles are inferred, or deductive - where principles are given, consequences and applications are deduced. Strengths and weakness of different learning styles are further discussed and a multi style approach is recommended.

"If professors teach exclusively in a manner that favours their students' less preferred learning style modes, the students' discomfort level may be great enough to interfere with their learning. On the other hand, if professors teach exclusively in their students' preferred modes, the students

may not develop the mental dexterity they need to reach their potential for achievement in school and as professionals.", [6].

The Myers-Briggs Type Indicator model [14] is mainly concerned with students' preferences from psychological point of view. They can be extraverts or introverts, sensors or intuitors, thinkers or feelers, judgers or perceivers. This leads to sixteen different learning style types, [14].

The Kolb's model is based on students' preferences for how to take information in and how to internalize information, [15]. The model consists of four types of learners concrete, reflective; abstract, reflective; abstract, active or concrete, active.

The Herrmann Brain Dominance Instrument model is concerned with preferences based on some brain functions, [12]. The four modes are analysis; methods and procedures; teamwork and communications; creative problem solving, systems thinking, synthesis, and design.

Learner-centered and teacher-centered teaching styles are discussed in [1].

Five teaching styles are described in [9] - expert, formal authority, personal model, facilitator, and delegator. Their advantages and disadvantages are also clearly formulated. Four teaching styles are identified in [19] - formal authority, demonstrator, facilitator, and delegator.

Our goal is to find a systematic way for detecting between students learning styles and lecturers teaching styles applying permutographs, [2] and relational concept analysis (RCA), [13], [10].

II. PRELIMINARIES

Let P be a non-empty ordered set. If $sup\{x, y\}$ and $inf\{x, y\}$ exist for all $x, y \in P$, then P is called a *lattice* [3]. In a lattice, illustrating partial ordering of knowledge values, the logical conjunction is identified with the meet operation and the logical disjunction with the join operation, [8].

Definition 1: [2] The permutograph on a set X is the graph, denoted by Σ_X, whose set of vertices is the set $\mathcal{L}_X$ of linear orders on X and whose edges are defined by the following adjacency relation, denoted Adj, between two linear orders: for $L, L' \in \mathcal{L}_X$, $L Adj L'$ if $|L \cap L'^d| = 1$.

For two linear orders L and L' on X representing preferences, $d(L, L') = |L \cap L'^d| = |L \backslash L'|$ is the number of disagreements on preferences between these two orders, [2].

The geodesic distance $\delta(L, L')$ between two linear orders L and L' in a permutograph is the minimum number of commutations to carry out in order to go from one to the other, [2].

In RCA, input data is organized as a pair made of a set of objects-to-attributes contexts $\mathbf{K} = \{\mathcal{K}_i\}_{i=1,...,n}$ and a set of objects-to-objects binary relations $\mathbf{R} = \{r_k\}_{k=1,...,m}$. Here, a relation $r \in R$ links two object sets from two contexts, i.e., there are $i_1, i_2 \in \{1,...,n\}$ (possibly $i_1 = i_2$) such that $r \subseteq O_{i_1} \times O_{i_2}$. Both contexts from $\mathbf{K}$ and relations from $\mathbf{R}$ are introduced as cross-tables,[11].

Definition 2: [11] (Relational Context Family (RCF)) An RCF is a pair $(\mathbf{K}, \mathbf{R})$ where:

- $\mathbf{K} = \{\mathcal{K}_i\}_{i=1,...,n}$ is a set of contexts $\mathcal{K}_i = (O_i, A_i, I_i)$

   and

- $\mathbf{R} = \{r_k\}_{k=1,...,m}$ is a set of relations r_k where

$$r_k \subseteq O_{i_1} \times O_{i_2} \text{ for some } i_1, i_2 \in 1,...,n.$$

Cluster analysis partition data into sets (clusters) sharing common properties, [2]. A frequently used tool in cluster analysis is a dissimilarity function d on a set of objects E, measuring the degree of dissemblance between the elements in E, [2].

III. SELECTIONS

Students are first properly introduced to the meaning of learning styles and are afterwards suggested to express their preferences via web based questionnaires.

We consider four groups of students formed according to gender and work experience. Criteria used under refinement order are based on learning styles models as in [4] and [5].

These four groups of students can be placed in seven sets due to application of one of the fore-mentioned criteria. Each of these seven sets has two subsets with 1 + 3 or 2 + 2 items in a subset.

After employing two of those criteria we obtain six sets following the refinement order. These six sets contain three subsets each with one or two items in a subset. All of them are placed in the 3rd row of the lattice in Fig. 1. In any of the six sets there is a couple of indiscernible elements (groups). Splitting these couples requires enforcement of yet another criterion. Set-valued functions developed in RCA are well suited for extracting knowledge from sets of students formed at different time periods.

Whether all criteria are to be applied or just some of them is up to a system modeling team. At the same time lattices as in Fig. 1 obtained from disjoint sets of students can be connected via RCA for extracting additional knowledge. The technical side of such processes is well explained in [11].

Four teaching styles described in [19] are ranked according to students' preferences as in permutograph in Fig. 2. The numbers in 1, 2, 3, 4 in Fig. 2 represent the four commonly understood teaching styles, i. e. formal authority (1), demonstrator (2), facilitator (3), and delegator (4). Teaching styles vary from topic to topic and students feedback can be followed by studying their responses, delivered via web

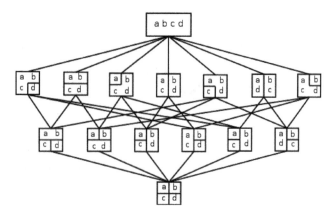

Fig. 1: Lattice of partitions

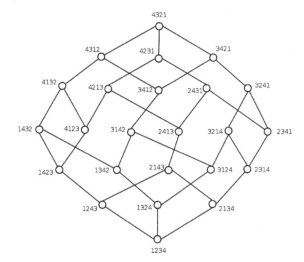

Fig. 2: Permutograph

based questionnaires. All possible orderings are summarized in Fig. 2, where two orderings connected by a strait line differ in positioning of two neighbor elements. This can be used while adjusting current teaching of new topics as well as performing further tuning of teaching the same course in the future.

Distances between vertices are applied while considering which group of users is effecting the order of preferences. Gender and age f. ex. are factors with a significant implication on preference orderings. This is to be incorporated in corresponding recommending processes. If users supply that type of information they would receive recommendations based on data from users with similar initial characteristics. If users do not provide such information they would receive recommendations based on the total collected data.

As an example about students' preferences one can look at the usual dilemma about orders in which problems are delivered by a course instructor: "learning how to apply a skill benefits more from blocked problem orders" while "learning when to apply a skill benefits more from interleaved problem orders", [16].

Another example is related to finding the degree to which students' preferences effect their progress in two consecutive

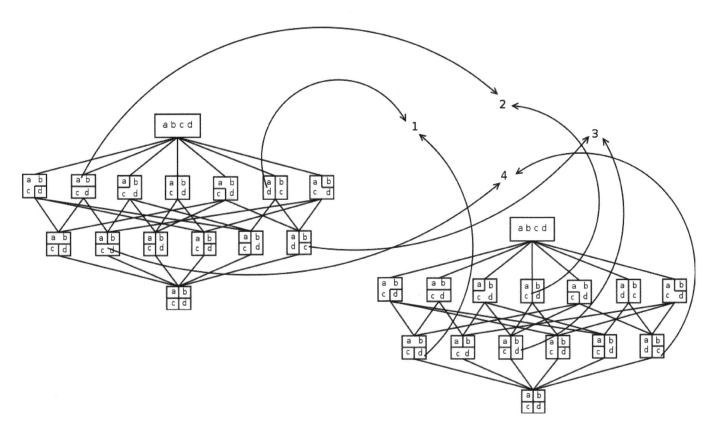

Fig. 3: Learning styles and preferences to teaching styles of two student groups

subjects when one of them is a prerequisite to the other. In a similar fashion one can work with related topics within a subject.

Distances between preferences can also be used to form clusters by joining existing clusters if they are within a predefined geodesic distance from a particular node, or creating new clusters with elements not within the geodesic distance. Once the clusters are formed an analysis of the reasons for their formations is to be performed.

Relational concept analysis is to be further applied for drawing conclusions about teaching different groups of students. Students groups, their learning styles and lectures teaching styles are to be collected in an information table as in [11]. The derived Hasse diagrams show correlations between learning styles and teaching styles, Fig. 3.

IV. CONCLUSION

There is no doubt about the existence of dependencies between students learning styles and lecturers teaching styles. Additional research has to be carried out in order to come up with meaningful recommendations to future instructors. Once a need for further tuning of a lecturer's teaching is established, additional efforts have to made for finding out what exactly has to be done. Both permutographs and relational concept analysis lend themselves very well to exploring compatibility between learning and teaching styles.

REFERENCES

[1] K. R. Barrett, B. L. Bower and N. C. Donovan, *Teaching Styles of Community College Instructors*, American Journal of Distance Education, vol. 21(1), pp. 37–49, 2007.

[2] N. Caspard, B. Leclerc, and B. Monjardet, *Finite Ordered Sets Concepts, Results and Uses*, Cambridge University Press, 2012.

[3] B. A. Davey and H. A. Priestley, *Introduction to lattices and order*, Cambridge University Press, Cambridge, 2005.

[4] R. M. Felder and L. K. Silverman, *Learning Styles and Teaching Styles in Engineering Education*, Engineering Education, vol. 78 (7), pp. 674–681, 1988.

[5] R. M. Felder, *Reaching the Second Tier: Learning and Teaching Styles in College Science Education*, Journal of College Science Teaching, vol. 23(5), pp. 286–290, 1993.

[6] R. M. Felder, *Matters of style*, ASEE Prism, vol. 6(4), pp. 18–23, 1996.

[7] R. M. Felder and J. Spurlin, *Applications, reliability, and validity of the index of learning styles*, International Journal of Engineering Education, vol. 21(1), pp. 103–112, 2005.

[8] B. Ganter and R. Wille, *Formal Concept Analysis*, Springer, 1999.

[9] A. Grasha, *Teaching with Style*, Pittsburgh, PA: Alliance Publishers, 1996.

[10] M. R. Hacene, M. Huchard, A. Napoli, and P. Valtchev, *A proposal for combining formal concept analysis and description logics for mining relational data.* In: Kuznetsov, S., Schmidt, S. (eds.) Proc. of the 5th Intl. Conf. on Formal Concept Analysis (ICFCA07). LNCS, vol. 4390, pp. 51–65, 2007.

[11] M. R. Hacene, M. Huchard, A. Napoli and P. Valtchev, *Relational concept analysis: mining concept lattices from multi-relational data*, Ann. Math. Artif. Intell., vol. 67, pp. 81–108, 2013.

[12] N. Herrmann, *The Creative Brain*, Lake Lure, NC, Brain Books, 1990.

[13] M. Huchard, M. R. Hacene, C. Roume, and P. Valtchev, *Relational concept discovery in structured datasets*, Ann. Math. Artif. Intell, vol. 49 (1-4), pp. 39–76, 2007.

[14] M. H. McCaulley, *The MBTI and Individual Pathways in Engineering Design*, Engineering Education, vol. 80, pp. 537–542, 1990.

[15] D. A. Kolb, *Experiential Learning: Experience as the Source of Learning and Development* Englewood Cliffs, NJ, Prentice-Hall, 1984.

[16] N. Li, W. W. Cohen, and K. R. Koedinger, *Problem Order Implications for Learning Transfer*, Lecture Notes in Computer Science, vol. 7315, pp. 185–194, 2012.

[17] T. A. Litzinger, H. L. Sang, J .C. Wise, and R. M. Felder, *A psychometric study of the index of learning styles*, Journal of Engineering Education, vol. 96(4), pp. 309–319, 2007.

[18] M. Lumsdaine and M. Lumsdaine, *Thinking Preferences of Engineering Students: Implications for Curriculum Restructuring*, Journal of Engineering Education, vol. 84(2), pp. 193–204, 1995.

[19] R. Wittmann-Price, M. Godshall, and L. Wilson, *Certified Nurse Educator (CNE) Review Manual*, Springer Publishing Company, 2 edition, 2013.

5

Wavelet Compressed PCA Models for Real-Time Image Registration in Augmented Reality Applications

Christopher Cooper
College of Engineering
North Carolina State University,
Raleigh, NC, 27695

John Cooper
Department of Chemistry and Biochemistry
Old Dominion University,
Norfolk, VA, 23529

Kent Wise
SGS Inc.
The Woodlands, TX, 77381

Makarand Deo*
Department of Engineering
Norfolk State University,
Norfolk, VA, 23504

Abstract—The use of augmented reality (AR) has shown great promise in enhancing medical training and diagnostics via interactive simulations. This paper presents a novel method to perform accurate and inexpensive image registration (IR) utilizing a pre-constructed database of reference objects in conjunction with a principal component analysis (PCA) model. In addition, a wavelet compression algorithm is utilized to enhance the speed of the registration process. The proposed method is used to perform registration of a virtual 3D heart model based on tracking of an asymmetric reference object. The results indicate that the accuracy of the method is dependent upon the extent of asymmetry of the reference object which required inclusion of higher order principal components in the model. A key advantage of the presented IR technique is the absence of a restart mechanism required by the existing approaches while allowing up to six orders of magnitude compression of the modeled image space. The results demonstrate that the method is computationally inexpensive and thus suitable for real-time augmented reality implementation.

Keywords—Image Registration; Principal Component Analysis; Wavelet Compression; Augmented Reality; Image Classification

I. INTRODUCTION

The utilization of augmented reality (AR) in the medical field provides multiple opportunities to enhance the access to and effectiveness of patient-specific medical information [1][2]. Using real-time AR systems allows the overlay, manipulation, and visualization of the various types of medical images acquired by MRI and tomography procedures (e.g., tissue, charge density, blood flow, etc.)[3][4]. Hence AR-based visualization techniques have been increasingly employed in safer medical practices for better understanding and accurate diagnostics. Creating an interactive 3D virtual model containing multiple dimensions of information, which can be manipulated and visualized in concert, provides immediate opportunities for high-quality medical training. Furthermore, the advanced AR-guided medical procedures have the potential to decrease the invasiveness and increase the safety and accuracy of a surgery by enhancing a surgeon's ability to utilize medical imagery during the operation [1][5]. The first step to achieving these goals however, is a robust and real-time registration of high resolution images [6]. This paper presents a novel registration method which is accurate and computationally inexpensive.

Image registration (IR) is the process of aligning two similar images, taken at different times or by different sensors, in order to correctly overlay an independent image [7]. IR techniques typically fall into two categories: feature-based [8] and intensity-based [9]. The former method relies on the detection and successful tracking of distinct image features, such as lines, corners, and contours, while the latter method determines a transformation using all of the image data. Each of these techniques relies on an optimization component, which determines the optimal spatial transformation, and a similarity metric, which compares the resemblance of the transformed scene image and the model image [7, 10]. Spatial transformations can be either rigid or non-rigid. Rigid transformations are composed of translation and rotation in three directions, for a total of six degrees of freedom. Non-rigid transformations account for these changes as well as those in the actual structure or anatomy of the object [11].

Initial IR techniques such as the iterative closest point (ICP) algorithm have produced incorrect transformations due to incoming image noise and prealignment errors [12]. One optimization approach to increase convergence range and avoid erroneous local optima is the use of hierarchical multi-scale, however down-sampling of images often suppresses key differences, leading to an absence of distinctive features in similar objects [11]. In response to these errors, evolutionary computation (EC) has been used to help alleviate the complex problems of image processing, most noticeably the need for a good initial estimation of the transformation. These models, included in the broader field of metaheuristics, rely on computational models of evolutionary processes to create populations of solutions [13]. One such example is the scatter search (SS) technique, which is a metaheuristic-based method

This work was supported in part by American Heart Association (AHA) Scientist Development Grant No. 12SDG11480010.

used in both feature-based and intensity-based methods [10]. Use of this technique provides noticeable advantages in the accuracy of transformations and eliminates prealignment error. Nevertheless, even IR techniques with metaheuristics rely on a restart mechanism when transformations become low quality. This is a result of the refinement process, which optimizes the previous transform in order to produce the new spatial transform for the incoming model image. Since optimizing a low-quality transformation is unlikely to produce a high-quality transformation, it is necessary to restart the algorithm and acquire a new initial transformation [10]. Recently, more advanced IR techniques have been proposed based on Speeded up Robust Features (SURF), optical flow method, and marker-free IR method [14][15]. However, these existing methods require extensive computations to achieve real-time IR which is a major concern that limits their use in real-time AR systems.

In this paper, a novel IR technique which is capable of achieving higher accuracy with substantially reduced computational time is presented. This was achieved by creating a database of compressed vectors from reference images of an object at all possible viewing angles and then constructing a corresponding principal component analysis (PCA) model prior to image registration. The proposed approach offers multiple benefits over existing methods. There is no need for an initial estimation or camera calibration, and furthermore, since the model operates independently for each incoming frame, there is no need for a restart mechanism. A systematic performance analysis of the proposed method is presented in this paper.

II. METHODS

A. Creation of Virtual Object Database

Current IR methods involve taking an existing image and transforming it in real time. With the use of high speed flash storage, it is now feasible and competitive to eliminate this transformation step, and simply recall previously generated high-resolution images. In this study, the virtual 2D images used in the registration process were generated from 3D imaging data prior to the real time registration process. A detailed 3D model of canine heart anatomy, derived from high resolution diffusion tensor magnetic resonance imaging (DTMRI) was used as the virtual object. A Virtual Object Database (VOD) was created by manipulating the orientation of the virtual heart model as a function of three orthogonal angles of rotation. Theoretically, there are an infinite number of possible orientations and hence 2D image views; however the ability to distinguish between similar orientations (2D images) decays as the change in an angle approaches zero degrees. Therefore, it is possible to represent an object within a pre-defined angular resolution with a limited number of image orientations. The degree of resolution, however, directly impacts the number of images needed in the VOD of 2D images. A total of 22,104 images were generated that fully represented all non-degenerate object orientations at 10° angle increments. At 5° resolution, a total of 186,624 images were

generated. Since the virtual object was implemented in digital form, a high-resolution database was programmatically created using MATLAB software.

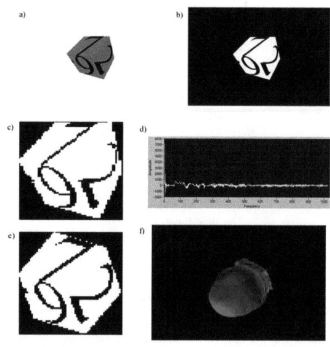

Fig. 1. This flow diagram shows the process of identifying the orientation of a reference object in the reference image and using it to correctly place a virtual heart image. The steps are: a) placement of 3D reference object in the scene, b) acquisition and processing of the reference 2D image, c) scaling of 2D image, d) wavelet transformation and compression of the 2D image, e) prediction of best match using PCA model, and f) registration and display of the appropriate virtual object image. A unique feature of this method is the absence of a restart mechanism

B. Creation of Reference Image Database

The first step in the proposed augmented reality method (shown in Figure 1) was registering an appropriate reference object in the scene to the appropriate 2D image from the VOD. Since the VOD was predetermined, the process was simplified as it only requires knowledge of the rotation of the reference object, its location within the scene, and the requisite scaling. In order to facilitate accurate determination of the three angles of rotation, the reference object must be appropriately designed. Although previous work has shown that this theoretically requires a reference object which has a minimum of four non-planar points [16], the results of this paper (described in Section 4) demonstrate that the accurate determination of the three orthogonal angles of rotation is highly dependent upon the asymmetry of the object such that as the asymmetrical complexity increases, the model accuracy increases as well.

Five reference objects (Objects I-V) with varying degrees of distinctive asymmetry (Figure 2) were created using a CAD software to study their effectiveness for precise object tracking based on optimal number of PCs required.

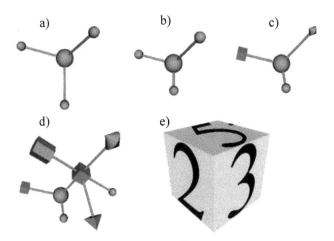

Fig. 2. The reference objects shown above are used to represent an increase in asymmetry for the reference object. The reference objects are a) Object I: Symmetric Spheres, b) Object II: Asymmetric Spheres, c) Object III: Three Shapes, d) Object IV: Multishape, and e) Object V: Dice

The first three reference objects possess only three distinct axes of asymmetry. Object I (Fig. 2a) contains bonds of equal length, resulting in a high degree of symmetry around the axes of rotation. It is worth noting that these symmetry elements are present in the 3D space of the real object. Object II (Fig. 2b) is similar, but all of the bond lengths are unique. In 3D space, this change results in the loss of all symmetry elements except for Identity. In Object III (Fig. 2c), two of the spheres have been replaced by unique shapes (a cube and a top). Object IV (Fig. 2d) possesses five axes extending from a cube where one axis has a highly complex structure attached to the end of the axis. Finally, Object V (Fig. 2e), a dice, has six perpendicular axes of asymmetry, created by a distinct number on each face. The numbers, which range from 2 to 7, were selected for their asymmetry, and hence the numbers 0, 1, and 8 were avoided since they possess symmetry elements other than the Identity.

The 3D reference objects were created as ".stl" files using Solid Edge, and reference images of the various rotations of each object were obtained in MATLAB in 10° increments for each of the three axes. The reference objects were also printed using a 3D printer and were used to acquire test images using a web camera at various object rotations and camera-to-object distances. Creating the reference object as a CAD file allows the Reference Image Database (RID) to be programmatically generated in the same manner as the VOD, while the ability to 3D print the object allows the generation of a physical reference object with high similarity to the RID. This approach ensures a high degree of correlation between the computer-generated images in the RID and the images of the reference object in the scenes that are acquired in real time.

C. Creation of Compressed Wavelet Vector Representation (CWVR) Database

In order to minimize the number of real-time calculations required for image registration, the resolution of the images in the RID was lowered and then the images were compressed further using a wavelet transform. This allowed each image in the RID to be represented by a compressed wavelet vector which was orders of magnitude smaller in number of pixels. These compressed wavelet vectors were arranged into a reference database which was used to construct the PCA model and to carry-out real-time calculations. Construction of the compressed wavelet vector reference (CWVR) database involved two steps: i) reference image scaling and ii) wavelet transformation and compression.

1) Image Scaling

Although the CAD file used to generate the RID provides high resolution, the results show that a lower resolution allows for faster processing while still allowing for sufficient information to maintain PCA model accuracy. Hence the resolution of the images in the RID was maintained at 64x64 pixels. Due to the calibration-free approach of the proposed model, there is no incoming information about the distance between the reference object and the camera. As such, a scene with a large camera-to-object separation will display a small reference object (low pixel resolution), while a scene with a small camera-to-object separation will display a larger one (high pixel resolution). In each case, however, the orientation of the reference object remains unchanged. Hence, if the object is appropriately scaled, and the scaled image possesses sufficient resolution, the numerical values of three distinct rotation angles can be determined from the PCA model. However, to preserve the scaling required for image registration, it is necessary that both the reference object images in the RID and in the scene be scaled in a similar manner. To achieve this, the incoming scene image was either up-sampled or down-sampled to the same resolution as the RID.

In both cases (scene and RID), the image was restricted to only the contents of a rectangular bounding box using the topmost, leftmost, rightmost, and bottommost points of the reference object. The horizontal scaling (S_h) required for image registration was defined as:

$$S_h = \frac{(i_R - i_L)}{n} \quad (1)$$

where i_R is the index of the rightmost side of the bounding box of the scene image, i_L is the index of the leftmost side of the bounding box of the scene image, and n is the horizontal pixel resolution in the RID. Similarly, the vertical scale (S_v) was defined as:

$$S_v = \frac{(i_T - i_B)}{n} \quad (2)$$

where i_T is the index of the topmost side of the bounding box of the scene image, i_B is the index of the bottommost side of the bounding box of the scene image, and n is the vertical pixel resolution in the RID (same as horizontal).

2) Wavelet Transform and Compression

Each scaled image was subsequently stripped into a single vector by unfolding the rows of the image. A wavelet transform with four wavelets and scaling functions [17] was applied to the resulting vector. The Daubechies family of orthogonal wavelets was chosen due to their extensive use in data compression. Specifically, the Daubechies wavelet filter with 8 taps and 4 vanishing moments was selected (also referred to as a D8 or db4 referring to the N=2A relationship between the number of taps, N, and the number of vanishing moments, A) as a reasonable compromise between image resolution and compression efficiency. The advantage of the wavelet transform is that it preserves both the frequency and the position information of the image vector (i.e., the function is not translationally invariant as is the case with most Fourier transform methods) [18]. Moreover, the use of discrete wavelet transform (DWT) is computationally efficient. Due to the nature of the reference object image, the dominant and requisite information was contained almost completely in the low frequency wavelet coefficients. Thus, the final step of compression involved truncating the high frequency wavelet coefficients to achieve image compression while preserving the lowest 1024 coefficients that retain essential information needed to determine the object orientation. As shown in Figure 3, the original input sample image (Panel A) and the reconstructed compressed image after wavelet transform (Panel B) were almost identical, but the latter required 75% less data for creation and storage. A slight blurring of sharp edges in the compressed images due to the loss of the high frequency components is evident in the figure but this had an insignificant impact on the accuracy of the PCA model.

D. Constructing the Principal Component Analysis (PCA) Model

Once the CWVR database was created, it was then used to construct a PCA Model [19]. The use of a PCA model provided two distinct advantages. First, it allowed the CWVR for all images of the reference object to be described using a multi-dimensional eigenvector space that accounted for the greatest variance of the underlying data structure. Second, it enabled each CWVR to be mapped into the eigenvector space by using a single scalar value (eigenvalue or score) for each eigenvector. Since the number of eigenvectors (more commonly referred to as principal components) required to account for the majority of variance was considerably less than the length of the CWVR, a significant further compression of the data dimensionality was achieved. For example, if the CWVR contained 1024 data points, and the variance was described by 10 principal components, then a compression of 100-fold was achieved since each CWVR could now be represented by only 10 eigenvalues within the space of the PCA model.

The PCA model was constructed by creating a data matrix X, where each row of the matrix corresponds to a CWVR, and the number of rows is equal to the number of reference images. The data matrix was then decomposed using a singular value decomposition algorithm:

$$X = U\Sigma V^T \tag{3}$$

where U is an $n \times m$ orthonormal matrix, Σ is an $m \times m$ eigenvalue matrix with all zero off-diagonal elements, $\mathbf{V}^T$ is an orthonormal $m \times m$ matrix, n is the number of reference images and m is the length of the CWVR. Each row of matrix V^T is an eigenvector or principal component (PC), thus the resulting number of principal components is equal to the length of the CWVR. Since V^T is an orthonormal matrix, all of the PC row vectors are orthogonal and define a multivariate space containing the compressed images. The coordinates of each compressed image within this multivariate space was given by the rows of a scores matrix, S, which is simply:

$$S = U\Sigma \tag{4}$$

where S is a $n \times m$ matrix. Thus the scores reflect where each sample lies on the PC axes. However, since the PCs were sorted in decreasing order of variance, the majority of the variance was described by the first few PCs and the higher order PCs were dominated mostly by noise. This allowed the PCA model to be constructed by truncating to the number of columns (k) in the scores matrix, S, and the loadings matrix, V:

$$X' = S'(V')^T \tag{5}$$

Where X' is an approximation of the compressed image data, and S' and V' have $k \times m$ dimension ($k \ll n$). T indicates the transpose. Figure 4 illustrates the matrix order reduction obtained while solving the Eqn. 5.

E. Image Registration

Since this was a rigid image registration method, there were six degrees of freedom which had to be determined in order to accurately display a virtual object in the scene image. These included three degrees of freedom in translation and three degrees of freedom in rotation.

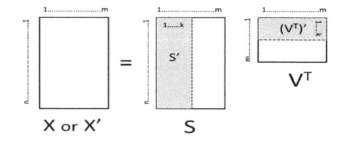

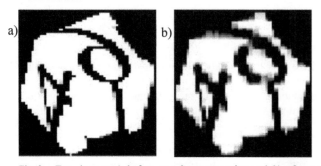

Fig. 3. Two images a) before wavelet compression and b) after wavelet compression are shown. The differences in the images are minimal, however, the amount of information required to reconstruct the image after compression is significantly smaller, and is used as the basis of the CWVR database

Fig. 4. The equation for creation of the PCA model is shown where n corresponds to the number of images in the PCA model and m corresponds to the length of the compressed wavelet vector representation (CWVR) of each image. The solid polygons correspond to the data matrix **X**, the scores matrix **S**, and the loadings (principal component) matrix **V**^{T}. The dashed polygon boxes correspond to the retained columns (k) of **S** and **V** which are used to construct an approximation (**X'**) of the data matrix to generate the PCA model

The translation that occurred between two images was determined by using the information acquired during the image scaling process. By comparing the geometric mean of the bounding box surrounding the reference object between the model and scene image, the translation on the two axes (which are parallel to the sides of the image) was determined. Translation along the third axis was accounted for by scaling the virtual object image according to the scale of the incoming reference object image. This was performed using the horizontal and vertical scale ratios given by Eqn. 1 and Eqn. 2, respectively.

The rotation around three orthogonal angles was simultaneously determined by using the PCA model as discussed earlier. An image of the reference object in an incoming scene was processed in the same way as that of the CWVR database. The PCs of the unknown object orientation were then calculated using the resulting CWVR (y) and the reference object PCA Model (i.e., the truncated matrix $(V')^T$) by solving Eqn. 6 for the scores vector s:

$$(V')^T s = y \qquad (6)$$

Since the PCs are orthogonal, $[(V')^T]^{-1} = V'$. This yielded a trivial solution for determining the scores for unknown compressed images:

$$s_{new} = y_{new} V' \qquad (7)$$

where s_{new} corresponds to the scores vector containing the coordinates of a new compressed image in PCA space and y_{new} is the CWVR for the new image. The significance of Eqn. 7 is that regardless of the size of the image database used to create the PCA model, the model of the image space was described by a $n \times k$ matrix, where k is the number of PCs and n is the length of the CWVR. Thus for an image database at 10 degree resolution (total 22,104 images) where each original image was defined by 1024x 1024 pixels ($n = 1024$), a PCA model containing 22 principal components ($k = 22$) yielded a compression of the database by six orders of magnitude.

Although it is possible for the new image to contain an exact match with the scores of the database, this probability was limited by the angular resolution of the database. For this reason, the best match was taken as the nearest neighbor (smallest distance) in PC model space, where the distance from nearest neighbors was calculated as the sum of the squares of the score differences:

$$d = \sum_{PC=1}^{k} (s_{PC}^{new} - s_{PC}^{model})^2 \qquad (8)$$

where d is the distance, k is the maximum number of PCs in the model, and s is the score for a particular PC for the new image being predicted and a neighboring images contained in the PCA model. Since each vector in the CWVR database was indexed to the angles of rotation of its reference image, identifying the best match also identified the corresponding values for rotation about three different axes.

Each reference object was tested using PCA Models constructed using a PC space ranging from one to 50 PCs in order to determine the optimal number of PCs for that object. The optimum number of PCs was chosen as the minimum PC number where adding an additional PC did not yield a statistically significant decrease in the standard error of the model predictions. This was determined by Malinowski's F-test [20]. Malinowski's F-test assumes that the sum of the eigenvalues can be decomposed into either parts which are significant or noise, and that the significant eigenvalues provide an estimate of the true number of principal components needed. If there are a maximum of p possible principal components (the minimum of either the number of samples or the number of variables), then the F-statistic for the sth eigenvalue (λ_s) is:

$$F_s = \frac{\lambda_s}{\sum_{j=s+1}^{p} \lambda_j / (p - s)} \qquad (9)$$

and the maximum PC is taken as that having the minimum F_s value.

The best match for an acquired image was restricted to a reference object image within the CWVR database (10° angle increments) and as such, an angle error of less than or equal to ±5° is considered accurate. Thus for an incoming sample image with a 45° rotation on a particular axis, a correct match is either 40° or 50° in that same angle. Figure 5 illustrates one example of the largest error for an incoming scene image with X-axis rotation =45°, Y-axis rotation =315°, Z-axis rotation =0° (henceforth written as: 45°, 315°,0°) (panel a) and the corresponding best RID correct match through the PCA

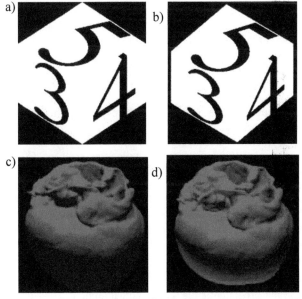

Fig. 5. Shown is an example of the largest error for a correct match given the selected angular resolution (10° increments) of the reference image database. The images shown are a) the scene image (45° 315° 0°), and b) the best RID match through PCA analysis (40° 310° 0°). The corresponding registered heart images are shown in c) and d). As seen, the difference between the two Object V images and their respective heart images is barely perceptible, but could be further reduced by using a higher angulation resolution (i.e. 5° increments)

analysis (40°, 310°, 0°). As seen, the difference between the two Object V images and their respective heart images is barely perceptible, but could be further reduced by using a higher angulation resolution (i.e. 5° increments).

The performance of the object matching algorithm was assessed by percent error (δ) calculated as:

$$\delta = \frac{i}{a} \times 100 \qquad (10)$$

Where i was the number of incorrect angle matches over a certain tolerance of error for a given PCA model and a was the total number of angles (or three times the number of images tested).

III. RESULTS

The percent standard errors for each reference object model as a function of the number of PCs are plotted in Figure 6. Each model reaches an apparent minimum percent error after reaching an optimum number of PCs, implying that the additional PCs are no longer providing significant additional information. For all objects, the error dropped significantly for the initial PCs after which the accuracy does not improve noticeably for higher order PCs. It was observed that the accuracy improved with extent of asymmetry in the objects with the most symmetric object (Object I) giving the highest error and the most asymmetric object (Object V) giving the lowest error regardless of the number of PCs included.

Table I lists the number of optimal PCs needed for each type of reference object. It was observe that as the asymmetry of the reference object increases, the number of PCs required to describe the variance of the CWVR database increases. Thus the highly symmetric object (Object I) required only 9 PCs, while the most asymmetric object (Object V) required 29 PCs. The table also provides percent errors for each reference object PCA model at the optimum PC number. The percent errors are broken down into three categories: 15° (the percent of incorrect predictions exceeding an error of 15°), 10° (percent of incorrect predictions exceeding an error of 10°), and 5° (percent of incorrect predictions exceeding an error of 5°). Also shown for each object PCA model is the root mean square of the distances between the reference and sample images where an error less than or equal to 5° is considered accurate. As can be observed, the results showed a significant increase in both accuracy and the optimum number of PCs as the complexity of the asymmetry in the reference objects increased.

The errors shown in Table I and Figure 6 both correspond to errors in the predicted angle vs. the actual angle of rotation. However, since each object possesses 3 orthogonal angles of rotation, it is possible for each image mismatch to correspond to one or more improper angles. The actual number of incorrect images and incorrect angles for each of the 672 tested samples is given in Table II.

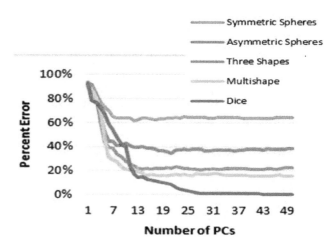

Fig. 6.   The graph shows the percent standard error of the PCA model for each reference object as a function of number of retained PCs. Initially the error decreases rapidly with each addition of a PC, however, after a certain number of PC's the resulting decrease in error becomes minimal, implying that the inclusion of further PC's is no longer providing significant information to the PCA model. Reference objects with higher degrees of asymmetry yield correspondingly lower errors once the minimal number of PCs is reached

TABLE I.   EFFECT OF REFERENCE OBJECT ASYMMETRY ON PCA OPTIMAL PC# AND PCA MODEL ERROR

Object	Optimal Number of PCs	Percent Error (5°)	Percent Error (10°)	Percent Error (15°)	Average RMS
Object I	9	63.79%	61.71%	56.35%	62.9964
Object II	12	39.63%	30.26%	28.47%	31.9440
Object III	14	21.58%	11.86%	9.62%	12.4449
Object IV	24	16.82%	8.48%	5.26%	6.7919
Object V	29	0.79%	0.00%	0.00%	0.1022

TABLE II.   THE NUMBER OF INCORRECT IMAGE AND ANGLE PREDICTIONS (AT 5°) WHEN PREDICTING THE 671 UNIQUE SAMPLES

Object	Optimal Number of PCs	Incorrect Images	Incorrect Angles	Average
Object I	9	498	1286	2.582
Object II	12	330	799	2.421
Object III	14	224	435	1.942
Object IV	24	189	339	1.794
Object V	29	9	16	1.778

As the asymmetry increased for the first four objects, the number of image mismatches decreased monotonically from 498 to 189. For the Object V PCA model, however, there was a precipitous decrease to only 9 incorrect images. Table II also gives the average number of incorrect angles per sample, and shows that as the average number of incorrect angles in each image mismatch decreased with an increase in asymmetry in the objects. One example of an image mismatch for Object II is shown in Figure 7. Panel a) corresponds to a test sample (60°, 50°, 210°) and Panel b) corresponds to the PCA predicted match (140°, 230°, 90°).

As can be seen, although the object possesses no symmetry in 3D space, there still exist combinations of distinct angle rotations which yield 2D views which are nearly identical. This predicted mismatch problem was further exacerbated upon image compression. Figure 8 shows one example for Object III after compression and scaling. The image with rotations (10°, 180°, 80°) (Panel A) was wrongly recognized as (0°, 0°, 40°) (Panel B), leading to an error of 230°. At the lower resolution, the image differences approach the noise limit.

For Object V, the highest error mismatch is shown in Figure 9 (reconstructed from their respective CWVRs). Despite the low resolution, the images are clearly distinguishable with the rotations of (90°, 90°, 285°) in Panel A, and that of (100°, 80°, 290°) in Panel B. The relatively small angle error combined with the *visually distinguishable features* of the two images suggests that the cause of the mismatch was rooted in the PC model itself. Indeed, a plot of the % variance attributable to each PC (Figure 10) indicated that the higher order PCs for object V constituted a significant amount of variance (Figure 10, inset) when compared to that of more symmetric object (Object II). This is consistent with the finding that more PCs are required to effectively describe the modeling space of objects with greater asymmetry (Figure 6). It also suggests that using an F-test may not be a reliable

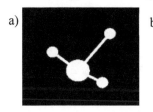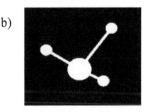

Fig. 7. An image mismatch resulting from PCA model of the Asymmetrical Spheres reference object. The images are shown at high resolution prior to compression. The left image (x-axis rotation = 60°, y-axis rotation = 50°, z-axis rotation 210°) is very similar, even at high resolution, to the right database image (x-axis rotation = 140°, y-axis rotation = 230°, z-axis rotation = 90°); however the angles are significantly different. This is a result of the loss of information as one moves from 3D to 2D space

method to determine the optimum number of PCs since the percentage of incorrectly identified object images beyond 29 PCs is relatively small (1.34%; 9 out of 672 images). When the Object V PC model is expanded to 45 PCs, the number of image mismatches dropped to 6 (0.89% incorrect) and no mismatches at 200 PCs. Thus the error in the Object V PCA models appears to originate in the ability of the PCA model to describe the asymmetry and not in the lack of distinguishable asymmetry. Since the truncation of PCs for a given model is part of the data compression, it suggests the need to balance compression in an effort to maintain the asymmetric properties required for accuracy. This is not noteworthy in case of more symmetric objects since the error remains high even for a significantly larger number of PCs (see Figure 6). Although a higher degree of asymmetry requires a higher number of PCs to maintain accuracy, the overall compression is still significant. For example, the original sample image contains over 1 million pixels, and conversion to a 64x64 image results into 4096 pixels. It is further reduced to 1024 pixels after wavelet compression and truncation. Thus even when using

200 PCs to represent the image, provides an additional factor of 5 in data compression and yields a total data compression of over 5000 per image. With regards to the image database space (22,103 images x 1024 x 1024 = 23 billion coordinates), the PCA model compression (200 PCs x 1024 length of CWV = 204,000 coordinates) still results in five *orders of magnitude* compression.

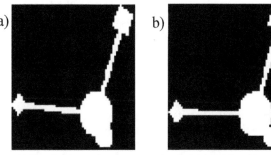

Fig. 8. The figure shows an example of a PCA predicted image-mismatch for the Object III PCA model. The images are shown after scaling and image compression. The differences between the top image (10°, 180°, 80°) and the bottom database image (0°,0°,40°) are almost indistinguishable at the lower resolution, where they approach the noise limit

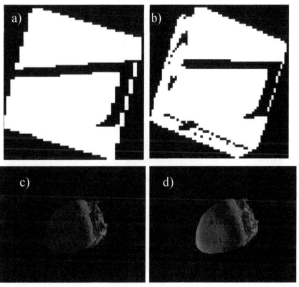

Fig. 9. An example of the maximum error obtained using a PCA model with 29 retained principal components for Object V is shown. The images are a) the sample image (90°, 90°, 285°), b) the best match (100°, 80°, 290°), c) the heart image corresponding to the sample image, and d) the heart image corresponding to the best match. Since the images are visually distinctive, it can be reasoned that the error lies in the PC model itself and indicates the need for additional PCs if more accuracy is required

The PCA models for each object were constructed using 22,103 images (i.e., 22,103 vectors in the CWVR database). Figure 11 shows the PCA space defined by these images at various orders of PCs for Object V (Fig. 11d-e) and Object II (Fig. 11a-c). The panels A and D show contents of the images extracted in the first three principal components (X, Y and Z axes). As can be seen, the plot for Object II (Panel a) exhibits a great deal of structure. Indeed, if the points in the plot are considered as a solid object, a 3-fold improper axis of rotation exists (S_3). For the Object V plot (Panel d), the defined space is less structured, and although not spherical, is significantly more homogeneous. As higher PC numbers are used to define the axes, the spread in the PC scores decreased for both objects, as expected.

IV. DISCUSSION

The paper presents a novel image registration method using pre-processed database of compressed image vectors spanning all possible image rotations and scaling. This method uses a combination of discrete wavelet transform to compress the images without losing any valuable information and principal component analysis to construct an accurate estimation model. This approach significantly reduced the computations and enabled real-time processing for seamless medical augmented reality applications.

The computational benefits of this approach are achieved by utilizing additional computational time prior to image registration for processing already acquired reference image database using a reference object with distinct and complex asymmetric properties. By acquiring reference object images

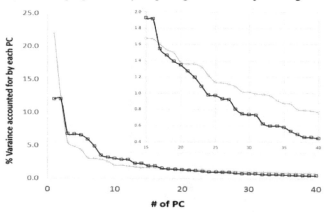

Fig. 10. The graph shows the relationship between %variance and the retained PC number for Object V and Object II. The graph inset shown in the top right shows that the Object V PCA model still contains a significant amount of variance at higher order PCs when compared to the PCA models of the more symmetric reference objects. This indicates that more PCs are required to accurately reflect the higher degree of asymmetry and is supported by the observation that inclusion of 200 PCs in the Object V PCA model eliminates all mismatches

from multiple viewpoints, a comprehensive model can be developed using principal component analysis which accurately matches an incoming image whose angles of rotation (with respect to the viewer) are not known, with a corresponding reference image. Since the reference image is indexed to the angles of rotation, image registration is straightforward and only requires scaling and positioning. Alternatively, these indexed angles can be used to determine an appropriate transform of the 3D virtual heart model into 2D image space. The advantages of the former are - 1) the bulk of the computations (generating high resolution heart images, building the reference image database, and constructing the PCA models) are carried out only once and in advance so that the real-time IR is computationally inexpensive; and 2) the current trend in computing storage is the use of high speed flash interfaced to the CPU via a high speed bus (e.g. PCIe) which allows extremely fast image recall. Thus the IR time of this algorithm remains superior to the existing method while

offering the advantage of a higher resolution rendering upon registration.

The robust nature of the model presented in this paper is created due to the use of a predefined, asymmetric reference object which is present in the incoming image. Unlike other methods, where tracked features are chosen in real time [8,9,21], in this method, the tracked features are predetermined. This significantly enhances the speed and accuracy of image registration at the cost of creating a more rigid technique which requires the presence and visibility of a specific reference object. Traditional IR methods rely on the creation of mathematical transformations to track features in scene images which are cumbersome to use with higher resolution images, and require "good" features which can be easily tracked [22] [23]. Intensity-based IR is one way to bypass this requirement, but these methods still employ a mathematical transformation, which ultimately increases the amount of real-time computation necessary for image registration [9]. Intensity-based IR methods also require a computationally expensive restart mechanism to obtain an optimal transform instead of trying to refine a bad transformation [10]. The proposed method, however, does not refine previous transforms, eliminating the need for a restart mechanism altogether.

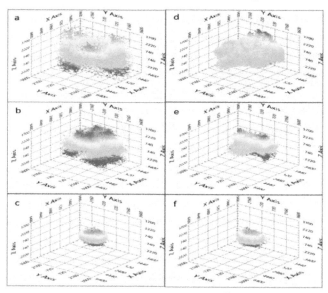

Fig. 11. The figure shows the PC space of the PCA model for Object V (right) and Object II (left) using different combinations of PCs for the axes of the space. The top images show the scores for the 1st (x-axis), 2nd (y-axis), and 3rd (z-axis) PCs. The middle images show the scores for the 4th, 5th, and 6th PCs; and the bottom images show the scores of 27th, 28th, and 29th PCs. A three-fold axis of improper rotation (S_3) can be seen in the Object II PCA space for the first 3 PCs (top left), while the similar PCA space for Object V is more homogenous (top right)

The significant difference of the proposed method compared to other IR techniques makes it difficult to draw a direct comparison. Nevertheless shape-based image retrieval techniques have often utilized PCA in order to reduce data dimensionality and decrease computation time. Image retrieval methods have demonstrated that more complex shapes are easier to use with PCA analysis [24], and the principal component descriptors are preferable to other methods of image identification and retrieval [25]. Content based image retrieval (CBIR) is a popular technique that utilizes such methods to search and retrieve images from large databases [26]. Typically employed to manage large volumes of digital images, this technique is similar to the proposed method which is repurposed as an IR method.

When uniform PCA space is encountered in a model, it is common to convert from a PCA classification method (as used in this paper) to a principal component regression (PCR) method [27] in order to derive more quantitative results (e.g., the ability to quantitatively interpolate between the modeled angles). Although it is beyond the scope of the present work, construction of a PCR model for this implementation of augmented reality would involve the multivariate regression of the scores of a PCA model against the angles of rotation. This could also be accomplished with a partial least squares model using a PLS-2 algorithm [28]. The more homogenous clustering for the Object V PCA model would suggest a greater likelihood of success of such quantitative modeling when using a higher degree of reference object asymmetry. This further exemplifies the importance of introducing complexity into the asymmetry of reference objects.

V. CONCLUSIONS

This paper presents a novel image registration technique involving image compression and PCA modeling based on the use of reference objects with complex asymmetry. The design provides a method to eliminate the real-time computational costs of performing geometric transforms and by using PCA classification, operates without the need for a restart mechanism. The method was validated using 672 object images to test a PCA model created from 22,103 reference images. The asymmetry of the reference objects was found to highly correlate with the accuracy of the image registration. In particular, for highly asymmetric objects, the accuracy was predominantly dependent upon the inclusion of enough principal component vectors to accurately describe the asymmetry of the objects described in the PCA model space. For higher symmetry objects, the inclusion of higher PC order models had little to no impact on the accuracy. Future studies could further investigate the properties and uses of complex asymmetry to enhance the accuracy of image registration methods. Since the majority of image processing in this method is done prior to the real-time process, and the data compression resulted in significant reduction in memory requirements, the proposed method is well suited for real-time medical augmented reality applications.

REFERENCES

[1] Azuma, R., Baillot, Y., Behringer, R., Feiner, S., Julier, S., & MacIntyre, B. "Recent advances in augmented reality. Computer Graphics and Applications," IEEE, vol. 21, no. 6, pp. 34-47. Nov, 2001.

[2] Van Krevelen, D., & Poelman, R. "A survey of augmented reality technologies, applications and limitations," International Journal of Virtual Reality, vol. 9, no. 2, pp. 1-19. Jun, 2010.

[3] Fritz, J., Paweena, U., Ungi, T., Flammang, A. J., Fichtinger, G., Iordachita, I. I., & Carrino, J. A. "Augmented reality visualisation using an image overlay system for MR-guided interventions: technical performance of spine injection procedures in human cadavers at 1.5 Tesla," European radiology, vol. 23, no. 1, pp. 235-245. Jul, 2013.

[4] Azagury, D., Ryou, M., Shaikh, S., San José Estépar, R., Lengyel, B., Jagadeesan, J., . . . Thompson, C. "Real-time computed tomography-based augmented reality for natural orifice transluminal endoscopic

surgery navigation," British Journal of Surgery, vol. 99, no. 9, pp. 1246-1253, Sep, 2012.

[5] Nakamoto, M., Ukimura, O., Faber, K., & Gill, I. S. "Current progress on augmented reality visualization in endoscopic surgery," Current opinion in urology, vol. 22, no. 2, pp. 121-126, Mar, 2012.

[6] Liao, H., Inomata, T., Sakuma, I., & Dohi, T. "3-D augmented reality for MRI-guided surgery using integral videography autostereoscopic zimage overlay," Biomedical Engineering, IEEE Transactions on, vol. 57, no.6, pp. 1476-1486, Jun, 2010.

[7] Zitova, B., & Flusser, J. "Image registration methods: a survey, Image and vision computing, vol. 21, no. 11, pp. 977-1000, Jun, 2003.

[8] Reddy, B. S., & Chatterji, B. N. "An FFT-based technique for translation, rotation, and scale-invariant image registration," IEEE transactions on image processing, vol. 5, no. 8, pp. 1266-1271, Aug, 1996.

[9] Kim, J., & Fessler, J. A. "Intensity-based image registration using robust correlation coefficients," Medical Imaging, IEEE Transactions on, vol. 23, no. 11, pp. 1430-1444, Nov, 2004.

[10] Valsecchi, A., Damas, S., Santamaría, J., & Marrakchi-Kacem, L. "Intensity-based image registration using scatter search," Artificial intelligence in medicine, vol. 60, no. 3, pp. 151-163, Feb, 2014.

[11] Markelj, P., Tomaževič, D., Likar, B., & Pernuš, F. "A review of 3D/2D registration methods for image-guided interventions," Medical image analysis, vol. 16, no. 3, pp. 642-661, Apr, 2010.

[12] Besl, P. J., & McKay, N. D. "Method for registration of 3-D shapes," Pattern Analysis and Machine Intelligence, IEEE Transactions on, vol. 14, no. 2, pp. 239-256, Feb, 1992.

[13] Glover, F., & Kochenberger, G. A. "Handbook of metaheuristics," Springer Science & Business Media, 2003.

[14] Li, H., Qi, M., & Wu, Y. "A Real-Time Registration Method of Augmented Reality Based on Surf and Optical Flow," Journal of Theoretical and Applied Information Technology, vol. 42, no. 2, pp. 281-286, Aug, 2012.

[15] Wang, J., Suenaga, H., Hoshi, K., Yang, L., Kobayashi, E., Sakuma, I., & Liao, H. "Augmented Reality Navigation with Automatic Marker-Free Image Registration Using 3D Image Overlay for Dental Surgery," IEEE Trans Biomed Eng., vol. 61, no. 4, pp. 1295-1304, Apr, 2014.

[16] Vallino, J. R. "Interactive augmented reality," University of Rochester, 1998.

[17] Mohamed, M., & Deriche, M. "An Approach for ECG Feature Extraction using Daubechies 4 (DB4) Wavelet," International Journal of Computer Applications, vol. 96, no. 12, pp. 36-41, Jun, 2014.

[18] Jensen, A., & Cour-Harbo, A. "Ripples in mathematics: the discrete wavelet transform," Springer, 2001.

[19] Wold, S., Esbensen, K., & Geladi, P. "Principal component analysis," Chemometrics and intelligent laboratory systems, vol. 2, no. 1, pp. 37-52, Aug, 1987.

[20] Malinowski, E. R. "Statistical F-tests for abstract factor analysis and target testing," Journal of Chemometrics, vol. 3, no. 1, pp. 49-60, Jan, 1989.

[21] Jin, H., Favaro, P., & Soatto, S. "Real-time feature tracking and outlier rejection with changes in illumination," Computer Vision, 2001. Proceedings. Eighth IEEE International Conference on, pp. 684-689, Jul, 2001.

[22] Lucas, B. D., & Kanade, T. "An iterative image registration technique with an application to stereo vision," IJCAI, vol. 81, pp. 674-679, Aug, 1981.

[23] Shi, J., & Tomasi, C. "Good features to track," Computer Vision and Pattern Recognition, Proceedings IEEE Computer Society Conference on, pp. 593-600, Jun, 1994.

[24] Tavoli, R., & Mahmoudi, F. "PCA-Based Relevance Feedback in Document Image Retrieval," International Journal of Computer Science Issues, vol. 9, no. 4, Jul, 2012.

[25] Wang, B., & Bangham, J. A. "PCA based shape descriptors for shape retrieval and the evaluations," Computational Intelligence and Security, International Conference on, pp. 1401-1406, Nov, 2006.

[26] Jain, A., Muthuganapathy, R., & Ramani, K. "Content-based image retrieval using shape and depth from an engineering database," Advances in Visual Computing, Springer,.pp.255-264, Nov, 2007.

[27] Geladi, P., & Esbensen, K. "Regression on multivariate images: principal component regression for modeling, prediction and visual diagnostic tools," Journal of Chemometrics, vol. 5, no. 2, pp. 97-111, Mar, 2005.

[28] Wold, S. "Nonlinear partial least squares modelling II. Spline inner relation," Chemometrics and intelligent laboratory systems, vol. 14, no. 1, pp. 71-84, Apr, 1992.

6

Sensitivity Analysis for Aerosol Refractive Index and Size Distribution Estimation Methods Based on Polarized Atmospheric Irradiance Measurements

Kohei Arai [1]

[1] Graduate School of Science and Engineering
Saga University
Saga City, Japan

Abstract—Aerosol refractive index and size distribution estimations based on polarized atmospheric irradiance measurements are proposed together with its application to reflectance based vicarious calibration. A method for reflectance based vicarious calibration with aerosol refractive index and size distribution estimation using atmospheric polarization irradiance data is proposed. It is possible to estimate aerosol refractive index and size distribution with atmospheric polarization irradiance measured with the different observation angles (scattering angles). The Top of the Atmosphere (TOA) or at-sensor radiance is estimated based on atmospheric codes with estimated refractive index and size distribution then vicarious calibration coefficient can be calculated by comparing to the acquired visible to near infrared instrument data onboard satellites. The estimated TOA radiance based on the proposed method is compared to that with aureole-meter based approach which is based on refractive index and size distribution estimated with solar direct, diffuse and aureole (Conventional AERONET approach). It is obvious that aureole-meter is not portable, heavy and large while polarization irradiance measurement instruments are light and small (portable size and weight).

Keywords—Degree of Polarization; aerosol refractive index; size distribution

I. INTRODUCTION

Earth observation satellites have a long history of being characterized by vicarious methods. These include the Marine Observation Satellite-1 [Arai, 1988], Landsat-7 Enhanced Thematic Mapper Plus [Barker, et al., 1999], SeaWiFS [Barnes, et al., 1999], SPOT-1 and 2 [Gelleman, et al., 1993], Hyperion [Folkman, et al., 1997], and POLDER [Hagolle, et al., 1999]. Vicarious approaches also provide a cross-comparison between sensors to characterize mission instruments onboard the same satellite [Arai, 1997] via the use of well-understood ground areas such as desert sites [Cosnefroy, et al., 1996]. Arai and Thome [2000] published an error budget analysis of solar reflectance-based vicarious calibration. The most dominant factor for vicarious calibration is surface reflectance measurement, followed by optical depth measurement, estimation of refractive index, aerosol size distribution, and identification error in test site pixels. Typical vicarious calibration accuracy is around 4%. Onboard calibrators cannot provide results of a higher accuracy than the preflight laboratory calibration. This means that the accuracy of the in-flight (absolute) calibration is inferior to the preflight results. This is because the preflight calibration source is used to calibrate the onboard calibrators. In addition, the uncertainty of the onboard calibrator typically increases with time. Hence, it makes good sense to include additional calibration approaches that are independent of the preflight calibration. Besides the normal and expected degradation of the onboard calibrators, they also run the risk of failing or operating improperly. Therefore, vicarious approaches are employed to provide further checks on the sensor's radiometric behavior. Given the understanding that the orbiting sensor's response will change over time, the ASTER science team developed a methodology, based on OBC results, to update preflight RCCs that are input to generate the Level-1B product [Thome, Arai et al., 2008]. The OBC results are also combined with vicarious calibration to produce the most accurate knowledge of ASTER's radiometric calibration.

The solar radiometers are relatively calibrated immediately prior to, during, or after each field campaign via the Langley method or Modified Langley method, and this allows for the determination of spectral atmospheric optical depths [Arai, et al., 2005]. The optical depth results are used as part of an inversion scheme to determine ozone optical depth and an aerosol size distribution. The aerosols are assumed to follow a power law distribution, also referred to as a Junge distribution. Columnar water vapor is derived from the solar extinction data using a modified-Langley approach. The atmospheric and surface data are used in a radiative transfer code. There are a variety of codes available that satisfy all the requirements of predicting the at-sensor radiance to the required accuracy. It has shown that similar conclusions are drawn for other code types such as doubling-adding, and the methods used in the 6S code [Lenoble 1985]. Besides these, another method takes into account polarizations in the calculation of down-welling and up-welling radiation [Arai et al., 2003]. It uses ground-based solar direct, diffuse and aureole radiance measurements as well as polarized radiance with several polarization angles [Arai and Liang, 2005, and Liang and Arai, 2005]. It is obvious that aureole-meter is not portable, heavy and large while polarization irradiance measurement instruments are light and small (portable size and weight). This study is based on a Lambertian view of the surface. The near-nadir view for the majority of the ASTER overpasses reduces the uncertainty of this assumption since the dominant direct-reflected solar irradiance is correctly taken into account. Strong gaseous absorption effects due to water vapor are determined using MODTRAN to compute transmittance for the sun-to-surface-to-satellite path for 1-nm intervals from 350 to 2500 nm. Also ozone absorption is taken into account based on MODTRAN with measured column ozone using atmospheric extinction measurements. This sun-to-ground-to-sensor transmittance is multiplied by the at-sensor radiance output from the radiative transfer code to correct the radiances for this strong absorption. While this approach is an approximation that excludes interactions between diffusely scattered radiances and absorption, it does not cause large uncertainties for ASTER applications because of the small absorption effect within most of the bands, and the typically high surface reflectance of the test sites used in this work.

For the multiple-scattering components calculation, it is easy to estimate Rayleigh scattering (molecule) with measured atmospheric pressure. Meanwhile Mie scattering (aerosol) is not so easy to estimate. Aerosol parameters, refractive index, size distribution, etc. have to be estimated. AERONET (Holben B.N. et al., 1998) and SKYNET Aoki, K. et al., 2005) allows for the estimation of aerosol parameters at the specific locations. They use aureole-meters and sky-radiometers which allow solar direct, diffuse and aureole irradiance. These ground-based instruments are heavy and large so that they equip them at the specific sites. Small and light portable polarization irradiance measuring instrument, on the other hand, is proposed by Arai (2009) for estimation of aerosol parameters. With a measured polarized irradiance at the specific observation angles (scattering angles) allows estimation of aerosol refractive index and size distribution. Estimated aerosol parameters are a little bit differing from those which are derived from AERONET as well as SKYNET. This paper describes at-sensor radiance of ASTER/VNIR with the estimated aerosol parameters derived from AERONET and SKYNET as well as the proposed method together with a sensitivity analysis.

II. Proposed Model

Reflectance based vicarious calibration method proposed here is based on MODTRAN with the following input parameters, Measured surface reflectance (Lambertian surface), Calculated molecule scattering based on a measured atmospheric pressure, Calculated aerosol scattering with the aerosol parameters, refractive index and size distribution which are estimated with measured polarized irradiance at several scattering angles (for instance seven scattering angles which ranges from 60 to 120 with 10 degree step) based on the proposed method, Calculate absorbance due to water vapor and ozone with measured column water and ozone.

At-sensor radiance is estimated based on MODTRAN and is compared to the actual ASTER/VNIR data derived radiance. The most influencing factor of the proposed method is

estimation accuracy of aerosol parameters, refractive index and size distribution.

Therefore, sensitivity of aerosol parameters on TOA radiance should be analyzed. Figure 1 shows the calculated TOA radiance in unit of $[W/cm^2/sr/\mu m]$ with the parameter of real and imaginary parts of refractive index while Figure 2 shows the calculated TOA radiance with the parameters of size distribution.

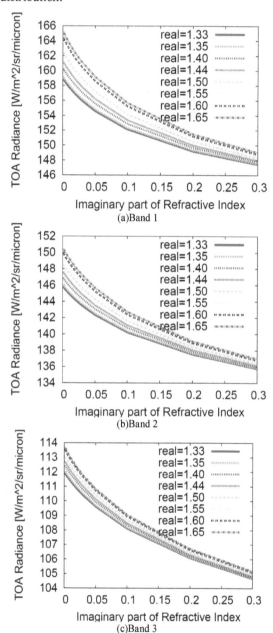

Fig. 1. Calculated TOA radiance derived from the field campaign which was conducted at Railroad valley on September 21 2008 with the parameters of real and imaginary parts of refractive index.

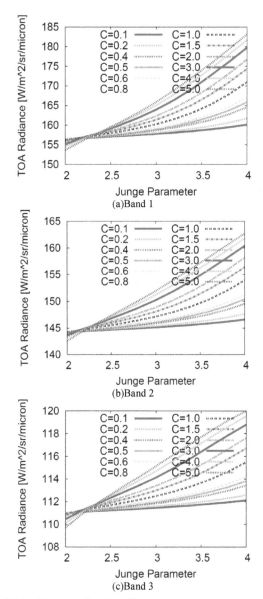

Fig. 2.   TOA radiance as a function of Junge parameter

These examples are derived from the field campaign which was conducted at Railroad valley on September 21 2008. Junge distribution, one of power low distributions is assumed as is expressed as the equation (1).

$$\frac{dN}{d\ln(r)} = Cr^{-\alpha}, \alpha = -(\upsilon+1) \qquad (1)$$

where α denotes the slope of the relation between volume and radius of aerosol particles and v is Junge parameter as are shown in Figure 3. Figure 3 also shows a typical aerosol density vertical profile. TOA radiance is increased with increasing of real part of refractive index and is decreased with increasing of imaginary part of refractive index. The TOA radiance-increasing ratio at shorter wavelength (Band 1) is much greater than that in the longer wavelength (Band 3N: Nadir view). Figure 1 and 2 also show that the calculated TOA radiance is changed below 2% when the estimated refractive index and Junge parameter are changed within a range of

±10% from the assumed typical values, 1.44 of real part of refractive index, 0.05 of imaginary part of refractive index and 3 of Junge parameter. This implies that required estimation accuracy of refractive index and size distribution is not so high; about ±10% would be enough if 2% were the required TOA radiance estimation accuracy.

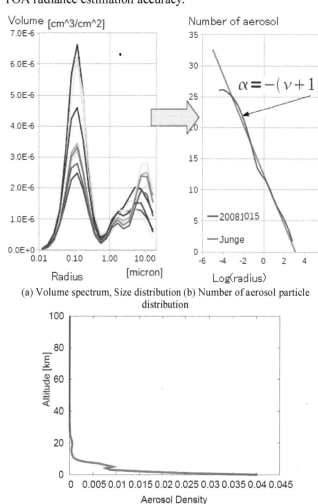

(a) Volume spectrum, Size distribution (b) Number of aerosol particle distribution

Fig. 3.   Volume spectrum and corresponding power low distribution, representation of aerosol size distribution with Junge parameter with corresponds to slope of the power low distribution together with aerosol density profile (an example of Saga (33:14.46N, 130:17.3E, 29m) field campaign which was conducted on October 15 2008).

III.   EXPERIMENTL

Field campaigns were conducted at Roach Lake on December 3 2008 and at Coyote Lake on December 10 2008, respectively. Table 1 shows the detailed information of the field campaigns.

Measured column ozone and water vapor are shown in Figure 4 (a) while relation between ln(wavelength) and ln(optical depth) are shown in Figure 4 (b), respectively. Also measured surface reflectance as well as estimated refractive index and size distribution are shown in Figure 4 (c) to (f). These are measured and estimated values for Roach Lake field campaign that was conducted on December 3 2008.

TABLE I.   DETAILED INFORMATION OF THE FIELD CAMPAIGNS
CONDUCTED.

Date and time (UTM)	December 3 2008, 18:38:34	December 10 2008, 18:38:34
Solar azimuth and zenith angles	154.48, 59.84	163.92, 59.61
Location	Roach Lake(38:30:18N,115:41:29W)	Coyote Lake(35:03:53N,116:44:50W)
Air-temperature, atmospheric pressure	22.5, 933hPa	22.1, 974hPa
Junge parameter(370/870, 500/870)	2.73, 3.15	5.89, 7.21
Ozone(DU), Water vapor(g/cm^2)	284.7, 0.24	271.6, 0.46

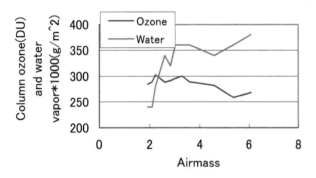

(a) Column ozone and water vapor

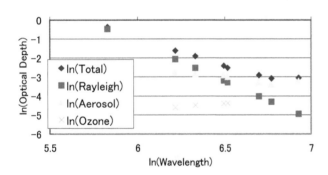

(b) Angstrome exponent and Junge parameter

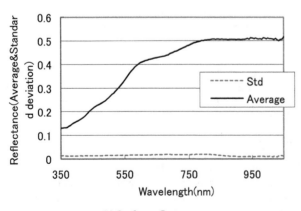

(c) Surface reflectance

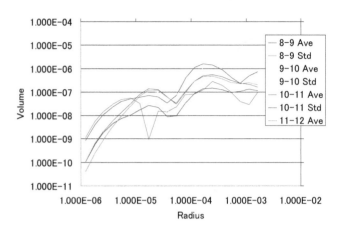

(d) Volume spectrum (Size distribution)

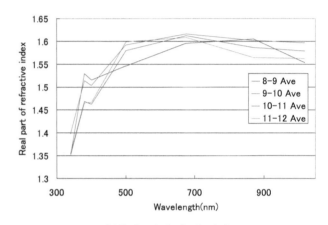

(e) Real part of refractive index

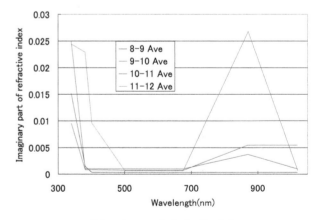

(f) Imaginary part of refractive index

Fig. 4. Atmospheric and surface characteristics of the test site at Roach Lake measured on December 3 2008.

Meanwhile those for Coyote Lake campaign are shown in Figure 5. Atmospheric optical depth for Coyote Lake campaign was very thin compared to Roach Lake campaign. In particular, Junge parameter for Coyote Lake campaign is twice much greater than that for Roach Lake campaign. This implies that small size of aerosol particles is dominant for Coyote Lake campaign in comparison to Roach Lake campaign.

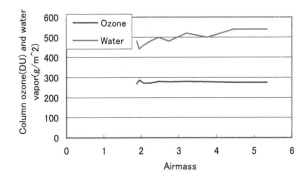

(a) Column ozone and water vapor

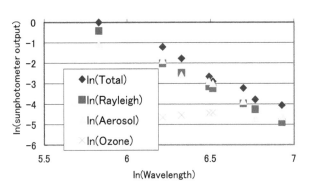

(b) Angstrome exponent and Junge parameter

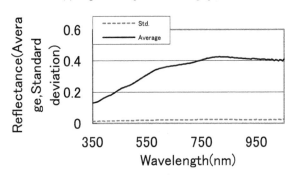

(c) Surface reflectance

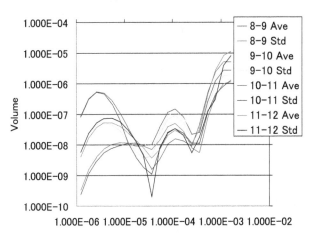

(d) Volume spectrum (Size distribution)

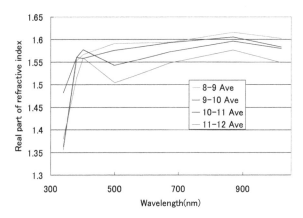

(e) Real part of refractive index

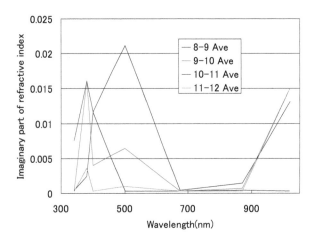

(f) Imaginary part of refractive index

Fig. 5. Atmospheric and surface characteristics of the test site at Coyote Lake measured on December 10 2008.

Refractive index and size distribution are estimated with skyradiometer data which allows measure solar direct, diffuse and aureole irradiance on the ground surface. Dr.Tsuchida and Dr.Kamei provided Skyradiometer data with their courtesy [Tsuchida and Kamei, 2009]. Using the modified skyrad.pack of software code, refractive index and size distribution are retrieved with these data. Although the original skyrad.pack provided by Dr.Nakajima (Nakajima et al., 2000) does not care about polarized radiance from the surface, the modified Arai-Ryo model takes p and s polarization of irradiance and radiance in the radiative transfer (Arai and Liang, 2005). On the other hand, measured scattering angle characteristics of Degree of Polarization (DP) are shown in Figure 6. Using curve-fitting algorithm of iterative method, most appropriate refractive index and size distribution (Junge parameter) is estimated. Through a comparison between estimated refractive index and Junge parameter by Arai-Ryo model with skyradiometer data and by curve fitting algorithm with seven scattering angles (60,70,80,90,100,110,120) of DP, it is found that both shows good coincidence (difference between both is within a range of ±5%. In accordance with the previous research, it is known that the estimation accuracy of refractive index and Junge parameter is approximately 6%. ±10% of refractive index and Junge parameter estimation accuracy causes ±2% of TOA radiance estimation accuracy so that 6% of accuracy of refractive index and Junge parameter would causes below 2% of TOA radiance estimation accuracy.

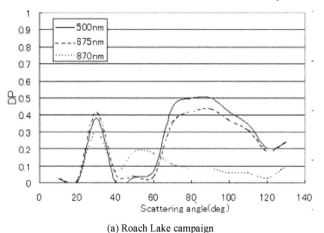

(a) Roach Lake campaign

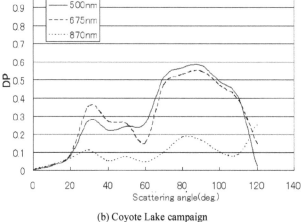

(b) Coyote Lake campaign

Fig. 6. DP measured for field campaigns at Roach Lake and Coyote Lake which were conducted on December 3 and 10 2008.

Using estimated refractive index and size distribution derived from skyradiometer data and DP data as well as surface reflectance, column ozone and water vapor, atmospheric pressure (Rayleigh scattering) TOA radiance is estimated based on MODTRAN. Table 2 shows the estimated refractive index and size distribution with skyradiometer data and DP data for Roach Lake and Coyote Lake campaigns. Both show a good coincidence, discrepancy of real part of refractive index ranges from -4.65 to 2.566%, difference of imaginary part of refractive index ranges from -5.86 to 3.846%, and discrepancy of Junge parameter ranges from 0.013 to 3.653%, respectively.

Meanwhile Table 3 shows the estimated TOA radiance with refractive index and size distribution derived from DP data and skyradiometer data. Also both show a good coincidence, below 15.22% of discrepancy. In particular, discrepancy at the shorter wavelength, 560 and 660nm of Bands 1 and 2 for Coyote Lake campaign is much greater than those for Band 3 for Coyote field campaign and Roach Lake campaign.

TABLE II. COMPARISON OF JUNGE PARAMETER AND REFRACTIVE INDEX DERIVED FROM SKYRADIOMETER DATA AND DP DATA

	Method	Junge	Real	Imaginary
08/12/03 Roach	Skyradiometer	3.372	1.582	0.0004
	DP	3.365	1.501	0.0003
08/12/10 Coyote	Skyradiometer	5.213	1.574	0.0068
	DP	5.214	1.541	0.0066

TABLE III. COMPARISON OF TOA RADIANCE DERIVED FROM SKYRADIOMETER DATA AND DP DATA

2008/12/3	L_DP	L_skyrad	% difference
B1(560)	111.75	110.47	1.145
B2(660)	114.4	113.65	0.656
B3N(810)	95.2	94.84	0.378
B3B(810)	94.58	94.56	0.021
2008/12/10	L_DP	L_skyrad	% difference
B1(560)	109.08	92.48	15.22
B2(660)	99.22	89.67	9.625
B3N(810)	76.7	73.22	4.537
B3B(810)	76.12	73.92	2.89

This is caused by relatively large Junge parameter, small size of aerosol particles are greater than large size of those for Coyote Lake campaign. Except these, the discrepancy between two methods for estimation of TOA radiance with skyradiometer data and DP data is below 4.5%. Due to the fact that aerosol optical depth increases in accordance with decreasing wavelength sharply for Coyote Lake field campaign, the discrepancy between estimated TOA radiance between two methods is greater than those in the longer wavelength regions. Also it is true that Junge parameter for Coyote Lake campaign is twice much greater than Roach Lake campaign. TOA radiance is sensitive to Junge parameter, in particular, greater Junge parameter regions as is shown in Figure 7 (which was derived from the field campaign which was conducted at Railroad valley on September 21 2008). In accordance with increasing of Junge parameter, the calculated TOA radiance is increased sharply. For these reasons, the discrepancy between two methods for Coyote Lake campaign is greater than that of Roach Lake.

IV. CONCLUSION

The estimated refractive index and size distribution using the proposed DP based method shows a good coincidence with the estimated those by the conventional skyradiometer (POM-01 which is manufactured by Prede Co. Ltd.), or aureole meter based method so that the proposed method does work well. The Junge parameter estimated by skyradiometer based method is derived from Angstrome exponent that is calculated with aerosol optical depth measured with skyradiometer while that by the proposed DP based method is derived from Angstrome exponent that is calculated with aerosol optical depth measured with polarized irradiance measuring instrument (MS720 which is manufactured by EKO Co. Ltd.).

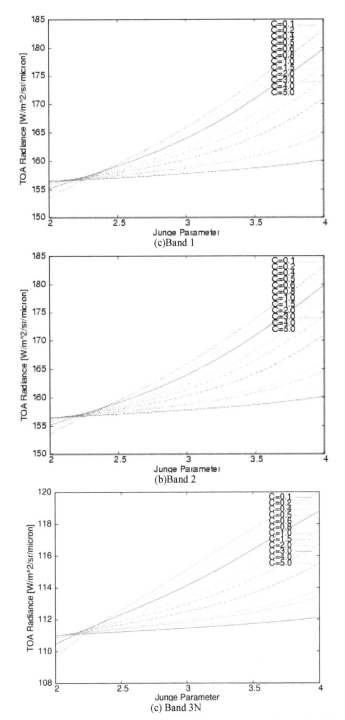

Fig. 7. Relation between TOA radiance and Junge parameter (Example of the calculated TOA radiance for the field campaign which was conducted at Railroad valley on September 21 2008)

The difference between both is caused by the difference of gain/offset of the two instruments, POM-1 and MS720. On the other hand, the differences of estimated refractive index between skyradiometer based and the proposed DP based methods are mainly caused by the estimation methods, inversion of radiance to refractive index for skyradiometer based method while least square method minimizing the discrepancy between the actual and simulated DP at the seven different scattering angles based on MODTRAN.

The difference of TOA radiance derived from the proposed DP based method and the conventional skyradiometer based method is within the range of 1.2% for relatively high reflectance and comparatively thin aerosol optical depth as well as small Junge parameter case (relatively large aerosol particles are dominant) and is within the range of 2.9 to 15.2% for relatively low reflectance and comparatively thick aerosol optical depth as well as large Junge parameter case (relatively small aerosol particles are dominant). Due to the fact that p and s polarized irradiance is relatively small for relatively high reflectance and comparatively thin aerosol optical depth as well as small Junge parameter case (relatively large aerosol particles are dominant), it is understandable.

It is obvious that skyradiometer and aureole meter is typically large and heavy in comparison to the polarized irradiance measuring instruments. It is possible to bring the polarized irradiance measuring instrument at anywhere easily. p and s polarized irradiance measurement at the seven different scattering angle takes around three minutes so that it has to be assumed that the atmosphere is stable for more than three minutes. p and s polarized irradiance is sensitive to the surface reflectance so that it is recommendable to use the proposed method for widely homogeneous ground cover targets.

ACKNOWLEDGMENT

The author would like to thank Mrs. Yui Nishimura for her efforts through experiments and simulations.

REFERENCES

[1]   Aoki, K., T. Takamura, and T. Nakajima, Aerosol optical properties measured by SKYNET sky radiometer validation network. *Proc. of the 2nd EarthCARE Workshop*, 133-134, 2005.

[2]   Arai K., Preliminary assessment of radiometric accuracy for MOS-1 sensors, International Journal of Remote Sensing, 9, 1, 5-12, 1988.

[3]   Arai, K., In-flight test site cross calibration between mission instruments onboard same platform, Advances in Space Research, 19, 9, 1317-1328, 1997.

[4]   Arai K., Post Launch Calibration of ASTER with MODIS data, Proceedings of the 3rd Annual IR Calibration Symposium in Utah State University, 1992.

[5]   Arai K., A.Ono and Y.Yamaguchi, Accuracy Assessment of the Interactive Calibration of ASTER/TIR with MODIS, Proceedings of the IGARSS'93, pp.1303-1305, 1993.

[6]   Arai K., H.Fujisada, ASTER Level 1 WG, ATBD: Analytical Theoretical Basis Document for Level 1 Products, 1995.

[7]   Arai K., and K.Thome, Error budget analysis of the reflectance based vicarious calibration for satellite-based visible to near infrared radiometers, Journal of the Japanese Society for Photogrammetry and Remote Sensing, 39, 1, 99-105, 2000.

[8]   Arai K. and X.Liang, Method for the top of the atmosphere radiance estimation taking into account the polarization in down and up welling radiance calculations, Journal of the Japanese Society for Photogrammetry and Remote Sensing, 44, 3, 4-12, 2005.

[9]   Arai and H.Tonooka, Radiometric performance evaluation of ASTER/VNIR, SWIR and TIR, IEEE Trans. on GeoScience and Remote Sensing, 43,12,2725-2732, 2005.

[10] Arai,K. and X.Liang, Characterization of aerosols in Saga city areas, Japan withy direct and diffuse solar irradiance and aureole observations, Advances in Space Research, 39, 1, 23-27, 2006.

[11] Arai, K., Vicarious calibration for solar reflection channels of radiometers onboard satellites with deserted area of data, Advances in Space Research, 39, 1, 13-19, 2006.

[12] Arai, Atmospheric Correction and Residual Errors in Vicarious Cross-Calibration of AVNIR and OCTS Both Onboard ADEOS, Advances in Space Research, 25, 5, 1055-1058, 1999.

[13] Barker, J.L., S.K. Dolan, et al., Landsat-7 mission and early results, SPIE, 3870, 299-311, 1999.

[14] Barnes, R.A., E.E.Eplee, et al., Changes in the radiometric sensitivity of SeaWiFS determined from lunar and solar based measurements, Applied Optics, 38, 4649-4664, 1999.

[15] Cosnefroy, H., M.Leroy and X.Briottet, Selection and characterization of Saharan and Arabian Desert sites for the calibration of optical satellite sensors, Remote Sensing of Environment, 58, 110-114, 1996.

[16] Folkman, M.A.,S.Sandor, et al., Updated results from performance characterization and calibration of the TRWIS III Hyperspectral Imager, Proc. SPIE, 3118-17, 142, 1997.

[17] Gellman, D.I., S.F. Biggar, et al., Review of SPOT-1 and 2 calibrations at White Sands from launch to the present, Proc. SPIE, Conf.No.1938, 118-125, 1993.

[18] Hagolle, O., P.Galoub, et al., Results of POLDER in-flight calibration, IEEE Trans. On Geoscience and Remote Sensing, 37, 1550-1566, 1999.

[19] Holben, B. N., et al., AERONET- A federated instrument network and data achieve for aerosol characterization, *Remote Sens.*, 12, 1147-1163, 1991.

[20] Holben, B.N., and Coauthors, AERONET-A federated instrument network and data archive for aerosol characterization. *Remote Sens. Environ.*, 66, 1-16. 1998.

[21] Lenoble, J., Edt. Radiative transfer in scattering and absorbing atmospheres: Standard computational procedures, A.Deepak Publishing Co.,Ltd.,

[22] Liang X. and K.Arai, Method for aerosol refractive index and size distribution with the solar direct, diffuse, aureole and polarization radiance, Journal of Remote Sensing Society of Japan, 25, 4, 357-366, 2005.

[23] Nakajima, T., M.Tanaka and T. Yamauchi, Retrieval of the optical properties of aerosols from aureole and extinction data, Applied Optics, 22, 19, 2951-2959, 1983.

[24] Ono.A., F.Sakuma, K.Arai, et al.,Pre-flight and Inflight Calibration for ASTER, Journal of Atmospheric and Ocean Technology, Vol.13, No.2, pp.321-335, Apr.,1996.

[25] Slater P., K.Thome, A.Ono, F.Sakuma, K.Arai, et al., Radiometric Calibration of ASTER, Journal of Remote Sensing Society of Japan, Vol.15, No.2, pp.16-23, June 1995.

[26] Thome, K., K. Arai, et.al., ASTER preflight and in-flight calibration and validation of level 2 products, IEEE Trans. on Geoscience and Remote Sensing, 36, 4, 1999.

[27] Thome, S.Schiller, J.Conel, K.Arai and S.Tsuchida, Results of the 1996 Earth Observing System vicarious calibration campaign at Lunar lake playa, Nevada (USA), Metrologia, 35, 631-638, 1998.

[28] Tsuchida, S. and A.Kamei, personal correspondence, 2009..

7

Evaluation of Cirrus Cloud Detection Accuracy of GOSAT/CAI and Landsat-8 with Laser Radar: Lidar and Confirmation with Calipso Data

Kohei Arai[1]

1Graduate School of Science and Engineering

Saga University

Saga City, Japan

Masanori Sakashita[1]

1Graduate School of Science and Engineering

Saga University

Saga City, Japan

Abstract—Cirrus cloud detection accuracy of GOSAT/CAI and Landsat-8 is evaluated with a ground based Laser Radar: Lidar data and sky view camera data. Also, the evaluation results are confirmed with Calipso data together with a topographic representation of vertical profile of cloud structure. Furthermore, origin of cirrus clouds is estimated with forward trajectory analysis. The results show that GOSAT/CAI de4rived cirrus clouds is not accurately enough due to missing of cirrus cloud detection spectral channel while Landsat-8 derived cirrus cloud.

Keywords—Cirrus cloud; GOSAT/CAI; Landsat; LiDAR; Sky view camera; Calipso; topogramphic representation of 3D clouds

I. INTRODUCTION

Cloud detection is one of though issues in satellite remote sensing in particular for cirrus clouds [1]-[16]. It is not so easy to detect cirrus clouds in particular for remote sensing satellite onboard instruments. In order to detect cirrus clouds, 1.38 micrometer wavelength channel is adopted for Moderate resolution of Imaging Spectrometer: MODIS 1 and Landsat-8 Operational Land Imager: OLI 2, etc. Green house gasses Observation Satellite / Cloud and Aerosol Imager: GOSAT 3 /CAI 4 is dedicated sensor for cloud and aerosol retrievals. Because that GOSAT/FTS (Fourier Transform Spectrometer 5) data is affected by clouds and aerosols, GOSAT/CAI is carried on the same platform of GOSAT satellite. Therefore, cloud flag and its confidence level are evaluated from the GOSAT/CAI and provide as Level 2 of GOSAT products together with Level 1B product as a source of Level 2 product. As mentioned above, it is not so easy to detect cirrus clouds. Although cirrus detection wavelength channel (1.38 micrometer) is required for detection of cirrus clouds, GOSAT/CAI does not have such channel. Therefore, it is not possible to detect cirrus cloud essentially. On the other hand, Landsat-8/OLI has cirrus detection channel. It is expected that cirrus detection can be done with Landsat-8 OLI data. Thus, cirrus cloud screening can be done for GOSAT/FTS observations.

In order to check the capability of cirrus cloud detection,

Light Detection and Ranging、 Laser Imaging Detection and Ranging: LiDAR data which allows measurement of back scattering ratio and depolarization ratio is used [17]-[29]. The ground based LiDAR is equipped at one of the GOSAT validation sites which is situated at Saga University, Japan. Therefore, vertical profile of aerosol particles as well as cloud particles are detected which results in detection of aerosols and clouds including cirrus clouds. Meantime, sky view camera observes hemispherical cloud conditions. Although it is possible to detect thick clouds, it is not easy to detect cirrus clouds with sky view camera. Vertical cloud structure can be retrieved with Cloud Aerosol Lidar and Infrared Pathfinder Satellite Observations: Calipso 6 data. Therefore, detected cirrus clouds can be validated with Calipso data. In this paper, a specific representation of vertical cloud structure is proposed. That is to representation of the retrieved structure on the topographic map which is projected on the globe. Forward trajectory analysis is also made for retrievals of the original source areas of the cirrus in concern through consideration of atmospheric conditions.

In the next section, the proposed method for evaluation of cirrus detection accuracy of GOSAT/CAI and Landsat-8 is described followed by experiments (method and procedure as well as the results from the experiments). Then validation of the evaluation results with sky view camera data and Calipso data is described followed by the specific representation of vertical cloud structure on the earth. Finally, conclusion and some discussion are followed.

II. METHOD AND PROCEDURE

A. GOSAT/FTS and CAI

GOSAT satellite is operating since January 23 2009 as the joint project among Ministry of Environment, JAXA and National Institute Environmental Science: NIES. GOSAT carries FTS and CAI as mission instruments as shown in Fig.1.

Major mission of GOSAT is to measure total column of carbon dioxide and methane which can be done with FTS instrument. In order to avoid influence due to aerosols and clouds, TANSO/CAI is also carried on GOSAT.

---

[1] http://modis.gsfc.nasa.gov/

[2] http://landsat.usgs.gov/landsat8.php

[3] http://www.gosat.nies.go.jp/

[4] http://www.gosat.nies.go.jp/eng/gosat/page2.htm

[5] https://en.wikipedia.org/wiki/Fourier_transform_infrared_spectroscopy

[6] http://www.icare.univ-lille1.fr/drupal/calipso

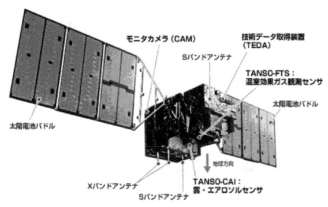

Fig. 1.   Mission instruments onboard GOSAT satellite

B. GOSAT Validation Site

There are TCCON validation sites in the world. One of these is Saga University site in Japan. The location is shown in Fig.2. Fig.3 shows the LiDAR site (Laser light and the container in which LiDAR is equipped).

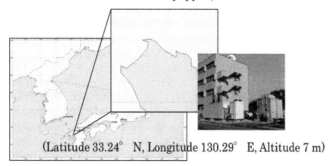

(Latitude 33.24° N, Longitude 130.29° E, Altitude 7 m)

Fig. 2.   Saga University TCCON site for GOSAT validation

Examples of the LiDAR data are shown in Fig.4 (a) together with PM2.5 data, CAI imagery data, the time series of PM2.5 data, and the sky view camera image. The right bottom graphs are the LiDAR data which is acquired at 14:00 Japan Standard time on May 29 2014. On the left, there is back scattering ratio data is situated while depolarization ratio is shown on the right. From these back scattering ratio and depolarization ratio, aerosol distribution and cloud vertical profile can be retrieved. Therefore, LiDAR data derived cloud vertical profile can be used for validation of CAI data derived clouds and Landsat-8 data derived clouds in particular, cirrus clouds. Meanwhile, Fig.4 (b) shows the LiDAR data which is acquired on April 26 2015. There is a peak of back scattered photon counts at around 10km of elevation (altitude above sea level). It is cirrus clouds. There is the ground based FTS for GOSAT validation which is situated just beside the LiDAR as shown in Fig.5. Other than these, there are sky view camera, sky radiometer which allows estimation of aerosol particle size distribution and refractive index retrievals for GOSAT validation. Examples of sky camera imagery data are shown in Fig.6 ((a) is for clear sky while (b) is for cloudy sky).

(a)Laser light

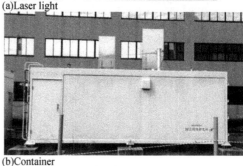

(b)Container

Fig. 3.   LiDAR at Saga University GOSAT validation site

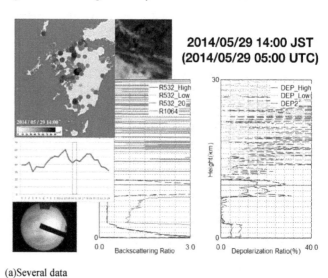

(a)Several data

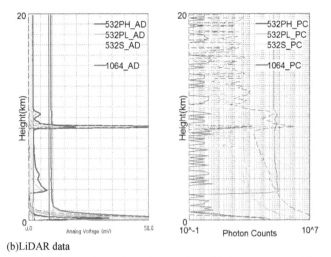

(b)LiDAR data

Fig. 4.   Examples of the LiDAR data together with the other measured data

Fig. 5.   Ground based FTS for GOSAT validation

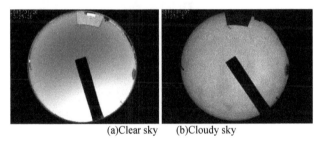

(a)Clear sky   (b)Cloudy sky

Fig. 6.   Examples of sky view camera data

C. Cloud Products Derived from CAI

There are two Level 2 cloud products, Cloud flag with 0 or 1 and Confidence level ranged from 0 to 1. Fig.7 shows lower and upper limits of clouds and clear sky. Using these definitions, four statistical tests are applied to the CAI data as shown in equation (1).

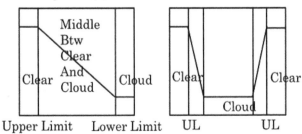

Fig. 7.   Definition of lower and upper limits

$$Q = 1 - \sqrt[n]{(1 - F_1)(1 - F_2)...(1 - F_n)} \qquad (1)$$

where F_i denotes statistical test results which are shown in Table 1.

TABLE I.   FOUR TESTS FOR CALCULATION OF CONFIDENCE LEVEL IN CLOUD DETECTION

Tests		Lower Limit	Upper Limit
Band2R		+0.195	+0.045
Band3R /Band2R	Smaller END	0.9	0.66
	LARGER END	1.1	1.7
NDVI	Smaller END	-0.1	-0.22
	LARGER END	0.22	0.46
Band3R /Band4R		1.06	0.86

D. GOSAT/CAI and Landsat-8 Imagery Data

Fig.8 shows examples of the acquired color images of GOSAT/CAI imagery data together with sky view camera images which area acquired at the Saga University validation site at the same time as satellite over pass time.

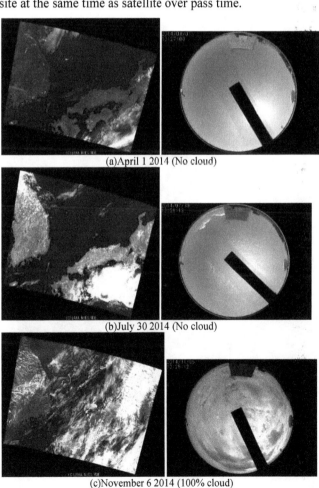

(a)April 1 2014 (No cloud)

(b)July 30 2014 (No cloud)

(c)November 6 2014 (100% cloud)

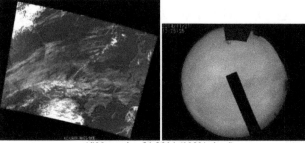

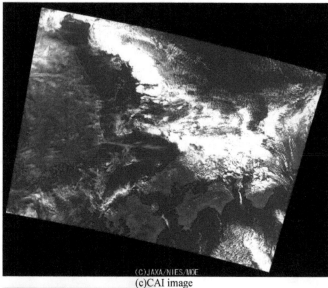

(d)November 21 2014 (100% cloud)

Fig. 8. Examples of GOSAT/CAI images and the sky view camera images at the Saga University validation site at the satellite over pass time

III. EXPERIMENTS

A. Match-Up Data Between CAI and LiDAR as well as Landsat-8 OLI

In order to evaluate cirrus detection capability of CAI and Landsa-8 OLI, match-up data have been searched for the term of the first half year of 2015. As the results from the search, the following two data are found,

January 20
April 26

CAI and Landsad-8 OLI images of the match-up data on January 20 2015 are shown in Fig.9 while those for April 26 2015 are shown in Fig.10.

(a)Cirrus detected image from Landsat-8 OLI

(c)CAI image

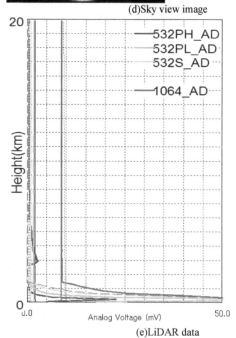

(d)Sky view image

(e)LiDAR data

Fig. 9. Match-up data on January 20 2015

(b)Landsat-8 OLI image

(a) shows cirrus detected image from Landsat-8 OLI imagery data. OLI band 9 is cirrus channel of which sensitive to 1.38 micrometer of wavelength. (b) shows color composite image of which red color is assigned to the near infrared channel, green color is assigned to band 9 of cirrus channel and blue color is assigned to mid-infrared wavelength channel, respectively. (c) shows natural color composite image of CAI.

(a)Cirrus detected image from Landsat-8 OLI

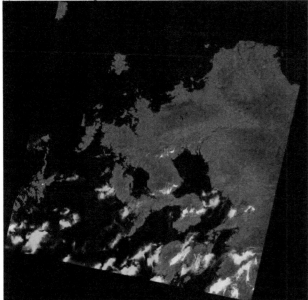

(b)Landsat-8 OLI image

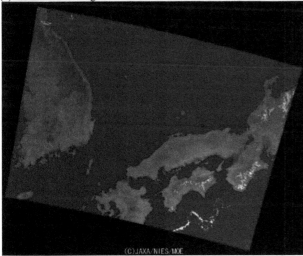

(c)CAI image

(d)Sky view camera image

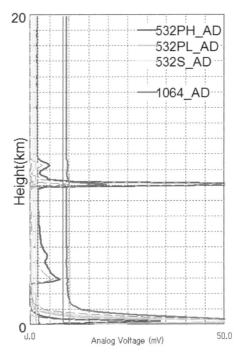

(e)LiDAR data

Fig. 10. Match-up data on April 26 2015

(d) shows sky view camera image while (e) shows LiDAR back scattered photon count data as the function of altitude. It is quite obvious that there is no cirrus cloud on January 20 2015 while there are cirrus clouds on April 26 2015. Green colored areas or pixels in (b) Landsat-8 OLI images indicate cirrus clouds.

Although Landsat-8 OLI image shows the cirrus clouds pixels in Fig.10 (b), Fig.10 (c) does not indicate any cirrus cloud at all. On the other hand, LiDAR data shows evidence of cirrus cloud existing as shown in Fig.10 (e). Therefore, it may say that Landsat-8 band 9 of cirrus channel does work to detect cirrus cloud while GOSAT/CAI does not work for detection of cirrus cloud due to missing cirrus channel.

B. Adjectment of Acquisition Time Difference Between Landsat-8 and GOSAT

Local mean times of the orbits of Landsat-8 and GOSAT are different each other for 30 minutes. Therefore, some adjustment of the time difference between both is required. By using forward trajectory analysis software tool provided by NOAA, original positions of cirrus cloud (30 minutes before the acquisition time) are estimated. The results from the forward trajectory analysis are shown in Fig.11. For January 20 2015, there is North-West wind while there is North-East wind for April 26 2015.

Therefore, the cirrus cloud locations are shifted for the distance which is shown in Fig.11 within 30 minutes in those directions. Thus the cirrus cloud detection accuracy can be done through comparisons between LiDAR data and Landsat-8 OLI data which is acquired at 30 minutes apart from the LiDAR acquisition.

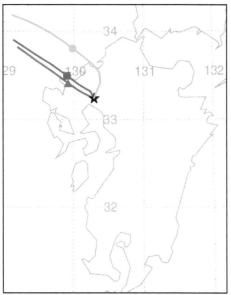

(a)January 20 2015

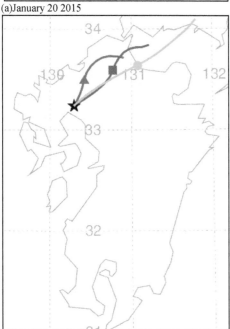

(b)April 26 2015

Fig. 11. Results from forward trajectory analysis

C. Summary of the Experimental Results

LiDAR data are acquired for 173 days within 518 days from April 1 2014 to August 31 2015 (Revisit cycle of the GOSAT satellite is 3 days). Within 173 days, LiDAR data are acquired 48 days (Acquisition ratio of LiDAR data to the total available days is just 33.01%). Cirrus clouds are observed for 11 days out of 48 days. Meanwhile, cirrus clouds are detected with CAI for just 8 days out of 11 days. On the other hand, cloud free situations are found with CAI for 18 days out of 37 days which is confirmed with LiDAR data. Due to the fact that the revisit cycle of Landsat-8 satellite is 16 days, just two match-up data between LiDAR and Landsat-8 OLI are collected for check cirrus detection accuracy. Two of match-up data show good coincidence between Landsat-8 OLI data utilized cirrus detection (no cirrus cloud and cirrus cloud existing situations).

D. Another Comparison Between Landsat-8 OLI data and GOSAT/CAI Imagery Data

Another match-up data between Landsat-8 OLI and GOSAT/CAI imagery data is found for April 7 2014 (Unfortunately LiDAR data is not acquired on that day). Fig.12 (a) shows GOSAT/CAI imagery data, (b) shows Landat-8 OLI imagery data on that day. Meanwhile, Fig.12 (c) shows the sky view camera image while (d) shows the results from the forward trajectory analysis for adjustment of the data acquisition time difference of 30 minutes between GOSAT/CAI and Landsat-8 OLI data.

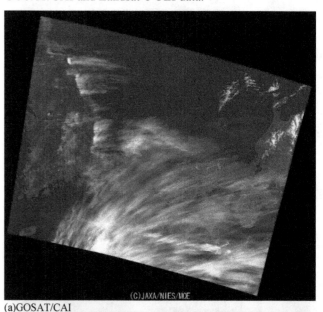

(a)GOSAT/CAI

(b)Landsat-8 OLI

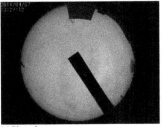

(c)Sky view camera

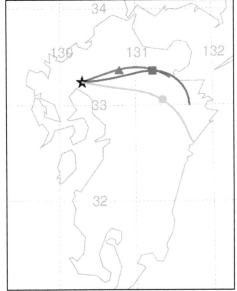

(d)Results from forward trajectory analysis

Fig. 12. Another match-up data between GOSAT/CAI and Landsat-8 OLI

Although GOSAT/CAI cloud product indicates there are cirrus clouds at the intensive study area of Saga University GOSAT Validation site, Landsat-8 OLI indicates there is no cirrus cloud at all. As shown in Fig.12 (c), there are thick clouds in the sky above the test site at the GOSAT satellite over pass time. These, however, are not cirrus clouds at all. Forward trajectory analysis result shows that there is West wind at that time.

E. Confirmation of Cirrus Cloud Detection Capability with Calipso Data

Cirrus clouds can be confirmed with Calipso data. As shown in Fig.13, vertical profile of the existing clouds are investigated with Calipso data.

Fig. 13. Calipso data derived vertical profile of cloud structure

The horizontal axis of the Fig.13 is sub-satellite track while vertical axis shows back scattered phone count from the cloud particles.

IV. CONCLUSION

Cirrus cloud detection accuracy of GOSAT/CAI and Landsat-8 is evaluated with a ground based Laser Radar: Lidar data and sky view camera data. Also, the evaluation results are confirmed with Calipso data together with a topographic representation of vertical profile of cloud structure. Furthermore, origin of cirrus clouds is estimated with forward trajectory analysis. The results show that GOSAT/CAI de4rived cirrus clouds is not accurately enough due to missing of cirrus cloud detection spectral channel while Landsat-8 derived cirrus cloud.

ACKNOWLEDGEMENTS

Authors would like to thank Dr. Shuji Kawakami of JAXA, Prof. Dr. Hirofumi Ohyama of Nagoya University, Dr. Isamu Morino and Dr. Osamu Uchino of NIES and their research staff for their efforts to conduct the experiments and valuable discussions.

REFERENCES

[1] Remote Sensing Society of Japan Edt. Kohei Arai, et al., Remote Sensing -An Introductory Textbook-, Maruzen Planet Publishing Co. Ltd., Chapter 9 Radiometric Correction and Cloud Detection, 155-172, p.301, ISBN978-4-86345-185-8, 2013.

[2] Kohei Arai, A Merged Dataset for Obtaining Cloud Free IR Data and a Cloud Cover Estimation within a Pixel for SST Retrieval, Asian-Pacific Remote Sensing Journal, Vol.4, No.2, pp.121-127, Jan.1992

[3] Kohei Arai, Yasunori Terayama, Yoko Ueda, Masao Moriyama, Cloud Coverage Estimation Within a Pixel by Means of Category decomposition, Journal of the Journal of Japan Society of Photogrammetry and Remote Sensing,, Vol.31, No.5, pp.4-10, Oct.1992.

[4] Kohei Arai, Tasuya Kawaguchi, Adjacency Effect Taking Into Account Layered Clouds Based on Monte Carlo Method, Journal of Remote Sensing Society of Japan, Vol.21, No.2, pp.179-185, (2001).

[5] Kohei Arai, Tatsuya Kawaguchi, Adjucency Effect of Layered Clouds taking Into Account Phase Function of Cloud Particles and Multi-Layered Plane Parallel Atmosphere Based on Monte Carlo Method, Journal of Japan Society of Photogrammetry and Remote Sensing, Vol.40, No.6, 2001.

[6] Kohei Arai, Adjacency effect of layered clouds estimated with Monte-Carlo simulation, Advances in Space Research, Vol.29, No.19, 1807-1812, 2002.

[7] K.Arai, Merged dataset with MOS-1/VTIR and NOAA/AVHRR enhancing cloud detectability, Proc. of the 28th COSPAR Congress, MA4-3.1, 1-8, 1990.

[8] K.Arai, Estimation of partial cloud coverage within a pixel, Proc. of the Pre-ISY International Symposium, 99-106, 1991.

[9] K.Arai, Sea surface temperature estimation taking into account partial cloud within a pixel, Proc. of the ISY conference on Earth and Space Information Systems, 10/13, 1992.

[10] K.Arai, Y.Ueda and Y.Terayama, Comparative study on estimation of partial cloud coverage within a pixel -Proposed adaptive least square method with constraints- Proc. of the European ISY Conference, 305/310, 1992.

[11] K.Arai, SST estimation of the pixels partially contaminated with cloud, Proc. of the Asian-Pacific ISY (International Space Year) Conference, 1992.

[12] Kohei Arai, Adjacency effect due to a box type of 3D clouds estimated with Monte Carlo simulation considering the phase function of cloud particles, Abstracts of the 33rd COSPAR Scientific Assembly, A1.2-0061, Warsaw, Poland, July 16-23, (2000).

[13] S.Sobue, K.Arai and N.Futamura, Development of a method of cloud detection in Japanese Earth observation satellites, Proceedings of the ISTS (International Space Science and Technology Symposium), N-6, 2002-n-21,2002.

[14] Shin-ichi Sobua and Kohei Arai, Development of method for cloud detection in ASTER image data, Proc. of the 24th International Symposium on Space Technology and Science (ISTS), n-01, (2004)

[15] T. Sakai, O. Uchino, I. Morino, T. Nagai, S. Kawakami, H. Ohyama, A. Uchiyama, A. Yamazaki, K.Arai, H. Okumura, Y. Takubo, T. Kawasaki, T. Akaho, T. Shibata, T. Nagahama, Y. Yoshida, N. Kikuchi, B. Liley, V. Sharlock, J. Robinson, T. Yokota, Impact of aerosol and cirrus clouds on the GOSAT observed CO2 and CH4 inferred from ground based lidar, skyradiometer and FTS data at prioritized observation sites,(2013), Proceedings of the 9th International Workshop on Greenhouse Gas measurements from Space, IWGGMS-9, 2013

[16] H.Okumura, Kohei Arai, Improvement of PM2.5 density distribution visualization system using ground-based sensor network and Mie Lidar, Proceedings of the Conference on Remote Sensing of Clouds and the Atmosphere, SPIE Remote Sensing, ERS 15-RS 104-50, 2015

[17] Osamu Uchino, Tetsu Sakai, Tomohiro Nagai, Masahisa Nakasato, Isamu Morino, Tatsuya Yokota, Tsuneo Matsunaga, Nobuo Sugimoto, Kohei Arai, Hiroshi Okumura, Development of transportable Lidar for validation of GOSAT satellite data products, Journal of Remote Sensing Society of Japan, 31, 4, 435-443, 2011

[18] O.Uchino, T.Sakai, T.Nagai, I.Morino, K.Arai, H.Okumura, S.Takubo, T.Kawasaki, Y.mano, T.Matsunaga, T.Yokota, On recent stratspheric aerosols observed by Lidar over Japan, Journal of Atmospheric Chemistry and Physics, 12, 11975-11984, 2012(doi:10.5194/acp-12, 11975-2012).

[19] Testu Sakai, Osamu Uchino, Isamu Morino, Tomohiro Nagai, Taiga Akaho, Kawasaki Tsuyoshi, Tetsu Sakai, Hiroshi Okumura, Kohei Arai, Akihiro Uchiyama, Akehiro Yamazaki, Tsuneo Matsunaga, Tatsuya Yokota, Vertical profile of volcanic prumes form Sakurajima volcano detected by Lidar and skyradiometer situated Saga and its optical property, Journal of Remote Sensing Society of Japan, 34, 3, 197-204, 2014

[20] 314. H. Okumura, S.Takubo, T.Kawsaki, I.N.Abdulah, T.Sakai, T.Maki, K.Arai, Web based data acquisition and management system for GOSAT validation Lidar data analysis, Proceedings of the SPIE Vol.8537, Conference 8537: Image and Signal Processing for remote Sensing , Paper #8537-43, system, 2012.

[21] 315. T.Sakai, H. Okumura, T.Kawsaki, I.N.Abdulah, O.Uchino, I.Morino, T.Yokota, T.Nagai, T.Sakai, T.Maki, K.Arai, Observation of aerosol parameters at Saga using GOSAT product validation Lidar, Proceedings of the SPIE Vol.8526, Conference 8526: Lidar Remote Sensing for Environmental Monitoring XIII, SPIE Asia-Pacific Remote Sensing, Paper #8295A-50,IP1, 2012.

[22] 332. Hiroshi Okumura, Shoichiro Takubo, Takeru Kawasaki, Indra Nugraha Abdulah, Osamu Uchino, Isamu Morino, Tatsuya Yokota, Tomohiro Nagai, Tetu Sakai, Takashi Maki, Kohei Arai, Improvement of web-based data acquisition and management system for GOSAT validation Lidar data analysis(2013), SPIE Electronic Imaging Conference, 2013.

[23] 335. Hiroshi Okumura, Kohei Arai, Observation of aerosol properties at Saga using GOSAT product validation LiDAR, Proceedings of the Conference on Image and Signal Processing fo Remote Sensing, SPIE #ERS13-RS107-38, 2013

[24] 337. T. Sakai, O. Uchino, I. Morino, T. Nagai, S. Kawakami, H. Ohyama, A. Uchiyama, A. Yamazaki, K.Arai, H. Okumura, Y. Takubo, T. Kawasaki, T. Akaho, T. Shibata, T. Nagahama, Y. Yoshida, N. Kikuchi, B. Liley, V. Sharlock, J. Robinson, T. Yokota, Impact of aerosol and cirrus clouds on the GOSAT observed CO2 and CH4 inferred from ground based lidar, skyradiometer and FTS data at prioritized observation sites,(2013), Proceedings of the 9th International Workshop on Greenhouse Gas measurements from Space, IWGGMS-9, 2013

[25] 338. Shuji Kawakami, Hirofumi Ohyama, Kei Shiomi, T.Fukamachi, Kohei Arai, C.Taura, H.Okumura, Observations of carbon dioxide and methane column amounts measured by high resolution of FTIR at Saga in 2011-2012, Proceedings of the International Symposium on Remote Sensing, ISRS-TCCOC 2013.(2013)

[26] 339. Osamu Uchino, T.Sakai, T.nagai, I.Morino, H.Ohyama, S.Kawakami, K.Shiomi, T Kawasaki, T.Akaho, H.Okumura, Kohei Arai, T.matsunaga, T.Yokota, Comparison of lower tropospheric ozone column observed by DIAL and GOSAT TANSO-FTS TIR, Proceedings of the AGU Fall Meeting 2013.(2013)

[27] 340. I Morino, T Sakai, T.Nagai, A.Uchiyama, A.Yamazaki, S Kawakami, H.Ohyama, Kohei Arai, H.Okumura, T.Shibata, T.Nagahama, N.Kikuchi, Y.Yoshida, Ben Liley, Vannessa Sherlock, John Robinson, O. Uchino, T.Yokota, Impact of aerosols and cirrus on the GOSAT onboard CO2 and CH4 inferred from ground based Lidar, skyradiometer and FTS data at prioritized observation sites, Proceedings of the AGU Fall Meeting 2013.(2013)

[28] 343. H.Okumura, K.Arai, Development of PM2.5 density distribution visualization system using gournd-level sensor network and Mie-Lidar, Proceedings of the SPIE European Remote Sensing Coference, ERS 14-RS107-97, 2014.

[29] 346. H.Okumura, Kohei Arai, Improvement of PM2.5 density distribution visualization system using ground-based sensor network and Mie Lidar, Proceedings of the Conference on Remote Sensing of Clouds and the Atmosphere, SPIE Remote Sensing, ERS 15-RS 104-50, 2015

8

Fuzzy Soft Sets Supporting Multi-Criteria Decision Processes

Sylvia Encheva

Stord/Haugesund University College

Bjørnsonsg. 45,

5528 Haugesund,

Norway

Abstract—Students experience various types of difficulties when it comes to examinations, where some of them are subject related while others are more of a psychological character. A number of factors influencing academic success or failure of undergraduate students are identified in various research studies. One of the many important questions related to that is how to select individuals endangered to be unable to complete a particular study program or a subject. The intention of this work is to develop an approach for early discovery of students who could face serious difficulties through their studies.

Keywords—Soft sets; Uncertainties; Decision making

I. INTRODUCTION

Exam failure is a serious problem for both students and the respective educational institutions where these students are enrolled inn. One of the important questions arising in such cases is related to early identification of students who are potentially in danger of exam failure.

Students experience various types of difficulties when it comes to examinations, where some of them are subject related while others are more of a psychological character. The former are usually more specific while the latter are more general. Some examples of the latter include anxiety, low level of concentration, increased stress level and sleep disorders, [14], [15]. A large number of factors influencing academic success or failure of university students is listed in [5]. Our intention is to identify students who might be in danger of not being able to complete a particular course at a very early stage of their enrollment and consequently provide them with individual recommendations. Since such processes are usually described in uncertain and unprecised ways handling them with methods from fuzzy soft set theory is proposed in this work. In the soft set theory [12], the initial description of the object has an approximate nature, [13]. Very useful group decision making methods based on intuitionistic fuzzy soft matrices are presented in [10]. In this paper one of their approaches is expended in a way that allows obtaining a set of interesting items that differ from the one with the highest score.

The rest of this work goes as follows. Definitions and statements are placed in Section II. The main results are presented in Section III, and a conclusion can be found in Section IV.

II. SOFT SETS

Let U be an initial universe set and E_U be the set of all possible parameters under consideration with respect to U. The power set of U (i.e., the set of all subsets of U) is denoted by $P(U)$ and $A \subseteq E$, [1]. A soft set is defined in the following way:

Definition 1: [12] A pair (F, A) is called a soft set over U, where F is a mapping given by

$$F : A \to P(U).$$

Definition 2: [3] Let U be an initial universe, $P(U)$ be the power set of U, E be the set of all parameters and X be a fuzzy set over E. An FP-soft set F_X on the universe U is defined by the set of ordered pairs

$$F_X = (\mu_X(x)/x, f_X(x)) :$$
$$x \in E, f_X(x) \in P(U), \mu_X(x) \in [0,1]\},$$

where the function $f_X : E \to P(U)$ is called approximate function such that $f_X(x) = \emptyset$ if $\mu_X(x) = 0$, and the function $\mu_X : E \to [0,1]$ is called membership function of FP-soft set F_X. The value of $\mu_X(x)$ is the degree of importance of the parameter x, and depends on the decision makers requirements.

Definition 3: [3] Let $F_X \in FPS(U)$, where $FPS(U)$ stands for the sets of all FP-soft sets over U. Then a fuzzy decision set of F_X, denoted by F_X^d, is defined by

$$F_X^d = \mu_{F_X^d}(u)/u : u \in U$$

which is a fuzzy set over U, its membership function $\mu_{F_X^d}$ is defined by $\mu_{F_X^d} : U \to [0,1]$,

$$\mu_{F_X^d}(u) = \frac{1}{|supp(X)|} \sum_{x \in supp(X)} \mu_X(x)\chi_{f_X(x)}(u)$$

where $supp(X)$ is the support set of X, $f_X(x)$ is the crisp subset determined by the parameter x and

$$\chi_{f_X(x)}(u) = \begin{cases} 1, & u \in f_X(x), \\ 0, & u \notin f_X(x). \end{cases}$$

Definition 4: [9] The union of two soft sets (F, A) and (G, B) over a common universe U is the soft set (H, C), where $C = A \cup B$, and $\forall\ e \in C$,

$$H(e) = \left\{ \begin{array}{ll} F(e), & if\ \ e \in A - B, \\ G(e), & if\ \ e \in B - A, \\ F(e) \cup G(e), & if\ \ e \in A \cap B. \end{array} \right.$$

It is denoted as $(F, A) \tilde{\cup} (G, B) = (H, C)$.

Definition 5: [9] The intersection of two soft sets (F, A) and (G, B) over a common universe U is the soft set (H, C), where $C = A \cap B$, and $\forall\ e \in C$, $H(e) = F(e)$ or $G(e)$ (as both are same set). It is denoted as $(F, A) \tilde{\cap} (G, B) = (H, C)$.

Definition 6: [9] Let (F, A) and (G, B) be soft sets over a common universe set U. Then

(a) $(F, A) \wedge (G, B)$ is a soft set defined by

$$(F, A) \wedge (G, B) = (H, A \times B),$$

where $H(\alpha, \beta) = F(\alpha) \cap G(\beta)$, $\forall (\alpha, \beta) \in A \times B$, and $\cap$ is the intersection operation of sets.

(b) $(F, A) \vee (G, B)$ is a soft set defined by

$$(F, A) \vee (G, B) = (K, A \times B),$$

where $K(\alpha, \beta) = F(\alpha) \cup G(\beta)$, $\forall (\alpha, \beta) \in A \times B$, and $\cup$ is the union operation of sets.

Soft set relations and functions are well presented in [2]. An intuitionsitic fuzzy soft sets based decision making is discussed in [8].

III. ATTRIBUTE SELECTION

Suppose three advisors are forming a committee that has to select attributes indicating potential exam failure. Advisors' opinions are to be taken with weights 0.5, 0.3 and 0.2 respectively, as in [10]. Weight distributions can also be determined by a decision making body that is in charge of that project. It is worth mentioning that in case the three advisors are assumed to have different influence, then there are not that many weight combinations that can actually effect attribute choice. Thus, if the lowest weight is 0.1 then the highest has to be at least 0.5 and if the lowest weight is 0.2 then the highest can be 0.4 (this implies two advisers with equal weight 0.4), 0.5 or 0.6 (this implies two advisers with equal weight 0.2).

In our case the set of attributes to be considered contains the following elements

A 1 - health related issues,

A 2 - last education relevant to this study has been

obtained at least five years ago,

A 3 - time consuming obligations outside of the study,

A 4 - preliminary test results,

A 5 - amount of time a student can devote to study that

subject weekly,

A 6 - student absences from classes, tutorials, etc.,

A 7 - insufficient preliminary knowledge,

TABLE I: Attributes significance

	O 1	O 2	O 3
A 1	(0.83, 0.1)	(0.6, 0.2)	(0.6, 0.1)
A 2	(0.3, 0.51)	(0.58, 0.4)	(0.8, 0.1)
A 3	*(0.6, 0.18)*	*(0.71, 0.24)*	(0.31, 0.5)
A 4	(0.9, 0.05)	(0.8, 0.13)	(0.4, 0.52)
A 5	(0.55, 0.3)	(0.9, 0.01)	(0.5, 0.36)
A 6	(0.47, 0.21)	(0.66, 0.3)	(0.7, 0.22)
A 7	(0.8, 0.08)	(0.8, 0.15)	(0.83, 0.1)
A 8	(0.58, 0.12)	(0.3, 0.64)	*(0.69, 0.3)*
λ_{med} (E)	(0.6, 0.18)	(0.71, 0.24)	(0.69, 0.3)

TABLE II: Values for all attributes

	O 1	O 2	O 3	Attributes values
A 1	1	0	0	1
A 2	0	0	1	1
A 3	1	1	0	2
A 4	1	1	0	2
A 5	0	1	0	1
A 6	0	0	1	1
A 7	1	1	1	3
A 8	0	0	1	1

A 8 - opportunities to work together with other students.

Each attribute is rated applying values from the set $0.1, ..., 1.0$, where 1.0 is the most important. Notations in Table I are as follows: O 1, O 2, O 3 represent opinions of first, second and third advisor with respect to attributes A 1, A 2, ..., A 8. A number in the first position of each couple describes a degree to which that attribute is important and the second number describes a degree to which that attribute is not important. A threshold vector λ_{med} (E) is based on median. Values for threshold vectors in Table I are emphasized.

The paper continues following mainly the work presented in [3]. Due to the specific nature of this investigation it is needed to tune a bit their approach. Thus instead of applying predefined degrees of attributes' importance values from Table II are taken. In [4], [8], and [10] they were referred to as choice values. For our study this seems more reasonable. Otherwise it will be necessary to ask every student to supply such degrees of importance. Most students would find such requests difficult and very few would be able to provide meaningful responses.

Next important difference is that the goal here is to identify all students in danger to fail their exam while following [3] one would find one student who seems to have more problems than the rest of his classmates. To achieve this the fuzzy decision set F_X^d is calculated and all students within the last quartile are selected.

To avoid confusions the terms 'classical set' and 'fuzzy soft set' are used in every single case without assuming that one or the other is understood by convention.

The classical set of proposed attributes is $\{A1, A2, ..., A8\}$ as described above. Students who answered a Web based inquiry are denoted by $\{St1, ..., St20\}$. Their responses are assumed to be binary in this case, see Table III. Nonbinary scale can be used if a finer grading is found to be more beneficial. Once again, the idea is to keep it simple. Students should not be overload with too many questions and too many

TABLE III: Responses from students

	A 1	A 2	A 3	A 4	A 5	A 6	A 7	A 8	F_{St}^d
St 1	×	×		×	×			×	0.75
St 2	×		×		×	×		×	0.75
St 3		×	×		×	×	×		1
St 4	×			×		×	×	×	1.25
St 5		×	×		×			×	0.625
St 6		×	×	×		×	×		1.25
St 7	×	×		×	×	×		×	1
St 8	×		×	×	×		×		1.25
St 9		×	×		×			×	0.625
St 10	×	×	×		×	×			0.875
St 11		×			×	×	×	×	0.875
St 12	×		×	×		×			0.75
St 13	×	×	×		×			×	0.75
St 14		×	×	×	×	×			0.875
St 15	×	×					×	×	0.75
St 16	×	×		×	×		×		1
St 17	×			×		×			0.5
St 18		×	×		×	×		×	0.875

options to choose from.

Members of the fuzzy decision set F_{St}^d are shown in Table III. After working on 4-quantiles of F_{St}^d we believe that students belonging to the forth quartile should be the ones to begin with. In other words, students $St3, St4, St6, St7, St8, St16$ should receive personal advises on what ought to be done in order to avoid exam failure.

Initially experience from previous courses is used. The intention is to tune the system after some time when new data has been collected. When it comes to handling situations requiring aggregation methods the approach in [10] is suggested. In case different datasets are to be used for drawing conclusions, applying statements presented in [9] seems to be quite appropriate.

A. Discussion

Another method that can be used involves formal concept analysis, [6], [7]. This is a method supporting data analysis among many other things. Once the sets of attributes and objects, and their relations are well presented in an information table, a corresponding concept lattice can be depicted. Each node in that lattice contains all the students that share the same attributes. A fuzzy function indicating degrees to which each node contents reflect danger in exam failure has to be build.

It seems that fuzzy soft sets are well equipped to handle problems presented in this work because every student is treated individually. The outcomes of formal concept analysis studies are beneficial for group of students and as a result some details concerning individuals might be omitted. Formal concept analysis based methods can be very helpful while dealing with new students and/or new advisors.

IV. CONCLUSION

Exam failure is most of the time a result of internal and external factors. Among the internal ones are lack of commitment and motivation, fear of exams, personal or financial problems, etc. The external ones are related to overloaded study programmes, supervision quality, inadequate requirements and so on. To determine which factors are of the highest importance one should study particular educational institutions and students groups. More research has to be done in order to determine the correct value of a proper q-quantile, as well.

REFERENCES

[1] M. I. Ali, F. Feng, X. Liu, W. K. Min, and M. Shabir, *On some new operations in soft set theory*, Computers and Mathematics with Applications, vol. 57, pp. 1547–1553, 2009.

[2] K. V. Babitha and J.J. Sunil, *Soft set relations and functions*, Computers and Mathematics with Applications, vol. 60, pp. 1840–1849, 2010.

[3] N. Çağman N., F. Çitak, and S. Enginoglu, *FP-soft set theory and its applications*, Annals of Fuzzy Mathematics and Informatics, vol. 2(2), pp. 219–226, 2011.

[4] F. Feng F. and Y. B. Jun, *An adjustable approach to fuzzy soft set based decision making*, Journal of Computer Applied Mathematics, vol. 234, pp. 10–20, 2010.

[5] W. J. Fraser and R. Killen, *Factors influencing academic success or failure of first-year and senior university students: do education students and lecturers perceive things differently?*, South African Journal of Education, vol. 23(4), pp. 254 – 260, 2003.

[6] B. Ganter and R. Wille, *Formal Concept Analysis*, Springer, 1999.

[7] B. Ganter, G. Stumme, and R. Wille, *Formal Concept Analysis: Foundations and Applications*, Lecture Notes in Artificial Intelligence, vol. 3626, Springer-Verlag, 2005.

[8] Y. Jiang, Y. Tang, and Q. Chen, *An adjustable approach to intuitionsitic fuzzy soft sets based decision making*, Applied Mathematical Modeling, vol. 35, pp. 824–836, 2011.

[9] P. K. Maji, R. Biswas, and A. R. Roy, *Soft set theory*, Computers and Mathematics with Applications, vol. 45, no. 4–5, pp. 555–562, 2003.

[10] J. Mao J., D. Yao D., and C. Wang, *Group decision making methods based on intuitionistic fuzzy soft matrices*, Applied Mathematical Modelling, vol. 37, pp. 6425–6436, 2013.

[11] N. Moha, J. Rezgui, Y-G. Gueheneuc, P. Valtchev, G. Boussaidi, *Using FCA to Suggest Refactorings to Correct Design Defects*, Concept Lattices and Their Applications, Lecture Notes in Computer Science, vol. 4923, pp. 269–275, 2008.

[12] D. Molodtsov, *Soft set theoryfirst results*, Computers and Mathematics with Applications, vol. 37, no. 4–5, pp. 19–31, 1999.

[13] M. M. Mushrif, S. Sengupta, and A. K. Ray, *Texture Classification Using a Novel, Soft-Set Theory Based Classification Algorithm*, ACCV 2006, Lecture Notes in Computer Science, vol. 3851, Springer-Verlag Berlin Heidelberg, pp. 246–254, 2006.

[14] S. S. Sazhin, *Teaching Mathematic to Engineering Students*, International Journal Engineering Education, vol. 14 (2), pp. 145–152, 1997.

[15] P. Vitasan, T. Herawanb, M. N. A. Wahabc, A. Othmana, and S. K. Sinnaduraic, *Exploring mathematics anxiety among engineering students*, Procedia - Social and Behavioral Sciences, vol. 8, pp. 482–489, 2010.

9

An Empirical Comparison of Tree-Based Learning Algorithms: An Egyptian Rice Diseases Classification Case Study

Mohammed E. El-Telbany
Computers and Systems Department
Electronics Research Institute Cairo,
Egypt

Mahmoud Warda
Computers Department
National Research Center
Cairo, Egypt

Abstract—Applications of learning algorithms in knowledge discovery are promising and relevant area of research. The classification algorithms of data mining have been successfully applied in the recent years to predict Egyptian rice diseases. Various classification algorithms can be applied on such data to devise methods that can predict the occurrence of diseases. However, the accuracy of such techniques differ according to the learning and classification rule used. Identifying the best classification algorithm among all available is a challenging task. In this study, a comprehensive comparative analysis of a tree-based different classification algorithms and their performance has been evaluated by using Egyptian rice diseases data set. The experimental results demonstrate that the performance of each classifier and the results indicate that the decision tree gave the best results.

Keywords—*Data Mining, Classification, Decision Trees, Bayesian Network, Random Forest, Rice Diseases.*

I. INTRODUCTION

Processing the huge data and retrieving meaningful information from it is a difficult task. Data mining is a wonderful tool for handling this task. The major components of the architecture for a typical data mining system are shown in Fig 1. The term Data Mining, also known as Knowledge Discovery in Databases (KDD) refers to the non trivial extraction of implicit, previously unknown and potentially useful information from data in databases [1]. They are several different data mining techniques such as *clustering, association, anomaly detection* and *classification* [2]. The classification process has been identified as an important problem in the emerging field of data mining as they try to find meaningful ways to interpret data sets. The goal of classification is to correctly predict the value of a designated discrete class variable, given a vector of predictors or attributes by produces a mapping from the input space to the space of target attributes [3]. There are various classification techniques each technique has its pros and cons. Recently, Fernandez-Delgado *et al.* [4] evaluate 179 classifiers arising from 17 families (e.g. statistics, symbolic artificial intelligence and data mining, connectionist approaches, and others are ensembles). The classifiers show strong variations in their results among data sets, the average accuracy might be of limited significance if a reduced collection of data sets is used [4]. For example, the largest merit of neural networks (NN) methods is that they are general: they can deal

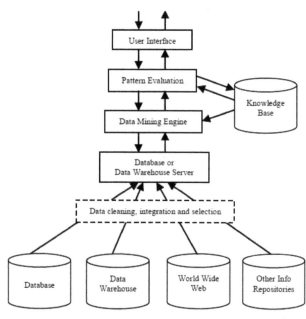

Fig. 1: Architecture of a Typical Data Mining System [1]

with problems with high dimensions and even with complex distributions of objects in the n-dimensional parameter space. However, the relative importance of potential input variables, long training process, and interpretative difficulties have often been criticized. Although the *support vector machine* (SVM) has a high performance in classification problems [5], the rules obtained by SVM algorithm are hard to understand directly and costly in computation. Due to the above-mentioned drawbacks of NN and SVM, the purpose of this paper, is to explore the performance of classification using various decision tree approaches which have the following advantages as follows [6]:

1) Decision trees are easy to interpret and understand;
2) Decision trees can be converted to a set of $if-then$ rules; and
3) Decision trees don't need priori assumptions about the nature of data, it is a *distribution-free*.

Since decision trees have the described advantages, they have proven to be effective tools in classification of Egyptian rice disease problems [7]. Specially, the transfer of experts from consultants and scientists to agriculturists, extends workers and farmers represent a bottleneck for the development of agriculture on the national. This information can be used as part of the farmers decision-making process to help to improve crop production. The aim of this paper is to evaluate the tree-based classifiers to select the classifier which more probably achieves the best performance for the Egyptian rice diseases which cause losses that estimated by 15% from the yield, malformation of the leaves or dwarfing of the plants. Discovering and controlling of diseases are the main aims and have a large effect for increasing density of Fadden and increasing gain for farmer then increasing the national income. Actually, the original contribution of this research paper is to measure and compare the performances of tree-based classification algorithms for Egyptian rice diseases. In particular, we have focused on the Bayesian network, random forest algorithms, comparing its performances with a decision tree using a variety of performance metrics. In this paper, four classification algorithms are investigated and presented for their performance. Section II, presents the related previous work. The proposed used classification algorithms are explained in section III. In section IV, our problem is formally described. Section V, describes data set used in this paper. In section VI an experimental results described for investigated types of classification algorithms including their performance measures. Finally, the conclusions are explained in section VII.

II. RELATED WORK

The objectives of applying data mining techniques in agriculture is to increase of productivity and food quality at reduced losses by accurate diagnosis and timely solution of the field problem. Using data mining classification algorithms, it become possible to discover the classification rules for diseases in rice crop [7], [8]. The image processing and pattern recognition techniques are used in developing an automated system for classifying diseases of infected rice plants [9]. They extracted features from the infected regions of the rice plant images by using a system that classifies different types of rice disease using self-organizing map (SOM) neural network. Feature selection stage was done using rough set theory to reduce the complexity of classifier and to minimize the loss of information where a rule base classifier has been generated to classify the different disease and provide superior result compare to traditional classifiers [9]. Also, SVM is used to disease identification in the rice crop from extracted features based on shape and texture, where a three disease leaf blight, sheath blight and rice blast are classified [10]. In another work, the brown spot in rice crop is identified using K-Means method for segmentation and NN for classification of disease [11]. The NN is used to identify the three rice diseases namely (i) Bacterial leaf blight, (ii) Brown spot, and (iii) Rice blast [12]. The fuzzy entropy and probabilistic neural network are used to identify and classifying the rice plant diseases. Developed a mobile application based on android operating system and features of the diseases were extracted using fuzzy entropy [13].

III. CLASSIFICATION ALGORITHMS

A total of four classification algorithms have been used in this comparative study. The classifiers in Weka have been categorized into different groups such as Bayes and Tree based classifiers, etc. A good mix of algorithms have been chosen from these groups that include decision tree, Naive Bayes net, random trees and random forest. The following sections briefly explain about each of these algorithms.

A. Decision Tree

The first algorithm used for comparison is a decision tree, which generates a classification-decision tree for the given data-set by recursive partitioning of data [14]. The algorithm considers all the possible tests that can split the data set and selects a test that gives the best information gain. For each discrete attribute, one test with outcomes as many as the number of distinct values of the attribute is considered. For each continuous attribute, binary tests involving every distinct values of the attribute are considered. In order to gather the entropy gain of all these binary tests efficiently, the training data set belonging to the node in consideration is sorted for the values of the continuous attribute and the entropy gains of the binary cut based on each distinct values are calculated in one scan of the sorted data. This process is repeated for each continuous attributes [15], [16]. In particular entropy, for an attribute is defined as in equation 1.

$$H(X) = -\sum_{j}^{m} p_j log_2(p_j) \tag{1}$$

Where p_j is defined as $P(X = V_j)$, the probability that X takes value V_j, and m is the number of different values that X admits. Due to their recursive partitioning strategy, decision trees tend to construct a complex structure of many internal nodes. This will often lead to over fitting. Therefore, the decision tree algorithm exhibits meta-parameters that allow the user to influence when to stop tree growing or how to prune a fully-grown tree.

B. Random Decision Tree

The second chosen algorithm for the comparison is the *random decision tree* presented by Fan *et al.* in [17]. The Random decision tree is an ensemble learning algorithm that generates many individual learners. It employs a bagging idea to produce a random set of data for constructing a decision tree. In the standard tree each node is split using the best split among all variables. The choice is bind on the type of the attribute, in particular if the feature can assume values in a finite set of options it cannot be chosen again in the sub tree rooted on it. However, if the feature is a continuous one, then a random threshold is chosen to split the decision and it can be chosen again several times in the same sub tree accordingly with the ancestor's decision. To enhance the accuracy of the method, since the random choice may leads to different results, multiple trees are trained in order to approximate the true mean. Considering k as the number of features of the dataset and N as the number of trees, then the confidence probability to have is:

$$1 - (1 - \frac{1}{k})^N \tag{2}$$

Considering the k features of the dataset, and the i classifying attributes, the most diversity among trees is with depth of

$$\frac{k}{2} \tag{3}$$

since the maximum value of the combination is

$$\binom{i}{k} \tag{4}$$

Once the structure is ready the training may take place, in particular each tuple of the dataset train all the trees generated in order to read only one time the data. Each node counts how many numbers of examples go through it. At the end of the training the leaves contain the probability distribution of each class, in particular for the tree i, considering $n[y]$ the number of instances of class y at the node reached by x, is:

$$P_i(y|x) = \frac{n[y]}{\sum_y n[y]} \tag{5}$$

The classification phase retrieves the probability distribution from each tree and average on the number of trees generated in the model:

$$P(y|x) = \frac{1}{N} \sum_{i=1}^{N} P_i(y|x) \tag{6}$$

C. Bayesian Network

Bayesian Networks encode conditional interdependence relationships through the position and direction of edges in a directed acyclic graph. The relationship between a node and its parent is quantified during network training. This classifier learns from training data the *conditional probability* of each attribute X_i given the class label C. Classification is then done by applying Bayes rule to compute the probability of C given the particular instances of $X_1 \ldots X_n$ and then predicting the class with the highest posterior probability. The goal of classification is to correctly predict the value of a designated discrete class variable given a vector of predictors or attributes [2]. In particular, the naive Bayes classifier is a Bayesian network where the class has no parents and each attribute has the class as its sole parent. Bayesian and neural network seem to be identical in their inner working. Their difference exist in the construction. Nodes in a neural network don't usually have clearly defined relationship and hidden node are more "discovered" than determined, whereas the relationships between nodes in Bayesian network are due to their conditional dependencies [18], [2].

D. Random Forest

The random forest classifier, described by Ho [19], [1], works by creating a bunch of decision trees randomly. Each single tree is created in a randomly selected subspace of the feature space. Trees in different subspaces complement each other's classifications. Actually, random forest is an ensemble classifier which consists of many decision tree and gives class as outputs i.e., the mode of the class's output by individual trees. Prediction is made by aggregating (majority vote for classification or averaging for regression) the predictions of the ensemble. Random Forests gives many classification trees without pruning [20]. The success of an ensemble strategy depends on two factors, the strength (accuracy) of individual base models and the diversity among them.

TABLE I: Possible value for each attribute from the Egyptian rice database

Attribute	Possible Values
Variety	gizal71,gizal77, gizal78
	sakhal0l, sakhal02, sakhal03 , sakhal04
Age	Real values
Part	leaves, leaves spot, nodes, panicles,
	grains, plant, flag leaves, leaf sheath, stem
Appearance	Spots, elongate, spindle, empty, circular, oval,
	fungal, spore balls, twisted, wrinkled, dray,
	short, few branches, barren, small, deformed,
	seam, few stems, stunted, stones, rot, empty seeding
Colour	gray, olive, brown, brownish, whitish, yellow,
	green, orange, greenish black, white, pale, blackish, blac k
Temperature	Real values
Disease	Blight, brown spot, false smut,
	white tipe, stem rot

IV. PROBLEM DEFINITION

The main aim of this work is to produce a comparison among different inductive learning the *optimal model* for a target function $t = F(x)$, given a training set of size n, $(x_1; t_1), \ldots, (x_n; t_n)$, an inductive learner produces a model $y = f(x)$ to approximate the true function $F(x)$. Usually, there exists x such that $y \neq t$. In order to compare performance, a loss function is introduced $L(t, y)$. Given the loss function $L(t, y)$, that measures the discrepancy between our function's class and reality, where t is the true class and y is the predicted class, an optimal model is one that minimizes the average loss $L(t, y)$ for all examples. The optimal decision y^* for x is the class that minimizes the expected loss $E_t(L(t, y^*))$ for a given example x when x is sampled repeatedly.

V. DATA SET DESCRIPTION

Rice is the worlds most common staple food for more than half of mankind. Because of its important, rice is considered a strategic resource in Egypt has been assigned as a high priority topic in its Agricultural Strategic Plans. Successful Egyptian rice production requires for growing a summer season (May to August) of 120 to 150 days according to the type of varieties as Gizal77 needs 125 day and Sakhal04 needs 135 day. Climate for the Egyptian rice is that daily temperature maximum = $30 - 35^o$, and minimum = $18 - 22^o$; humidity = 55%-65%; wind speed = $1 - 2m$. Egypt increase productivity through a well-organized rice research program, which was established in the early eighties. In the last decade, intensive efforts have been devoted to improve rice production. Consequently, the national average yields of rice increased by 65% i. e., from $(2.4t/fed.)$ during the lowest period $1984 - 1986$ to $(3.95t/fed.)$ in 2002 [21]. Many affecting diseases infect the Egyptian rice crop; some diseases are considered more important than others. In this study, we focus into the most important diseases, which are five; blight, brown spot, false smut, white tip nematode and stem rot sequence. Each case in the data set is described by seven attributes. We have a total of 206 samples and the attribute and possible values are listed in Table I.

VI. EXPERIMENTAL EVALUATION

To gauge and investigate the performance on the selected classification methods or tree-based learning algorithms, many experiments are implemented within the WEKA framework

[22]. The Weka is an open source data mining workbench software which is used for simulation of practical measurements. It contains tools for data preprocessing, classification, regression, clustering, association rule and visualization. It does not only supports data mining algorithms, but also data preparation and meta-learners like bagging and boosting [22]. In order to test the efficiency of tree-based classification algorithms, training and test sets are used. Usually disjoint, subsets, the training set to build a classification tree(s) and the test set so as to check and validate the trained model. Also, cross-validation process applied where same sized disjoint sets are created so as to train the model fold wise. n-fold cross-validation, (usually $n = 10$) is used to divide the data into equally sized k subsets/ folds. In such case the model is trained using $(k - 1)$ folds and the kth fold is used as a test set. The whole process is repeated n times in an attempt to use all the folds for testing thus allowing the whole of the data to be used for both training and testing. In our data, ten cross-validation bootstraps, each with 138 (66%) training cases and 68(34%) testing cases, were used for the performance evaluation. The simulation results are partitioned into two parts for easier analysis and evaluation. On the first part, correctly and incorrectly classified instances will be partitioned in percentage value and subsequently Kappa statistics, mean absolute error and root mean squared error will be in numeric value only. We also show the relative absolute error and root relative squared error in percentage for references and evaluation. The results of the simulation are shown in Tables II and III. Table II mainly summarizes the result based on accuracy and time taken for each simulation. Meanwhile, Table III shows the result based on error during the simulation.

TABLE II: Evaluation results of different classification algorithms

Alg.	Correctly %	Incorrectly %	time (sec.)	Kappa statistics
Decision Trees	97.57	2.42	0.01	0.97
Random Trees	94.66	5.33	0.07	0.92
Bayes Net	93.68	6.31	0.06	0.93
Random Forest	95.63	4.36	0.07	0.94

TABLE III: The errors of different classification algorithm

Alg.	Mean Abs. Error	Root Mean Squ. Error	Relative Abs. Error(%)	Root Relative Squ. Error(%)
Decision Tree	0.04	0.12	12.8	30.7
Random Trees	0.06	0.133	19.61	33.7
Bayes Net	0.129	0.199	41.31	50.4
Random Forest	0.036	0.124	11.44	31.4

Figure 2 shows the evaluation of different classification algorithms which are summarized in Table III. From the confusion matrix to analyse the performance criterion for the classifiers in disease detection accuracy, precision, recall and Mathews correlation coefficient (MCC) have been computed for the dataset as shown in Table IV. MCC is a special case of the linear correlation coefficient, and therefore also scales between +1 (perfect correlation) and -1 (anti correlation), with 0 indicating randomness. Accuracy, precision (specificity), recall (sensitivity) and MCC are calculated using the equations (7), (8), (9) and (10) respectively, where T_p is the number

of true positives, T_n is the number of true negatives, F_p is the number of false positives and F_n is the number of false negatives.

$$accuracy = \frac{T_p + T_n}{T_p + T_n + F_p + F_n} \qquad (7)$$

$$specificity = \frac{T_p}{T_p + F_p} \qquad (8)$$

$$sensitivity = \frac{T_p}{T_p + F_n} \qquad (9)$$

$$MCC = \frac{T_p * T_n - F_p * F_n}{\sqrt{(T_p + F_n)(T_p + F_p)(T_n + F_n)(T_n + F_p)}} \qquad (10)$$

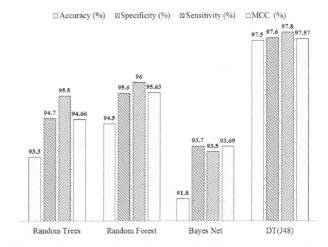

Fig. 2: The Roote Mean Square (RMSE) of each algorithm

TABLE IV: Accuracy, Specificity, Sensitivity and MCC of different classification algorithm

Alg.	Accuracy (%)	Specificity (%)	Sensitivity (%)	MCC
Decision Tree	97.57	97.8	97.6	0.95
Random Tree	93.69	93.5	93.7	0.92
Bayes Net	95.63	96.0	95.6	0.95
Random Forest	94.66	95.5	94.7	0.068

VII. CONCLUSIONS AND FUTURE WORK

Data mining in agriculture is a very interesting research topic and can be used in many applications such as yields prediction, disease detection, optimizing the pesticide usage and so on. There are many algorithms that have been presented for classification in diagnosing the Egyptian rice diseases data set. However, we have choose four algorithms the J48 decision tree, Bayes net, random trees and random forest that belongs to the Tree-based category which are easy to interpret and understand. we conduct many experiments to evaluate the four classifiers for Egyptian rice diseases. The above analysis shows that for the J48 decision tree achieves highest sensitivity, specificity and accuracy and lowest RMS error, than Bayes net, random trees and random forest. J4.8 gave the best results due to the pruning process which simplify the tree and remove unrelevant branches. Moreover, the random forest superior over random trees due to boosting process [23], [24].

Lastly, it should be mentioned that the predictive accuracy is the probability that a model correctly classifies an independent observation not used for model construction. A tree that involves irrelevant variables is not only more cumbersome to interpret but also potentially misleading. selecting an informative features and removing irrelevant/redundant features drastically reduces the running time of a learning algorithm and yields a more general classifier [25], [26]. So, in future works we intend to apply relevant methods for *feature selection* in classification to improve our results as a preprocessing stage before the classification process.

ACKNOWLEDGEMENT

The authors would like to thank the anonymous reviewers for their revision and help to enhancement the paper writing. Also, we are indebted to Central Laboratory for Agricultural Expert Systems staff for providing us with their experiences and data set. And above all, God for His continuous guidance.

REFERENCES

[1] J. Han, M. Kamber and J.Pei, Data mining: concepts and techniques, Elsevier Inc., 3rd edition, 2012.

[2] Y. Nong, Data mining: theories, algorithms, and examples, CRC Press, 2014.

[3] M. Zaki and W. Meira, Data mining and analysis: foundations and algorithms, Cambridge University Press, 2014.

[4] M. Fernandez-Delgado, E. Cernadas, S. Barro and D. Amorim, "Do we need hundreds of classifiers to solve real world classification problems?," in Machine Learning Research, 15, pp. 3133-3181, 2014.

[5] C. Bishop, Pattern recognition and machine learning, Springer New York, 2006.

[6] Y. Zhao and Y. Zhang, *Comparison of decision tree methods for finding active objects*, arXiv:0708.4274v1, 2007.

[7] M. El-Telbany, M. Warda and M. El-Borahy, "Mining the classification rules for Egyptian rice diseases," in International Arab Journal of Information Technology (IAJIT), Jordan, Vol. 3, No. 4, 2006.

[8] A. Nithya, V. Sundaram, "Classification rules for Indian Rice diseases," in International Journal of Computer Science (IJCSI), Vol. 8, Issue 1, 2011.

[9] S. Phadikar, J. Sil and A. Das, "Rice diseases classification using feature selection and rule generation techniques," in Comput. Electron. Agric., 90, pp. 7685, 2013.

[10] Q. Yao, Z. Guan, Y. Zhou, J. Tang, Y. Hu and B. Yang, "Application of support vector machine for detecting rice diseases using shape and color texture features," in International Conference on Engineering Computation, pp.79-83, 2009.

[11] D. Al-Bashish, M. Braik, S. Bani-Ahmad, "A Framework for Detection and Classification of Plant Leaf and Stem Diseases," in International Conference on Signal and Image Processing, pp. 113-118, 2010.

[12] J. Orillo, J. Cruz, L. Agapito, P. Satimbre and I.Valenzuela, "Identification of diseases in rice plant (oryza sativa) using back propagation Artificial Neural Network." in 7th International Conference on Humanoid, Nanotechnology, Information Technology, Communication and Control, Environment and Management (HNICEM), 2014.

[13] K. Majid, Y. Herdiyeni and A.Rauf, "I-PEDIA: Mobile application for paddy disease identification using fuzzy entropy and probabilistic neural network," in International Conference on Advanced Computer Science and Information Systems (ICACSIS), 2013.

[14] T. Hastie, R. Tibshirani and J. Friedman, The elements of statistical learning: data mining, inference and prediction, the Mathematical Intelligence, 27(2): pp. 83-85, 2005.

[15] J. Quinlan, "Induction of decision trees," in Machine Learning, 1(1), pp. 81-106, 1986.

[16] T. Mitchell, Machine Learning, McGraw Hill, 1997.

[17] W. Fan, H. Wang, P. Yu, and S. Ma, "Is random model better? On its accuracy and efficiency," in 3rd IEEE International Conference on Data Mining, 2003.

[18] N. Friedman, D. Geiger and M. Goldszmidt, "Bayesian network classifiers," in Machine Learning, 29, pp. 131-163, 1997.

[19] K. Ho , "Random decision forests," in IEEE Proceedings of the 3^{d} International Conference on Document Analysis and Recognition, pp. 278-282, 2005.

[20] L. Breiman, Random Forests, Machine learning, Springer, pp. 5-32, 2001.

[21] Sakha Research Center, "The results of rice program for rice research and development," Laboratory for Agricultural Expert Systems, Ministry of Agriculture, Egypt, 2002.

[22] M. Hall, E. Frank, G. Holmes, B. Pfahringer, P. Reutemann and I. Witten,", The WEKA data mining software an update," in ACM SIGKDD Explorations Newsletter, 2009.

[23] L. Rokach and O. Maimon, Data mining with decision trees: theory and applications, World Scientific Publishing, 2nd ed. 2015.

[24] T. G. Dietterich, "An experimental comparison of three methods for constructing ensembles of decision trees: Bagging, Boosting, and Randomization," in Machine Learning, pp. 1-22, 1999.

[25] C. Aggarwal, Data mining: the textbook, Springer, 2015.

[26] S. Garca, J. Luengo and F. Herrera, Data preprocessing in data mining, Springer, 2014.

10

Innovative Processes in Computer Assisted Language Learning

Khaled M. Alhawiti

Computer Science Department, Faculty of Computers and Information technology
Tabuk University, Tabuk, Saudi Arabia

Abstract—**Reading ability of an individual is believed to be one of the major sections in language competency. From this perspective, determination of topical writings for second language learners is considered tough exam for language instructor. This mixed i.e. qualitative and quantitative research study aims to address the innovative processes in computer-assisted language learning through surveying the reading level and streamline content of the ESL students in the classrooms designed for students. This study is based on empirical research to measure the reading level among the ESL students. The findings of this study have revealed that using the procedures of language preparing such as shortened text as well as assessed component tools used for automatic text simplification is profitable for both the ESL students and the teachers.**

Keywords—*Natural Language Processing; Computer Assisted Language Learning; Syntactic Simplification Tools*

OUTLINE

This paper will encompass various sections such as introduction to the topic, materials and methods, results, discussion, conclusions, and future work. In the first part named as "introduction," it will describe the different levels of English as Second language (ESL) learners in the United States as well as their specific needs to be successful immigrant students. The second part of this paper will encompass the prior researches conducted relevant to the topic of interest. The subsequent part i.e. results and discussion will focus upon the provision of results of the data gathered through primary sources. The final part of this paper will conclude the study along with the provision of future work.

I.   INTRODUCTION

The Educational system of the United States is confronted with the testing assignment of instructing developing quantities of understudies for whom English is a second language. Washington had 72,215 understudies (7.2% of all understudies) between the school year 2001 and 2002. These understudies were related to the LEP, known as Limited English Proficient. From a year, 2003 onwards more than 2.9 million understudies got English language learner (ELL) administrations, including 19% of all government funded school understudies in California and 20% of all understudies in Texas [1]. In any case, in 2001-2002, 21% of LEP understudies had been in the project for more than three years.

On the other hand, reading is considered to be the basic piece of language and educational advancement, yet discovering appropriate reading material for LEP understudies

is considered frequent upsetting. To help the learners who lie below the evaluation level, the educators with bilingual instruction search out "high investment level" writings at "low reading levels." For example, writings at a first or 2nd grade level back the 5th grade science educational program. Evaluated course readings and different materials are accessible, yet these do not meet the high investment/low reading level model.

In addition, learners also need to be engaged in supplemental reading outside of evaluated reading material for class ventures. Educators additionally require finding material, which consists of a blend of levels, since understudies need distinctive writings to peruse autonomously vs. with assistance from the instructor. This study address the issue by creating computerized instruments to help instructors and understudies discover reading-level fitting writings matched to a specific point to assist furnish these understudies with more intelligible reading material [1].

The term "Natural Language Processing" (NLP) refers to the innovation that is a perfect asset for computerizing the errand of selecting fitting material for reading understudies for bilingual. Data recovery systems effectively find "topical materials" and many of them answers complex questions in content databases on the World Wide Web. A compelling mechanized approach used for evaluating the reading level of the recovered content is still required [2]. Notwithstanding understudies in bilingual instruction, these devices will likewise be valuable for individuals with inabilitiy for learning and education understudies for adults. In both of these circumstances, the understudy's level of reading does not match their educated level as well as investments.

A. BILLINGUAL EDUCAITON

The term "Bilingual education" refers to the different approaches of teaching to the students who have the ability to communicate in multiple languages. From this perspective, "English as a Second Language" (ESL) programs are designed to prepare immigrant students for English-speaking classrooms. Some schools also offer bilingual programs recognized as "dual language immersion" in which throughout the school day two languages are used. In various ESL classrooms, students are ranging from various levels. They require various text levels to review with the assistance and individually involving instructors to locate or generate a great text variety. It would not be wrong to state that teachers working for dual language programs have to face the challenge. This is mainly due to that, students may learn a few subjects in either of the language.

B. COMPUTER ASSISTED LANGUAGE LEARNING

Combined with advanced information technologies (IT), communication systems are considered key to the information society. IT is a baked information revolution, as it gives new human intelligence and vast capabilities. It would not be wrong to state that IT provides resources, and change the way people work and live. It is a new way of living and working together, a new means for communication and interaction in 21st century. The training needs of citizens extend beyond the first studies leading to a degree and extend throughout their lives. The introduction of computers and ICT in the classrooms is mainly due to three perspectives. These perspectives may include initially that students learn about computers because they focus their interest on the technological components. Secondly, they learn computers to employ a range of tailored programs for teaching. Finally, learning with computers and using them as tools give the students benefits of all their applications and connectivity [3].

The computer-assisted language learning is described as incorporation of unit CPU or process and peripherals (monitor, keyboard, mouse, microphone, speaker, video camera, and printer). In the same way, it is also described as a series of texts-processor software, browsers, and educational games. It is an educational application of Information and Communications Technology (ICT). These new technologies help the student to learn more effectively, by allowing them to learn or practice the target language at any time and communicate with the teacher remote, enabling e-learning [3]. Modern multimedia programs often offer an attractive presentation to the user/student to combine text with animated sequences of images and sound. Today the computer is already part of the service of instrumental ELE student at the University, in language schools in self-learning centers [4].

The language learning and computer-assisted instruction (computer-assisted language) are the two complementary facets of the same phenomenon, one from the perspective of the learner and the other from the perspective of the teacher. Computer programs and materials designed for ALAO have some drawbacks. By its very nature, it can hardly pick up all the nuances of human language and the negotiation of meaning that occurs in verbal interaction, such as gestures and intonation [4]. Moreover, these programs do not always cover the various language skills as naturally happens in the human language. Yet, ALAO offers the students a number of advantages, such as:

1) The student chooses a menu among various options for the item, the level of language difficulty, etc.

2) Each student chooses their study schedule; work at their pace, repeating a difficult activity and overlooking another who is not interested

3) The student is frequently the center of the learning process; this helps the student to take responsibility for his or her learning, while contributing to the formation as an independent learner

4) Students get help and correction. A machine usually afflicts to correct the error except when an individual edits it. The computer encourages self-evaluation

C. IMPACT OF TECHNOLOGY ON LEARNING PROCESS

Technology has positively affected the learning process of students in today's society. It has shown that there are many benefits of education in our society. The application of information technology is a feasible and necessary activity. These new technologies are affecting the educational world, particularly the students in the field of training. It is particularly because the media plays a powerful role in the learning process of educational training in term of multiplication. On the other hand, the notion known as "natural language processing" serves as an essential tool to teach the students, specifically the ones with special needs [4].

In a world where distance shrinks and borders disappear, growing mobility of people has a great significance, as they provide an opportunity to work together and to solve problems. The field of education enjoys this scenario with new forms. It appears that the inclusive schools try to solve and address the educational needs of different social groups or sectors of informal education. They find the ICT as a general contributor to the solution of such problems [5].

D. NLP TOOLS FOR LANGUAGE LEARNERS

Currently, there exist few Natural Language Processing (NLP) tools to support teachers as well as language learners. In this way, the existing systems concentrate mainly on constructing vocabulary and involve a large amount of human intervention. The observation of Horst et.al reveals that whereas the reading is a useful technique for vocabulary building, it also helps the readers to recognize new words in accordance to the context. Text Ladder system of Ghadirian classifies selected articles by teachers for the optimum arrangement of vocabulary. At the same time as functional, it is a requisite for the teacher to locate all the stories physically; the tool mainly sorts them [5].

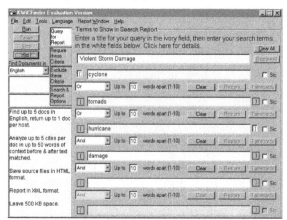

Fig. 1.   Fletcher's KWiC Finder

Fletcher has successfully developed a concordance tool to be used for "foreign language teaching" (FLT). He often makes use of web in the EFL classes in order to find examples related to the FLT.

In this way, he easily checks the questionable usage of various words. On the other hand, he also finds new words, particularly, which have not yet made it into dictionaries. His tool named as "KWiCFinder" finds examples of "keywords in context" (KWiC) and routinely constructs synopsis documents. It was intended mainly as a filter to pace up the job of locating examples of particular words. The research conducted at the "Carnegie Mellon University" contributed in the development of REAP. It is an intelligent system meant for tutoring and constructing to identify online reading material particularly for ESL students. It is particularly based on grade level unigram models, curriculum, and the level of reading of students individually with the incorporation of grammar rules. For example, the rule known as handcrafted is present in the most up-to-date version of the system [6].

For the adult ESL students at grade 6-8, REAP is mainly targeted. A pre-processed database of articles is constructed by the system that entails several words from the Academic World List (570 words). It is a fact that the exact choice of word list is not a primary element of the system. The approach of REAP mainly focuses on students acquisition at an individual level, facilitated through a gram model usage. Our Study considering the same context believes that it would be useful to revisit the complex syntactic structure within the test as well as the short phrases.

In contrast, acquisition for vocabulary remains the vital part of language learning, the content material however needs to be learned by the students. Brunelle & Boonthum-Denecke (2012) explain how their work differentiates from previous studies where they have focused on more features. These features are structural in nature of reading level used particularly to permit the specification for the users regarding the topic discretely from the level [20].

1) Goals and Contribution

This study underlines two main objectives. Firstly, the application of natural language processing and its existing technology to the issues faced in bilingual education. These issues can be related to either teachers or students. Secondly, enabling and enhancing the modern approaches in the area of NLP [18].

The mentioned purpose is also used for the development of the tool that would prove to be helpful in assisting instructors and teachers to locate appropriate level. This assessment can also be done through the development of new techniques of assessment at the reading level and simplification of text to make significant contributions. Such contributions would be mentioned further in this study [7].

2) Reading level detection

Reading level detection is the primary example that would be constructed for corpus of articles that are clean text. It thereafter extends these detectors to be applied to web retrieved pages by the use of a standard search engine. That is how the developing such reading level detectors that are trainable for plain text [7]. These detectors act as SVM classifiers (Support vector machine. They include the usual grade level features including "parser" functions, "n gram" language model scores, and so on.

3) Extension of Read in Level Detectors for Web Pagesdetection

The static group of excellent quality test to the dynamic recourses of test was found to enable to produce additional challenges. It was eventually found that those "web pages" returned inclusive of several pages that trained through detectors on "clean text" merely are not constructed to hold [8]. To discriminate the "web pages" along with text, narrative in nature from those that chiefly have advertisements, links, or other unnecessary content, that would substantially reduce the amount of discarded pages by approximately 50% [8].

4) Investigation of Extension of Algorithm for adjustment Detectors for Individuals

The assessment at the reading level is a variable and subjective issue. Various annotators have diverse insights of the suitability of "articles for a particular grade level." This is partially due to the inconsistency between the students while working. One of the major aims is to develop the detectors that may be modified in the active learning style or in the significance feedback. In this way, one can learn the application of existing processes of SVM to this task. In order to meet the requirements of each user, it is necessary to adapt the reading level detectors via observations from each instructor [9]. It was impossible to attain developments using existing techniques what were known as SVM adaptation. Nevertheless, these were created on non-text-based tasks.

5) Theory/Calculation

The main purpose of the study was to present an investigation of corpus of manual and original news articles that are simplified. The main goal of the analysis is to gain an insight about the practices of people to simplify text to frame enhanced development tools. Burstein et.al approached to choose synonyms for the words they saw challenging. They had the opinion that the use of synonyms could be used to simplify the vocabulary [7]. Instead of the concept of synonym, this section aims to be presenting the analysis of corpus that is related to the manually simplified and original news-articles. The research focused on gaining answer to following questions:

- Suggest the differences in usage of phrase types and POS, found in simplified in comparison to original sentences?

- Identify the traits of the dropped sentences, in resultant of simplified article?

- Identify the traits of split sentences, in resultant of simplified article?

Unlike other studies, our study laid emphasis on corpus of manual and original sentences that are simplified. This study incorporated corpus of paired articles, however it is to be noted that each sentence selected may not have resultant simplified sentence. The corpus of this study makes it possible to discover where the sentences have been dropped and simplified by the rewriters.

6) Aligned Corpus of News Articles

This study took 108 authentic news articles that had parallel-abridged editions framed by literacy works. These literacy works consists of websites for instructors and learners. The target audience selected for articles was native speakers accompanied with poor reading skills.

TABLE I.   CORPUS OF 108 PAIRS(ARIBDGED/ORIGNAL)

	Original	Abridged
# of Sentences(Total)	2439	2359
Words (Total)	40282	28584
Length of Sentence (Avg. Words)	15.5	13.0

7) Corpus Statistics

Upon analysis, it was deduced that number of abridged sentences was nearly equal to original sentences. It was also deduced that there were 29% fewer words in the abridged article set, and the average length of the sentences was 16% shorter in the used set as shown in table 1. In order to explore the differences among abridged and original sentence the study made use of automatic parser. The main purpose of the automatic parser was to acquire tags for parts of speech and parses for sentences.

Table 2 signifies the average length of abridged sentences was 16% shorter; therefore, fewer POS tags and words per sentence were fewer. It was noticed that there was a percentage decrease in the average frequency for adjectives, coordinating conjunctions, and adverbs. There was a 31% decrease for nouns, and 45% for pronouns, which denoted that nouns are unlikely to be replaced with pronouns and deleted less often.

TABLE II.   AVERAGE FREQUENCY, SELECTED POS TAGS (ORIGNAL/ABRIDGED/DIFFERENCE)

Tag	Original	Abridged	Difference (%)
Adjective	1.3	0.8	38%
Adverb	1.1	0.5	55%
CC	0.6	0.4	33%
Determiner	1.8	1.3	27%
IN	1.7	1.4	18%
Noun	3.2	2.2	31%
Proper Noun	1.4	1.0	28%
Pronoun	1.1	0.6	45%
Verb	2.1	1.5	28%

8) Original and Aligned Sentences

The original sentences were distributed in categories based on alignment explained in the above section. This categorization allowed us to drop or align sentences to "one or more abridged sentences". The sentences, that were aligned to precisely the other sentence. In this way, the study calculated the length of the abridged sentence. The study calculated whether the abridged sentence is 19.5% shorter, longer, or roughly equal to the length of the original sentence.

The Sentence is hypothesized to be split that is associated to more than single sentence that is abridged [10]. Similarly, sentences that are aligned to a single and a shorter sentence are assumed as split with one part of the sentence that is dropped. However, it is to note that the average length of these sentences is longer than that of sentences in other categories. Nevertheless, the standard sentence length in such categories is comparatively longer [22].

TABLE III.   ALIGNMENT (ORIGINAL TO ABR. SENTENCES)

Category	# of Sentences (%)	Avg. length
Total	2439 (100%)	16.5
1 - 0 (dropped)	663 (30%)	14.1
1 - >=2 (split)	370 (19%)	24.6
1 - 1 (total)	988 (47%)	15.8
1 - 1 (shorter abr.)	320 (14%)	21.0
1 - 1 (same length abr.)	525 (29%)	14.4
1 - 1 (longer abr.)	103 (4%)	9.1
2 - 1 (merged)	127 (7%)	14.6

9) Annotating True Split Sentences

In this study, around 20% original sentences are adjusted to more than single abridged sentence. The study also assumed that sentences with one part dropped that are aligned to the shorter abridged sentences could be split. On the other hand, sentences having no split points were categorized as "edited," and the sentences conveying same information were marked as "different" [11]. As shown in Table 4, the original sentences were spread among 3 categories i.e. the hypothesized, one-to one splits, and one-to-many splits. In addition, it is not surprising that making a new sentence that is somewhat shorter seems more plausible as compared to the sentence that is changed into two new sentences having no obvious split points [12].

TABLE IV.   HYPOTHESIZED SPLIT- SENTENCE, (DISTRIBUTION)

Category	# of Sentences	
	1 to Many (%)	1 to 1 (%)
Total	441 (100%)	365 (100%)
True split	356 (80%)	198 (54%)
Edited	16 (4%)	162 (45%)
Different	69 (16%)	5 (1%)

10) Analysis of Split vs. Unsplit Sentences

The first and foremost step in simplification that is automatically done via sentence selection to split. The study selected long sentences to be split, incorporating other characteristics as well. For the purpose of in-depth analysis, 1675 sentences were used. The can be identified as 356 "true splits" to 1319 un split sentences mentioned in Table 4. For this study, the different sentences as well as the edited sentences are assumed split sentences as they are measured unsplit. As a matter of fact, the average amount of phrases "identified by the parser S, NP, etc." as well as the average length of the phrases is longer. Therefore, it is assumed that the split of sentence depends upon the syntactic features in addition to the length of a particular sentence. [13]

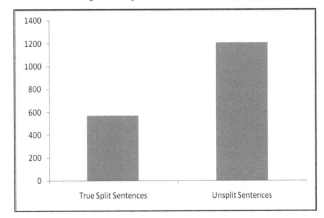

Fig. 2.   True Split Sentences vs Unsplit Sentences

In order to examine the most important feature for splitting sentence the study made use of "C4.5 decision tree learner" to build a classifier for unsplit and split sentences. From this perspective, the rule generators of C4.5 were selected for this study considering the easiness in the emerging results. This can be described in other words that in this section the major focus was on analysis instead of classification. From this perspective, the study incorporated a few features of sentence. The features, on the other hand, may include "Sentence length in words," "POS", and "Phrase." In other words, POS include number of adverbs, adjectives, CC, determiners, IN, nouns, pronouns, proper nouns, and verbs. On the other hand, the term "Phrase" may include average length and number of SBAR, S, PP, NP, and VP.

11) Analysis of Dropped Sentences
The researcher tends to attempt, in this section, to evaluate the "dropped sentences" in the comparison with the other original sentences. As "C4.5 rule generator" was used in the previous section in order to see the most important feature of sentence, "the basis for analyze the sentences to drop" seems more credible to see the dropped sentences. On the other hand, it is more credible to be content-based as compared to the syntactic. In this way, content-based is considered to be a quite different group of feature in the present section. It is forecasted instinctively that the replicated sentences can be dropped as well.

II.   RESULTS AND DISCUSSION

Based on the prior work above, the study decomposed the problem, what is known as simplification, into the following four component problems: sentence selection, sentence splitting, sentence compression, and lexical replacement.

The decoupling of such processes is useful for recognition their role in simplification, and for leveraging existing text processing tools. The key aspect is Sentence selection, furthermore, component of summarization systems and major research addresses this problem. Summarization systems targets to extract the most vital information from an article but majority select longer and more complicated sentences since those sentences convey major information [14]. By the same token, the articles that are simplified in the Literacy works corpus analyzed in paper above are seventy percent as long as the original articles.

In this study, the Literacy works corpus hand-aligned was studied to train the models for the selection of sentence and then splitting. The use of selected set developed from "Literacy works corpus" as well as "two test sets selected from two sources on the web" were used to evaluate the mentioned models as well as the processing tools of the existing language. The assessment of "automatic simplification" is considered to be quite similar to the compression tasks and closely-related summarization, as it generally involves the evaluation that is to be handwritten. This way is also associated with the human decision-making or gold standard as it is related to the "quality of the results." On the other hand, a huge variety of articles "from the Literacy works corpus" has been selected for the present study in order to use the development set as well as for the experiments.

The study created two additional test sets. The initial consists of usage of 5 articles from the online Principles of Aeronautics textbook written by Cislunar Aerospace, Inc.

These articles selected were about the same length as the Literacy works articles, and pointed out the factor that the "average sentence length" is longer for the corpus comparatively. Nevertheless, the "average sentence length" that is longer will represent the different challenge for the purpose of simplification though the mentioned articles are included in the similar "news domain" as the Literacy works corpus are.

Several of other systems do not use the quotation feature, rather than that it appears to be useful for the mentioned corpus. It is believed that it is helpful in general for such application. The content features represent the related purpose of the "tf-idf features used by Nobata and Sekine". Nevertheless, such features are to offer "information about content at the level of a single document instead of requiring a corpus of related documents." On the other hand, some features are mainly used in the extractive summarization. Nevertheless, these features are not quite relevant to the unimpeded domain.

Using the IND "decision tree package", the instructor may train "the classifier" with the mentioned features that are mainly based on dropped and not dropped sentences in the interpreted corpus. The IND usage in the place of "C4.5" is thus because of the package that provides probability to estimate the classifications. Eventually, to apply the classifier to the new feature known as "vector x for a sentence", all of the B trees 7) may be applied to "the vector and the resulting posterior probabilities" that are initially averaged and then regularized to the account for "re-sampling", i.e.

$$p(drop|x) = \frac{q_d}{0.5B} \sum_{i=1}^{B} p_i(drop|T_i(x))$$

where B = 48 and = 0.28 is the previous prospect of the "dropped" class. The application of the classifier to developing the set gives way performance, which is not much better than probability. From this perspective, the present paper did not follow any additional evaluation of the classifier on the set test. As a matter of fact, the present study is intended to compare to the set results of the development along with the "summarizer results" occurring in the subsequent section. The manual review related to the "various decision trees" identify that "all of the types of feature used, with the "quote" feature that often occur near to the top of tree [15]. The both feature categories i.e. redundancy and position emerge significant. It is unclear that what category is used consistently than the other one. The presence of complexity in using commands a language that is more rewarding than GUI [17]. However, they are still being used in various applications, typically without the users' intentions. People, however, do not notice simple control line interfaces that have been integrated into systems, which include a search box on the Web and page range box in Microsoft Word's printing options. Unnoticeable interface is a sign of a good interface. An extensive variety of functions can be fluently provided by a text interface in comparison to GUI, which lack scalability. The major cause is that the text is so much lighter than graphics [16]. The aptitude for pure texts to be effortlessly view, copy, paste, edit, stored, and share is accessible in almost every user interface and application

III.   CONCLUSIONS

The purpose of the present research was to apply as well as to extend the existing "NLP technology" to be used for the problems faced by students as well as teachers within the context where bilingual education takes place. In this way, the study advanced "the state of the art in the relevant NLP areas". The paper has discussed in detail about the tools that support the instructors in order to choose appropriate-level and topical texts to be used for their students by the use of innovations and techniques of reading assessment. This study also explored the "characteristics of abridged text" as well as it "assessed component tools for automatic text simplification". It would not be wrong to state that this study contributed to the area where "reading level assessment" takes place and includes the "development of reading level detectors for clean text and extending them to a most varied text found on the World Wide Web". The researcher found that the combining "SVM classifiers" within the traditional grade-level, the scores of the models named as n-gram language, and the features that are parser-based are based on the LMs alone in form of n-gram.

Through the study, it was established that various methods of SVM adaptation have been developed for the "non-text-based tasks with well-separable components do not essentially apply to this task". Nevertheless, the study finds that "other classifier/adaptation combinations will provide better results in the future". With the use of Literacy works as well as "news article corpus", it was suggested that the further studies will characterizes the texts to expand insight into what people generally do when performing the type of text-adaptation. The study concluded that the hand-aligned sentences were to illustrate that how sentences in the original description and abridged description relate to one another other. From this perspective, the "split points" were marked at the places where an original sentence is mapping to two "abridged sentences". The resulting corpus in such observations will be the contributing factor to the "field of NLP research." Therefore, it could be used for the purpose of research on the topic that includes sentence alignment, simplification, and summarization [19].

IV.   FUTURE WORK

There are various possible paths for the work that will be conducted in the future that might be based on this research. In this section, the author discusses the four most important directions such as future work for simplification, future work for adaptation, creating the system that can be used for interaction between teachers and students, and application to languages that are other than target language. The basic focus of the author was to develop the tools that can be used for English language. Nevertheless, the research would be equally helpful for students as well as teachers to find as well as simplify the texts available in the target language. On the other hand, the other then English languages English may have advantage from "the use of additional features" as well. For example, it can be used to capture the richer morphology. In the same way, extending and modifying the tools that have been developed in the research is the additional area for future work, as it is to create a system that may work in real-time.

REFERENCES

[1]   J. Amaral, L. A., & Meurers, D. (2011). On using intelligent computer-assisted language learning in real-life foreign language teaching and learning. ReCALL,23(01), 4-24.

[2]   Chapelle, C. A. (2010). The spread of computer-assisted language learning.Language Teaching, 43(01), 66-74.

[3]   Cohen, K. B. (2013). Biomedical Natural Language Processing and Text Mining.Methods in Biomedical Informatics: A Pragmatic Approach, 141.

[4]   Tyagi, D., Joshi, T., Ghule, D., & Joshi, A. (2014). An Interactive Answering System using Template Matching and SQL Mapping for Natural Language Processing. International Journal, 2(2).

[5]   Kamath, R. S. (2013). Development of Intelligent Virtual Environment by Natural Language Processing. Special issue of International Journal of Latest Trends in Engineering and Technology.

[6]   Field, D., Richardson, J. T., Pulman, S., Van Labeke, N., & Whitelock, D. (2014). An exploration of the features of graded student essays using domain-independent natural language processing techniques. International Journal of e-Assessment, 4(1).

[7]   G. Eason, B. Noble, and I. N. Sneddon, "On certain integrals of Lipschitz-Hankel type involving products of Bessel functions," Phil. Trans. Roy. Soc. London, vol. A247, pp. 529–551, April 1955. (references)

[8]   Reeves, T., & McKenney, S. E. (2013). Computer-assisted Language Learning and Design-based Research: Increased Complexity for Sure, Enhanced Impact Perhaps.

[9]   Song, P., Shu, A., Zhou, A., Wallach, D., & Crandall, J. R. (2012). A pointillism approach for natural language processing of social media. arXiv preprint arXiv:1206.4958.

[10]   Denny, J. C., Choma, N. N., Peterson, J. F., Miller, R. A., Bastarache, L., Li, M., & Peterson, N. B. (2012). Natural language processing improves identification of colorectal cancer testing in the electronic medical record.Medical Decision Making, 32(1), 188-197.

[11]   Nakata, T. (2011). Computer-assisted second language vocabulary learning in a paired-associate paradigm: a critical investigation of flashcard software.Computer Assisted Language Learning, 24(1), 17-38.

[12] Esit, O. (2011). Your verbal zone: an intelligent computer-assisted language learning program in support of Turkish learners' vocabulary learning. Computer Assisted Language Learning, 24(3), 211-232.

[13] Kennewick, R. A., Locke, D., Kennewick, M. R., Kennewick, R., & Freeman, T. (2011). U.S. Patent No. 8,015,006. Washington, DC: U.S. Patent and Trademark Office.

[14] Gorjian, B., Moosavinia, S. R., Ebrahimi Kavari, K., Asgari, P., & Hydarei, A. (2011). The impact of asynchronous computer-assisted language learning approaches on English as a foreign language high and low achievers' vocabulary retention and recall. Computer Assisted Language Learning, 24(5), 383-391.

[15] Litman, D., Moore, J. D., Dzikovska, M., & Farrow, E. (2010). Using natural language processing to analyze tutorial dialogue corpora across domains and modalities.

[16] Meurers, D. (2012). Natural language processing and language learning. The Encyclopedia of Applied Linguistics.

[17] Jarvis, S. (1984). Language Learning Technology and Alternatives For Public Education. CALICO Journal, 1(4), 11-16.

[18] Michos, S. E., Fakotakis, N. & Kokkinakis, G. (1996). Towards an adaptive natural language interface to command languages.. Natural Language Engineering, 2, 191-209.

[19] Lewis, D. D., & Jones, K. S. (1996). Natural language processing for information retrieval. Communications of the ACM, 39(1), 92-101.

[20] Kerr, D., Mousavi, H., & Iseli, M. (2013). Automatic Short Essay Scoring Using Natural Language Processing to Extract Semantic Information in the Form of Propositions. CRESST Report, 831.

[21] Brunelle, J. F., & Boonthum-Denecke, C. (2012). Natural Language Processing Tools. Cross-disciplinary Advances in Applied Natural Language Processing: Issues and Approaches, 9.

[22] Murff, H. J., FitzHenry, F., Matheny, M. E., Gentry, N., Kotter, K. L., Crimin, K., ... & Speroff, T. (2011). Automated identification of postoperative complications within an electronic medical record using natural language processing. Jama, 306(8), 848-855.

11

Speed and Vibration Performance as well as Obstacle Avoidance Performance of Electric Wheel Chair Controlled by Human Eyes Only

Kohei Arai [1]
Graduate School of Science and Engineering
Saga University
Saga City, Japan

Ronny Mardiyanto
Department of Electric and Electronics
Institute Technology of Surabaya
Surabaya, Indonesia

Abstract—Speed and vibration performance as well as obstacle avoidance performance of the previously proposed Electric Wheel Chair: EWC controlled by human eyes only is conducted. Experimental results show acceptable performances of speed vibration performance as well as obstacle avoidance performance for disabled persons. More importantly, disabled persons are satisfied with the proposed EWC because it works by their eyes only. Without hands and finger, they can control EWC freely.

Keywords—Human Computer Interaction; Gaze; Obstacle Avoidance; Electric Wheel Chair Control

I.   INTRODUCTION

The Electric Wheel Chair: EWC controlled by human eyes only is proposed previously. It works well in principle. Some experiments show acceptable performances previously. This paper describes the performance of speed control and vibrations as well as obstacle avoidance performance.

The proposed system consists of forward and backward looking Web cameras mounted glass and pocket PC that allows Bluetooth communications. Thus users can be moved using the system. Pocket PC can be communicated with not only with Input and Output devices but also the other pocket PCs mounted on the other Electric Wheel Chairs: EWCs so that created and updated map information can be shares with many EWCs. The system provides obstacle finding with forward-looking camera so that EWCs can avoid obstacles. Location information of obstacles is uploaded to the other EWCs through Bluetooth communications. Thus all the EWCs can be controlled safely avoiding obstacles with the shared map information.

The following section describes the proposed system followed by experiments for the proposed system in terms of control performance of EWC, in particular, obstacle avoidance performance.

II.   THE PROPOSED COMPUTER INPUT SYSTEM WITH HUMAN EYES-ONLY

A. Hardware configuration

Hardware configuration is shown in Figure1. The proposed system consists of (1) two Web cameras mounted glass, (2) Pocket PC, (3) Ultrasonic Sensor. One of two Web cameras looks forward and the other one looks backward (acquires users' eye image). The web camera used 1.3 Mega pixel OrbiCam (Visible camera) and The Pocket PC used Sony VAIO UX180P with Intel Solo Processor U1400 1.20GHz and 512 MB RAM. First camera is used for acquired eye behavior and other camera is used for detect the obstacle. The Electric Wheel Chair used Yamaha JW-I type.

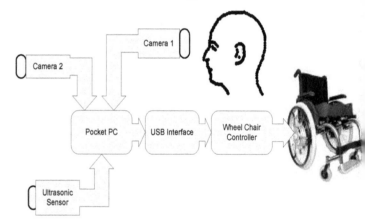

Fig. 1.   Hardware configuration

B. The method used and system parameters

In order to control EWC, at least four keys, move forward, turn right, turn left and stop are required. For the safety reason, users have to look forward so that the key layout that is shown in Figure 2 is proposed.

Stop		
Turn left	Move forward	Turn right
Stop		

Fig. 2.   3 by 3 of key layout for EWC control

Namely, key consists of 9 keys (3 by 3). Move forward and turn left/right are aligned on the middle row. Stop key is aligned on the other rows, top and bottom rows. Users understand the location of desired key so that it can be selected with users' eye-only. The backward looking Web camera whose resolution is 640 by 480 pixels acquires users' eye and its surrounding. Using OpenCV[14] of eye detection and tracking installed on the Pocket PC, users' eye is detected and tracked. If the OpenCV cannot detect users' eye, then EWC is stopped for safety reason. EWC is also stopped when users look at the different location other than the three keys aligned on the center row. When users are surprised human eyes used to be large. Such a situation can be detected with acquired image with backward looking Web camera so that EWC is stopped.

Intentional blink can be detected if the eye is closed for more than 0.4 seconds because accidental blink is finished within 0.3 seconds. In this connection, it is easy to distinguish between intentional and accidental blink. Also, key selection can be done every 0.4 seconds. Thus the system recognizes user specified key every 0.4 seconds. In order to make sure the user specified key, 10 frames per seconds of frame rate is selected for backward looking camera.

C. Eye Detection and Tracking[17]

Figure 3 shows the process flow of eye detection and tracking. Eye is detected by Viola-Jones classifier. The viola-Jones classifier employs ADABOOST at each node in the cascade to learn a high detection rate the cost of low rejection rate multi-tree classifier at each node of the cascade. To apply the viola-Jones classifier on the system, we use viola-Jones function in OpenCV[15]. Before use the function, we should create xml file data. The training sample (face or eye image) must be collected. There are two sample types: negative and positive sample. Negative sample correspond to non-object images. Positive sample correspond to object image. After acquired image, OpenCv will search the face center location and continue with search the eye center location. Advantage this function is fast and robust.

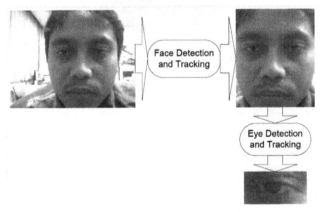

Fig. 3. Eye detection and tracking

D. Template Matching

Eye behavior is detected by template matching. Template matching which used is not based on histograms; the function matches an actual image patch against an input by sliding the patch over the input image. There are several template matching methods:

1) Square difference matching methods

These methods match the square difference, so perfect match will be 0 and bad matches will be large.

$$R_{sq\_diff}(x, y) = \sum_{x',y'}[T(x', y') - I(x+x', y+y')]^2 \tag{1}$$

This method will obtain good result only if both images have same pixel intensity. Because of output result only 0 and 1, it not sophisticated for the proposed system.

2) Correlation matching methods

These methods multiplicatively match the template against the image, so a perfect match will be large and bad matches will be small or 0.

$$R_{ccorr}(x, y) = \sum_{x',y'}[T(x', y'), I(x+x', y+y')]^2 \tag{2}$$

3) Correlation coefficient matching methods.

These methods match a template relative to its mean against the image relative to its mean, so a perfect match will be 1 and a perfect mismatch will be -1; a value of 0 simply means that there is no correlation.

$$R_{ccoeff}(x, y) = \tag{3}$$
$$\sum_{x',y'}[T'(x', y').I'(x+x', y+y')]^2$$
$$T'(x', y') = \tag{4}$$
$$T(x', y') - \frac{1}{(w.h)}\sum_{x'',y''}T(x'', y'')$$
$$I'(x+x', y+y') =$$
$$I(x+x', y+y') - \frac{1}{(w.h)}\sum_{x'',y''}I(x+x'', y+y'') \tag{5}$$

In this system we used Correlation coefficient methods and give additional normalized to reduce the effects of lighting difference between the template and the image.

$$z(x, y) = \sqrt{\sum_{x',y'}T(x', y')^2 . \sum_{x',y'}I(x+x', y+y')^2} \tag{6}$$

Result values for this method that give the normalized computation are:

$$R_{ccoeff_{norm}}(x, y) = \frac{R_{ccoeff}(x, y)}{z(x, y)} \tag{7}$$

In this system, four template images are used to determine eye gaze. These images are acquired during calibration step. The best result will obtain eye gaze. The threshold is set to 90% match.

E. Calibration Step

Eye gaze measurements that apply template matching methods will does work on fix illumination and fix condition. Illumination changes, shadow, different of eye shape when it is used by other users, and others difference condition will obtain mismatch result.

This problem can be solved by calibration step. The calibration step consists of acquiring template image and self evaluation as is shown in Figure 4.

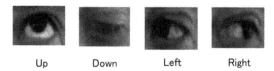

Up   Down   Left   Right

Fig. 4. Result of calibration step. These steps obtain up, down, left, and right image. Up image is used for move forward, down is used for stop, left is used for turn left, and right is used for turn right

System will acquire eye image when looking at down, up, left, and right keys. Next, system will evaluate template image. If template image is good template, it will be used. If not, system will acquire eye image again until good template are obtained. When system is started, it will check that template images are proportional. Illumination, shadow, and eye shape will compare with current eye image. If there are not proportional image, calibration step will be ran.

F. Custom Microcontroller

Yamaha JW-I Electric wheel chair type is used. To control EWC using Pocket PC, custom microcontroller circuit is used. USB interface on pocket PC is used to connect with other peripheral. The custom microcontroller circuit is use RS232 communication. Microcontroller AT89S51 type is used. To connect between pocket PC and custom microcontroller circuit, USB to Serial communication is used. The microcontroller will drive relay to move the EWC. Microcontroller connection is shown in Figure 5.

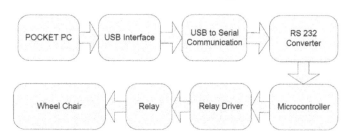

Fig. 5. Microcontroller AT89S51 connects to other peripheral through serial communication. Serial communication type should be converted to USB communication using USB to serial converter.

G. Obstacle Avoidance

In order to safety reason, obstacle avoidance system is implemented in our system. Obstacle avoidance system will able to identify the obstacle in front of EWC and avoid it. This system consists of two approaches: (1) Obstacle detection, and (2) Best Path Finding.

H. Obstacle Detection

Obstacle detection is consisting of image processing based and ultrasonic sensor based. Image processing based utilizes background subtraction between current image I(x, y) and background image B(x, y). Background subtraction method will obtain black-white image S(x, y) which represent obstacle. On this image, obstacles appear as white pixel. By using searching of outer line from white pixels, we can determine position and size of obstacle.

$$S(x,y) = 0, \text{ if } |I(x,y) - B(x,y)| < threshold \quad (8)$$

$$S(x,y) = 255, \text{ if } |I(x,y) - B(x,y)| <= threshold \quad (9)$$

Weakness of background subtraction method is working only if two images have same position of translation, scale, shear, and rotation. To solve these problems, we utilize affine transformation. This transformation requires 3 points that appears on both images as is shown in Figure 6.

Fig. 6. Background subtraction method, left-top is background image and right-top is current image. Background subtraction method will obtain black-white image (bottom)

Translation, scale, share, and rotation parameter can be determined from these points. Affine transformation required at least three noticeable important points. These points should be appears on both images. It can be detected by several ways: corner, edge, specific object, and etc. Corner and edge have many points and its will create computation problem. In our system, we decide to use specific object. The specific object can be an easily recognized object, text character, chessboard wall, and etc. Flow process of obstacle detection using affine transformation is shown in Figure 7.

The specific object will obtain one coordinate from center of area. Obstacle detection using affine transformation requires identifying at least three kind of object, so it will obtain three noticeable important points.

First step of obstacle detection is convert source image into gray image. By using template matching, system will find specific object position. Using six noticeable points, system calculate translation, share, scale, and rotation parameter. These parameters will used for creating affine matrix.

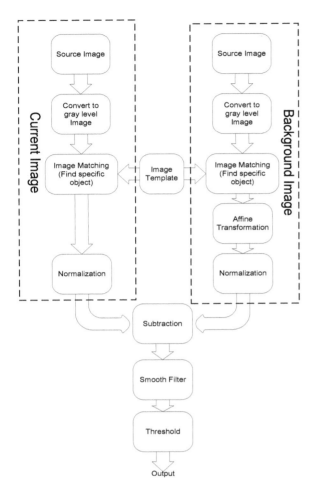

Fig. 7. Obstacle Detection. This methods rely on subtraction between two images and affine transformation.

Affine matrix apply on background image will obtain affine image. Normalization is used to eliminate disturbance such shadow, noise, and etc. Next, subtract between output form current image and background image. Smooth filter is used to reduce noise which is caused by subtraction process. Last step is applying threshold on image and it will obtain black and white image. Obstacle will be signed as group of white pixel. After black white image is obtained, we should return center of white pixel area into current coordinate and coordinate of obstacle is founded. Because of so many type of specific object, we got best performance of specific object by using chessboard wall. The advantage of the chessboard (Figure 8) is easily detected and robust on distance changes. We use 3 types of chessboard: 3 by 4, 3 by 5, and 3 by 6 as are shown in Figure 9. The other obstacle detection is use ultrasonic sensor. This sensor has advantage when visual system does not work. In case EWC move into surrounding glass door, smoke condition, and minimum lighting will caused visual system of obstacle detection fail. Ultrasonic sensor consists of transmitter and receiver part. We use ultrasonic sensor with 40 kHz frequency resonance.

Fig. 8. Chessboard as specific point

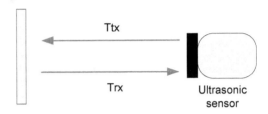

Fig. 9. Ultrasonic sensor, distance is measured by converting half of Ttx transmitting time and Trx receiving time

I. Best path finding

When EWC detect the obstacle, it must understand where the best path should be chosen if EWC want to go into specific place. Best path will be chosen based on floor layout map and image map. Image map is created by acquiring background images in every 1 m as is shown in Figure 10. In every location (x, y) have one image. Instance, if the location has size 10 x 10 m, it will have 100 images. After acquire background images, floor layout map will be added manually as initial condition. This map is setup manually based on room layout. Example is shown in Figure 11.

Combination between background images and layout path will obtain main map. EWC will move consider the main map. After obstacle is detected in current path, EWC will switch to another path which have same destination. Obstacle avoidance methods are also useful when user is not confident to pass the obstacle. EWC will take over and pass the obstacle. Best path is chosen based on Dijkstra algorithm. Furthermore, System also able to renew existing map.

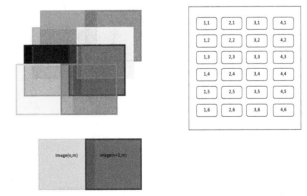

Fig. 10. Image map, created by acquire image in every 1m.

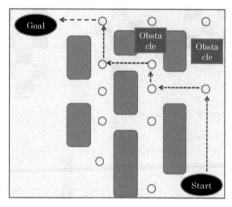

Fig. 11. Best path finding using Dijkstra, best path is chosen based on shortest distance

III. EXPERIMENTS

A. Key selection

The proposed method experiments by measuring success rate eye gaze detection as hit keys selection. Success rate of eye gaze detection is measured by change user distance to camera. Experiments say that minimum distance of success rate is 13 cm and maximum distance is 38 cm. The distance beyond these, OpenCV will not detect the eye and caused the system fail to measure eye gaze. Hit keys selection experiment was done by using real time video 640 by 480 pixels. Data experiments when measure hit keys selection on range distance 24 – 30 cm is shown in Table 1.

TABLE I.   HIT KEYS MEASUREMENT

Distance (cm)	Success Rate			
	Stop (%)	Move Forward (%)	Turn Left (%)	Turn Right (%)
24	100	100	100	100
25	100	100	100	100
26	100	100	100	100
27	100	100	100	100
28	100	100	100	100
29	100	100	100	100
30	94.12	100	100	100

The result of hit key measurement is shown almost perfectly, Error which cannot understand the eye input is caused by the system fail to verify the template image. Error will zero if good quality template is used. To make sure that good quality template is used; System will always verify every template and analyzed it. If templates have poor quality, system will conduct calibration step again. This step will repeat until good quality template is obtained.

B. Obstacle avoidance performance.

Experiment of obstacle avoidance is conducted by acquiring image in corridor with distance 1m per images. Location which image was acquired, is set with (x, y) coordinate. So, in every (x, y) location will have 1 background image. This image will be used as background reference. Obstacle is detected by subtract background image with current image. To eliminate problem which caused by different position between these images, affine transformation is used to transform background image to affine image which

have same position with current image. Applying affine transformation will does work if only if three noticeable important points are appears on both images. These important points are represented by three types of chessboard. If the chessboard is successful detected, then by using affine transformation and subtraction between background image and current image, obstacle is founded. Experiments of chessboard detection is conducted by measuring maximum location which still detected by system. Chessboard is put in fix location. After that, EWC is move around the chessboard. Maximum location which detected by system is recorded. Data experiments are shown in Figure 12.

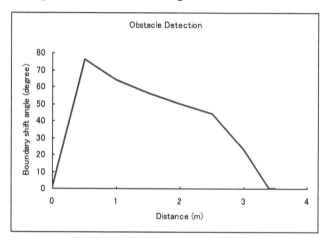

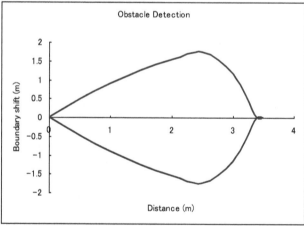

Fig. 12. measuring obstacle avoidance performances

Data show that boundary shift will decrease when distance between camera and chessboard is increase. This experiment is equal to obstacle detection. After three types of chessboard are detected, affine transformation will use these points (three chessboard center of areas) to subtract with current image and obstacle position will be found.

C. Ultrasonic Sensor

Objective of this experiment is measure accuracy of ultrasonic sensor before used in the system. This experiment is conducted by measure sensor output for measuring distance and comparing with real data. Some object is put on front of sensor with varies distances and measure it. The experiment is conducted on distance 0 cm until 3 m.

Speed and Vibration Performance as well as Obstacle Avoidance Performance of Electric Wheel Chair Controlled...

71

Ultrasonic sensor use PING type parallax product and microcontroller AT89S51 as processing data (convert from time value to distance output). Graph of accuracy sensor is shown in Figure 13.

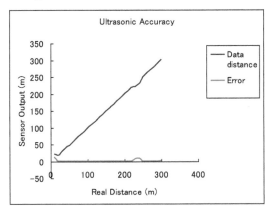

Fig. 13. Experiment of ultrasonic accuracy. This shown that minimum distance is around 3 cm and maximum distance is 3 m. This range is appropriate for detect the object.

Elevation angle is require to know how width of the beam sensor. Ultrasonic sensor with width beam is not benefit to our system. Narrow beam will obtain good result because it will not influence with any disturbance. This experiment is conducted by measure elevation angle from 0 cm until 3 m. Graph of elevation angle is shown in Figure 14.

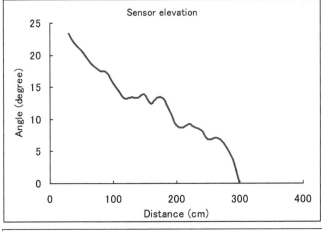

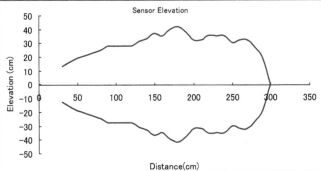

Fig. 14. Experiment of Elevation angle. Top side is distance versus angle, and bottom side is distance versus distance elevation.

D. Performance of Electric Wheel Chair

This experiment is conducted for measure EWC performance on starting up acceleration, forward and backward breaking deceleration. Also conducted speed measurement when EWC move forward, backward, turn left, and turn right. EWC is drive by user who has weight is 73 kg. We record the duration time and convert it into speed value. Experiment data of speed measurement is shown in Table 2. Graph of EWC acceleration and deceleration when start and stop duration is shown in Figure 15.

TABLE II.   SPEED MEASUREMENT

Moving	Speed (m/s)
Forward	0.29
Backward	0.26
Turn Left	0.11
Turn Right	0.11

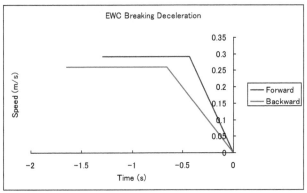

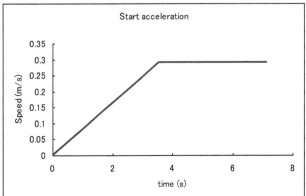

Fig. 15. Experiment of EWC acceleration and deceleration

E. Processing time each process

In order to apply whole method into EWC application, processing time should be measured to identify performance of our real time system. Figure 16 shows transient time of eye detection and tracking, n the beginning of chart, it seem this method take long time around 300 ms. In this time, system still process face detection, eye detection and others process before running template matching method. After eye location is founded, then system bypass previous step and cause process working fast.

Meanwhile, Figure 17 shows processing time of eye detection and tracking on steady state condition, it looks faster than transient condition.

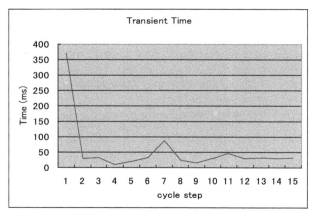

Fig. 16. Transient time of eye detection and tracking, n the beginning of chart, it seem this method take long time around 300 ms. In this time, system still process face detection, eye detection and others process before running template matching method. After eye location is founded, then system bypass previous step and cause process working fast.

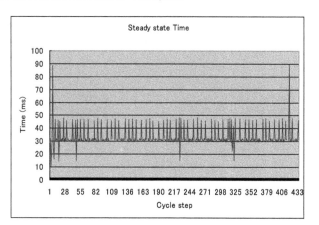

Fig. 17. Processing time of eye detection and tracking on steady state condition, it looks faster than transient condition.

F. Eye detection and tracking

This experiment is conducted using Optiplex 755 Dell computer with Intel Core 2 Quad CPU 2.66 GHz and 2G of RAM. We use NET COWBOY DC-NCR131 camera as visual input. Experimental result show average steady state processing time is 32.625 ms. it also shows difference processing time between transient and steady state condition. Transient time require more time than steady state time.

G. Eye gaze measurement

Objective of this experiment is measure processing time on Eye gaze identification. It is conducted by using ACER computer ASPIRE 5572 Series Core Duo T2050 1.6 GHz CPU

and 1G of RAM. Result data show average processing time of this method is 342.379 ms. Figure 18 shows processing time of Eye gaze method.

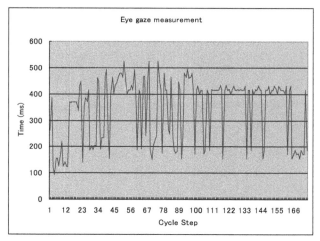

Fig. 18. processing time of Eye gaze method

H. Obstacle detection

This experiment also was conducted using Optiplex 755 Dell computer with Intel Core 2 Quad CPU 2.66 GHz and 2G of RAM. NET COWBOY DC-NCR131 camera as visual input is also used. Experimental result show average processing time is 32.625 ms. Figure 19 shows processing time of obstacle detection.

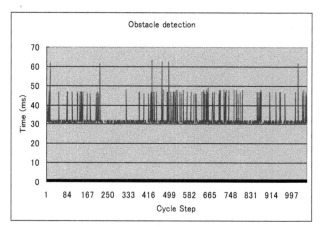

Fig. 19. processing time of obstacle detection

I. Ultrasonic sensor

We implemented ultrasonic sensor parallax PING type. This sensor is controlled by using custom microcontroller AT89S51. Data was stored into computer by using USB communication. Result data show average processing time is 568.658 ms. Figure 20 also shows processing time of ultrasonic sensor, it look take longer time than others.

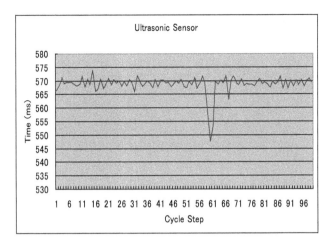

Fig. 20. processing time of ultrasonic sensor, it look take longer time than others.

IV. CONCLUSION

It is concluded that the proposed key-in system with human eyes only works 100% perfectly for the number of keys is four, start, stop, turn right and left. Also it is concluded that the proposed EWC control system does work in a real basis avoiding obstacles on range bellow 3.4 m using image processing method and bellow 3 m using ultrasonic sensor. By the proposed system, EWC is able to identify obstacle and avoid them. Obstacle avoidance can assist user into interest place although undesired condition such as dark areas, glass wall or door, smoke area, and etc. By implemented this system, EWC will move more safely and comfortable.

ACKNOWLEDGMENT

The author would like to thank Dr.Djoko of Institute of Technology Surabaya, Indonesia for his efforts through experiments and simulations.

REFERENCES

[1] Park, K.S., and Lee K.T., Eye controlled human/computer interface using the line of sight and intentional blink, Computer Ind. Eng., 30, 3, 463-473, 1996.

[2] Ito, Sudo, the Ifuku part: "The look input type communication equipment for serious physically handicapped persons", the Institute of Electronics, Information and Communication Engineers paper magazine, J83D1, 5, 495.503, 2000.

[3] Yamada, Fukuda: "The text creation and the peripheral equipment control device" by eye movement, the Institute of Electronics, Information and Communication Engineers paper magazine , J69D, 7, 1103-1107, 1986.

[4] Mitsunori Yamada, the research trend of the latest eye movement, an electric information-and-telecommunications academic journal, MBE95-132, NC95-90, 145-152, 1995.

[5] K.Abe, S.Ohiamd M.Ohyama: "An Eye-gaze Input System based on the Limbus Tracking Method by Image Analysis for Seriously Physically Handicapped People", 7th ERCIM Workshop "User Interface for All" Adjunct Proc., 185-186, 2002.

[6] http://www.creact.co.jp/jpn/por.pdf

[7] Kuno, Yagi, Fujii, Koga, Uchikawa: "Development of the look input interface using EOG", the Information Processing Society of Japan paper magazine, 39, 5, 1455-1462, 1998.

[8] D.A.Robinson: A method of measuring eye movement using a sclera search coil in a magnetic field, IEEE Trans. on Biomedical Electronics, 10, 137-145, 1963.

[9] http://webvision.med.utah.edu/

[10] Ito, Nara: "Eye movement measurement by picture taking in and processing via a video capture card", an Institute of Electronics, Information and Communication Engineers technical report, 102, 128 , 31-36, 2002.

[11] Kishimoto, Yonemura, Hirose, Changchiang: "Development of the look input system by a cursor move system", and the Institute of Image Information and Television Engineers, 55, 6, 917-919, 2001.

[12] Corno, L.Farinetti, I. Signorile: "A Cost-Effective Solution for Eye-Gaze Assistive Technology", Proc. IEEE International Conf. on Multimedia and Expo, 2, 433-436, 2002.

[13] Abe, Ochi, Oi, Daisen: "The look input system using the sclera reflection method by image analysis", and the Institute of Image Information and Television Engineers, 57, 10, 1354-1360, 2003.

[14] Abe, Daisen, Oi: "The multi-index look input system which used the image analysis under available light", and the Institute of Image Information and Television Engineers, 58, 11, 1656-1664, 2004.

[15] Abe, Daisen, Oi: "The look input platform for serious physically handicapped persons", Human Interface Society Human interface symposium 2004 collected papers, 1145-1148, 2004.

[16] http://homepages.inf.ed.ac.uk/rbf/CVpnline/LOCAL_COPIES/GONG1/cvOnline-skinColourAnalysis.html.

[17] Gary Bradski, Andrian Kaebler, "Learning Computer Vision with the OpenCV Library", O'REILLY, 214-219, 2008.

12

Bidirectional Extraction of Phrases for Expanding Queries in Academic Paper Retrieval

Yuzana Win
Graduate School of Engineering
Nagasaki University
Nagasaki, Japan

Tomonari Masada
Graduate School of Engineering
Nagasaki University
Nagasaki, Japan

Abstract—This paper proposes a new method for query expansion based on bidirectional extraction of phrases as word *n*-grams from research paper titles. The proposed method aims to extract information relevant to users' needs and interests and thus to provide a useful system for technical paper retrieval. The outcome of proposed method are the trigrams as phrases that can be used for query expansion. *First*, word trigrams are extracted from research paper titles. *Second*, a co-occurrence graph of the extracted trigrams is constructed. To construct the co-occurrence graph, the direction of edges is considered in two ways: *forward* and *reverse*. In the forward and reverse co-occurrence graphs, the trigrams point to other trigrams appearing after and before them in a paper title, respectively. *Third*, Jaccard similarity is computed between trigrams as the weight of the graph edge. *Fourth*, the weighted version of PageRank is applied. Consequently, the following two types of phrases can be obtained as the trigrams associated with the higher PageRank scores. The trigrams of the one type, which are obtained from the forward co-occurrence graph, can form a more specific query when users add a technical word or words before them. Those of the other type, obtained from the reverse co-occurrence graph, can form a more specific query when users add a technical word or words after them. The extraction of phrases is evaluated as additional features in the paper title classification task using SVM. The experimental results show that the classification accuracy is improved than the accuracy achieved when the standard TF-IDF text features are only used. Moreover, the trigrams extracted by the proposed method can be utilized to expand query words in research paper retrieval.

Keywords—word n-grams; Jaccard similarity; PageRank; TF-IDF; query expansion; information retrieval; feature extraction

I. INTRODUCTION

In these days, it is an important but complex task to get valuable information by searching the Web. With the rapid increase of information, users often perceive the difficulty of accessing the rich information resource effectively and of obtaining the information associated with their needs accurately. When users want to find the information relevant to their needs, they are required to find appropriate query words or phrases. However, the search results may not be relevant due to the inability of the queries to represent the needs accurately. Especially in academic paper retrieval, in many cases, users also want to find the papers focusing on specific and precise research topics, not general and vague topics. It can be considerably difficult for users to formulate a query for retrieving the papers discussing clear and specific topics. If the query contains only a single word, the search result consists of papers discussing a wide range of topics. That is, while the recall is high, the precision is low. If the query contains too many words, users may get only a limited number of academic papers as a search result. That is, while the precision is high, the recall is low. To overcome the above problems, the solution of this paper is to provide users with help in extracting from a large text set phrases that can be used to expand a less specific query. By expanding queries with the extracted phrases, users may get a search result containing a sufficient number of papers talking about specific research topics.

This paper proposes a new method for extracting important phrases as word *n*-grams from research paper titles. The extracted phrases are expected to be fruitful in query expansion for academic information retrieval. The proposed method is special in the following sense. The method extracts two types of phrases, each of which realizes a different query expansion, i.e., the expansion to the left and the expansion to the right. For example, the proposed method gives "a framework for" and "in sensor networks" as its outcome. The phrase "a framework for" can expand queries like "clustering", "classification", etc., to the left and give more specific queries like "a framework for clustering", "a framework for classification", etc. The phrase "in sensor networks" can expand queries like "clustering", "classification", etc., to the right and give more specific queries like "clustering in sensor networks", "classification in sensor networks", etc.

A brief explanation of the proposed method is given as follows. *First*, the proposed method extracts word trigrams as phrases that can be used for query expansion from a large number of research paper titles. There are two reasons why we focus on trigrams. The one reason is that, while word *n*-grams will be useful for text analysis, longer *n*-grams may cause data sparseness problem. Because the *n*-grams longer than three may be too long to obtain a sufficient large number of technical papers as a search result. The other is that unigrams and bigrams are too short to make a single word query express a specific and precise topic. *Second*, the proposed method builds a co-occurrence graph of the extracted trigrams. To construct the co-occurrence graph, the extracted word trigrams are used as nodes and the co-occurrence relations of trigrams appearing in the same paper titles as edges. Here, both the forward and reverse directions of edges are considered. In the forward co-occurrence graph, the trigram points to other trigrams appearing after it in a paper title. In the reverse co-occurrence graph, the trigram points to other trigrams appearing before it in a paper title. *Third*, the proposed method evaluates the Jaccard similarity for all co-occurring pair of trigrams and utilizes the similarity as the edge weight. And *fourth*, the proposed method applies a weighted version of PageRank on the forward and reverse co-occurrence graphs. As a result, we can get the top-ranked trigrams with reference to PageRank scores. Many of the top-ranked trigrams given from these two co-occurrence graphs can be regarded as important phrases. Details will be explained later.

Our first paper [18] describes a method for exploring technical phrase frames by extracting word *n*-grams. However, this paper introduces a new approach that applies weighted PageRank algorithm on the forward and reverse co-occurrence graphs of trigrams. The distinction between these two types of co-occurrence graphs does not appear in [18]. As a result, the two types of top-ranked trigrams are obtained. The performance of the extracted trigrams are evaluated as additional features in paper title classification using SVM. This evaluation is also not included in [18].

The remainder of this paper is divided into four sections. Section 2 describes the related work. Section 3 explains the proposed method. Section 4 contains the results of the evaluation experiment. The final section concludes the paper with discussion on future work.

II. RELATED WORK

The extraction of important word sequences, e.g. keyphrases and key sentences, is relevant to our problem. There are two types of extraction, i.e., supervised [2], [6], [7], [9] and unsupervised methods [1], [3], [4], [8], [10], [11]. Natural language processing techniques [12], [13], [14] have also been used for keyphrase extraction.

Mihalcea [15] proposed an unsupervised method for automatic sentence extraction using graph-based ranking algorithms. The author used a text graph to represent the interconnection of words or other text entities with meaningful relations, ranked the entire sentences in weighted graphs manner, sorted in reversed order of their scores and selected the top ranked sentences for summary. The author evaluated the method in text summarization task. The experimental results show that graph-based ranking algorithms (HITS and PageRank) are useful for sentence extraction when applied to graphs extracted from texts.

Litvak et al. [17] analyzed two graph-based approaches, i.e., unsupervised and supervised ones, which enhance to extract keywords to be used in summarizing documents. The researchers built a graph to represent the co-occurrence in a window of a fixed number of words. They used HITS algorithm to get the top-ranked keyword and identified the keywords in order to generate the summarization. As a result, they argued that if a large number of summarized documents were available then supervised classification was the most accurate to identify the keywords in a document graph. Unless the number of summarized documents are large, unsupervised classification is better to extract the keywords in a graph.

Wan et al. [16] proposed CollabRank, a collaborative approach to single-document keyphrase extraction from multiple documents. They implemented the CollabRank to obtain

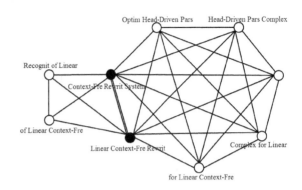

Fig. 1. A small portion of the co-occurrence graph

document clusters by using the clustering algorithm. They used the graph-based ranking algorithm to extract the keyphrases within each document cluster. They built a graph based on all candidate words in the documents of the given cluster and evaluated the candidate phrases in the document based on the scores of the words contained in the phrases. Finally, they chose a few phrases with highest scores as the keyphrases of the document.

Contribution. This paper proposes a method that applies weighted PageRank algorithm on the forward and reverse co-occurrence graphs of trigrams. Consequently, the method can extract two different types of trigrams that can be used for query expansion: 1) Many of the trigrams obtained from the forward co-occurrence graph can form a more specific query when users add a word *before* them (e.g. **clustering** for web search"); 2) Many of the trigrams obtained from the reverse co-occurrence graph can form a more specific query when users add a word *after* them (e.g. "automatic extraction of **clustering**"). This kind of bidirectional nature of extraction was not achieved by any of the PageRank-type methods described above.

III. THE PROPOSED METHOD

In this section, the four steps of the proposed method are explained.

A. Word Trigrams

First, the proposed method extracts trigrams from a large set of research paper titles after applying stemming. For example, the proposed method extracts from the paper title "Recognition of Linear Context-Free Rewriting Systems" the following trigrams: "Recognit of Linear","of Linear Context-Fre", "Linear Context-Fre Rewrit", and "Context-Fre Rewrit System". Word trigrams are extracted by using the natural language toolkit for python (NLTK).

B. Co-occurrence Graph

The next step of the proposed method is to construct a co-occurrence graph of the extracted trigrams. In order to build the co-occurrence graph, the extracted word trigrams are used as nodes. When two trigrams appear in the same title, they are connected by an edge. Fig. 1 shows a small portion of the co-occurrence graph. This portion is obtained from the

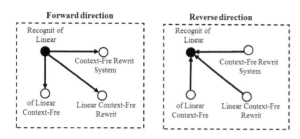

Fig. 2.   Co-occurring pairs of trigrams according to forward and reverse directions

TABLE I.   DATA SETS

Fields	Venue
NLP	ACL, EACL, COLING, CICLing, NAACL, IJCNLP, EMNLP, NLDB, TSD
DM	SIGMOD, VLDB, PODS, SIGIR, WWW, KDD, ICDE, ISWC, CIDR, ICDM, ICDT, EDBT, SDM, CIKM, ER, ICIS, SSTD, WebDB, SSDBM, CAiSE, ECIS, PAKDD
ALG	STOC, FOCS, ICALP, STACS, ISAAC, MFCS, FSTTCS, FCT, COCOON, CSR, WoLLIC
PRG	POPL, PLDI, ECOOP, OOPSLA, ISMM, ICLP, ICFP, CGO, ESOP, FOSSACS, CP, CC, LOPSTR, FLOPS, HOPL, AOSD

following two paper titles: "Recognition of Linear Context-Free Rewriting Systems" and "Optimal Head-Driven Parsing Complexity for Linear Context-Free Rewriting Systems".

Further, the direction of edges is specified according to the order of trigrams. The direction of edges is determined in two ways: *forward* and *reverse* directions, as shown in Fig. 2. On the left panel of Fig. 2, the trigram "Recognit of Linear" points to the trigrams "of Linear Context-Free", "Linear Context-Fre Rewrit", and "Context-Fre Rewrit System", because the latter three trigrams appear *after* the trigram "Recognit of Linear" in the paper title "Recognition of Linear Context-Free Rewriting Systems". This direction is called *forward* direction. In contrast, on the right panel of Fig. 2, the same trigram "Recongnit of Linear" is pointed by the other three trigrams. In this case, each trigram points to the trigrams appearing *before* it. This direction is called *reverse* direction. According to the forward and reverse directions of edges, the two co-occurrence graphs, i.e., forward co-occurrence graph and reverse co-occurrence graph, can be constructed.

C. Jaccard Similarity

In the third step, the Jaccard similarity is evaluated for all co-occurring pairs of trigrams and the similarity is utilized as the edge weight. Let (t_1, t_2) denote a pair of trigrams whose similarity is to be calculated. Let $S(t_i)$ denote the set of paper titles that contain the trigram t_i. The Jaccard similarity is computed between two trigrams t_1 and t_2 as follows:

$$sim(t_1, t_2) = \frac{|S(t_1) \cap S(t_2)|}{|S(t_1) \cup S(t_2)|} \quad (1)$$

After assigning the Jaccard similarity to each edge, a weighted version of PageRank algorithm is applied. The survey paper [5] analyzed many binary similarity measures. There are two reasons why we compute the Jaccard similarity. The first one is that it is simple to compute. The second one is that the Jaccard similarity is measured with the exclusion of *negative matches* [5]. In our approach, negative matches are related to the research paper titles where both of the trigrams under consideration do not appear and are not that important.

D. Weighted PageRank Algorithm

The last step of the proposed method applies weighted PageRank algorithm on both forward and reverse co-occurrence graphs of the extracted trigrams. Let $P(t_i)$ denote the PageRank scores of the trigram t_i. Let w_{ji} denote the weight assigned to the edge connecting the two co-occurring pairs of nodes, t_i and t_j. w_{ji} is set to the corresponding

Jaccard similarity. Then the PageRank score of the trigrams is calculated t_i by applying the Eq. (2) as below:

$$P(t_i) = \frac{1-d}{N} + d \times \sum_{t_j \in M(t_i)} \frac{w_{ji}}{\sum_{t_k \in M(t_j)} w_{jk}} P(t_j) \quad (2)$$

where $M(t_i)$ denotes the set of nodes which point to t_i and N is the total number of extracted trigrams. The parameter d is the damping factor that is usually set to 0.85. $\sum_{t_k \in M(t_j)} w_{jk}$ is the sum of the weights assigned to each neighbor t_k in $M(t_j)$. Intuitively, if a node is pointed by many high-scored neighbors, the node may get a high score. However, the proposed method combines the Jaccard similarity and weighted PageRank algorithm. Therefore, if a node is pointed by many high-scored neighbors with large Jaccard similarities, then the node may obtain a high score.

IV.   EXPERIMENTAL RESULTS

A. Evaluation in Text Classification

The trigrams extracted by the proposed method were evaluated as additional features in the paper title classification task. We used SVM (Support Vector Machine) for classification and checked whether the trigrams extracted by the proposed method improved the classification accuracy.

The proposed method was tested in the binary classification of the paper titles obtained from DBLP (Digital Bibliography & Library Project) [1]. Each DBLP record included a list of authors, title, conference name or journal name, year, page numbers, etc. Academic conferences were chosen in the four fields: Natural Language Processing (NLP), Data Management (DM), Algorithms and Theory (ALG), and Programming Languages (PRG). We only selected top conferences and used the research paper titles presented in the conferences shown in Table I. As a result, the total number of paper titles contained in NLP, DM, ALG, PRG data sets are 10,666, 27,573, 16,468, and 9,434, respectively. In the preprocessing, stop words were removed and Porter Stemmer [2] was used to stem words to their root forms.

Classification was conducted on the four data sets, i,e., NLP paper titles, DM paper titles, ALG paper titles and PRG paper titles. From these four data sets, six different pair of data sets were obtained as ALG_PRG, DM_ALG, DM_PRG, NLP_ALG, NLP_DM and NLP_PRG. For each pair, the data was randomly split into 90% of the paper titles for training and 10% for testing, and the classification was performed

---

[1] http://www.dblp.com/

[2] http://www.tartarus.org/martin/PorterStemmer/

with SVM. The accuracies were averaged over the ten results obtained from the 10-fold cross-validation.

TF-IDF term weighting was used to compose a feature vector for each paper title based on the formula: $\text{tf\_idf}(t, d) = \text{tf}(t, d) \times \log(N/\text{df}(t))$, where $\text{tf}(t, d)$ is the frequency of term t in document d, and $\text{df}(t)$ is the document frequency of t, i.e., the number of documents where t appears. N is the total number of documents in the corpus. In the experiment, the TF in TF-IDF was modified by using the trigrams obtained by the proposed method to improve the classification accuracy. First, we find the trigram having the largest PageRank score in each paper title. For each paper title d, the set of the three words of the trigram is denoted having the largest PageRank score by $W(d)$. Then, weight(t, d) is used, defined by Eqs. (3) and (4) in place of $\text{tf}(t, d)$:

$$\text{weight}(t, d) \equiv \alpha \times \text{tf}(t, d) + 1 \text{ for } t \in W(d) \quad (3)$$

$$\text{weight}(t, d) \equiv \alpha \times \text{tf}(t, d) \text{ for } t \notin W(d) \quad (4)$$

In Eqs. (3) and (4), α is the term reweighting parameter and is chosen as an integer. All word counts are increased by a factor of α and then the word counts are increased by one only for the words of the trigram having the largest PageRank score (cf. Eq. (3)). For example, we assume that the trigram "probabilist inform flow" has the largest PageRank score among the trigrams appearing in the paper title "Decidability of Parameterized Probabilistic Information Flow" after stemming. Then only the counts of the three words "probabilist", "inform", and "flow" are increased by one after we increase the counts of all words by a factor of α. If the trigrams extracted by the proposed method are important in the sense that they are closely related to a particular research topic and thus help discriminating the research topic from other topics, the reweighting described above may improve the classification accuracy.

SVM was trained with linear kernel by setting $C = 1.0$ and the classification accuracy was obtained in terms of Area Under the ROC curve (AUC). The term reweighting parameter α was varied from 3 to 27, and the mean and standard deviation of AUC in the 10-fold cross validation were recorded.

Tables II and III summarize the p-values obtained by comparing the standard TF-IDF (i.e., TF-IDF without modification of TF) and the TF-IDF based on the TF modified by Eqs. (3) and (4) in terms of AUC. The p-values are obtained in a paired two-sided t-test. If the classification accuracy of the proposed method is not as high as the frequency-based method, the p-value is assigned with a minus symbol. When the p-value is less than 0.05, we can say that the improvement is statistically significant and thus give the p-value in bold in Tables II and III. The results of Table II are given by using the trigrams obtained from the forward co-occurrence graph. On the other hand, the results of Table III are given by using trigrams obtained from the reverse co-occurrence graph. Tables II and III show the term reweighting factor α yielding the best p-values on each data set. Only for the two pairs, i.e., DM_PRG and NLP_PRG, in Table III, we could not get a statistically significant improvement. For all remaining cases in Tables II and III, we could get a significantly better accuracy than the standard TF-IDF. Based on these results, it can be said that the classification accuracy is improved by modifying the TF in TF-IDF by the trigrams the proposed method gives. So we claim that the proposed method can extract the features that

TABLE II. p-VALUES FOR PAIRED T-TEST ON ROC CURVE AUC (FORWARD)

Data sets	Standard TF-IDF	Modified TF-IDF	p-value
ALG_PRG	0.942075	0.942493 ($\alpha = 3$)	**0.009**
DM_ALG	0.978106	0.978225 ($\alpha = 8$)	**0.021**
DM_PRG	0.971507	0.971669 ($\alpha = 10$)	**0.029**
NLP_ALG	0.989345	0.989452 ($\alpha = 8$)	**0.003**
NLP_DM	0.954356	0.954432 ($\alpha = 27$)	**0.048**
NLP_PRG	0.985577	0.985633 ($\alpha = 19$)	**0.047**

TABLE III. p-VALUES FOR PAIRED T-TEST ON ROC CURVE AUC (REVERSE)

Data sets	Standard TF-IDF	Modified TF-IDF	p-value
ALG_PRG	0.942075	0.943616 ($\alpha = 9$)	**0.013**
DM_ALG	0.978106	0.978163 ($\alpha = 8$)	**0.020**
DM_PRG	0.971507	0.971566 ($\alpha = 23$)	0.052
NLP_ALG	0.989345	0.989422 ($\alpha = 12$)	**0.016**
NLP_DM	0.954990	0.954883 ($\alpha = 20$)	-0.053
NLP_PRG	0.985899	0.985959 ($\alpha = 22$)	**0.041**

are useful in discriminating different research topics as the trigrams having large PageRank scores.

B. Comparing with Frequency-based Trigram Extraction

To discuss the special nature of the trigrams extracted by the proposed method, we compared the proposed method with a simple method for the extraction of trigrams, i.e., the frequency-based extraction. In the frequency-based method, the same data sets were used and the same preprocessing were applied as in the proposed method. Then, the number of occurrences, i.e., frequency, were counted for every trigram, and the higher-ranked trigrams based on their frequencies were obtained. The difference between two methods are clarified by displaying examples.

Tables IV and V summarize the trigrams obtained by the frequency-based method and by the proposed method for ALG and DM data sets, respectively. For example, "the complex of" and 381 in the top cell of the left column of Table IV mean that the frequency is 381 for the trigram "the complex of". Moreover, "and relat problem" and 6.05×10^{-4} in the top cell of the middle column of Table IV mean that the PageRank is 6.05×10^{-4} for the trigram "and relat problem" in the forward co-occurrence graph. Furthermore, "on the complex" and 19.59 $\times 10^{-4}$ in the top cell of the right column of Table IV mean that the PageRank is 19.59 $\times 10^{-4}$ for the trigram "on the complex" in the reverse co-occurrence graph.

We can observe that many trigrams obtained from the forward co-occurrence graph can expand queries to the right. For example, the trigram "in web search" can expand the queries like "ranking" and "queries" to give more specific queries like "ranking in web search" and "queries in web search". On the other hand, many trigrams obtained from the reverse co-occurrence graph can expand queries to the left. For example, the trigram "efficient algorithm for" can expand the queries like "computing" and "mining" to give more specific queries like "efficient algorithm for computing" and "efficient algorithm for mining". This is a remarkable feature of the proposed method. In contrast, the frequency-based method

TABLE IV.   Top-10 (stemmed) trigrams of ALG

Frequency		PageRank			
		Forward ($\times 10^{-4}$)		Reverse ($\times 10^{-4}$)	
the complex of	381	and relat problem	6.05	on the complex	19.59
lower bound for	259	in polynomi time	5.52	the complex of	19.23
approxim algorithm for	225	in linear time	5.46	lower bound on	8.89
algorithm for the	209	and it applic	4.10	approxim algorithm for	8.69
on the complex	162	term rewrit system	3.75	lower bound for	8.12
the power of	103	in the plane	3.72	a note on	7.20
with applic to	100	constraint satisfact problem	3.24	bound on the	7.09
bound on the	91	in planar graph	3.05	effici algorithm for	6.84
lower bound on	81	of complex class	2.37	the power of	6.13
effici algorithm for	73	and their applic	2.34	on the power	5.76

TABLE V.   Top-10 (stemmed) trigrams of DM

Frequency		PageRank			
		Forward ($\times 10^{-4}$)		Reverse ($\times 10^{-4}$)	
a case studi	229	on the web	4.71	the impact of	8.06
the role of	219	for inform retriev	2.91	the effect of	6.56
the impact of	216	in inform retriev	2.68	the role of	4.41
a framework for	201	a case studi	2.62	a framework for	3.39
the effect of	184	in web search	2.50	a comparison of	3.04
for inform retriev	140	in social network	2.33	the influenc of	2.83
the case of	140	an empir studi	2.03	a studi of	2.71
in inform system	137	inform retriev system	2.00	the use of	2.06
on the web	134	an exploratori studi	1.94	effici process of	1.85
of inform system	123	in social media	1.90	an analysi of	1.84

TABLE VI. p-values for paired t-test on ROC curve AUC (forward)

Data sets	Frequency-based	Proposed	p-value
ALG_PRG	0.94904 ($\alpha = 5$)	0.94928 ($\alpha = 3$)	0.058
DM_ALG	0.98194 ($\alpha = 4$)	0.98202 ($\alpha = 2$)	0.546
DM_PRG	0.97613 ($\alpha = 2$)	0.97622 ($\alpha = 2$)	0.333
NLP_ALG	0.99140 ($\alpha = 5$)	0.99148 ($\alpha = 4$)	**0.031**
NLP_DM	0.96251 ($\alpha = 2$)	0.96276 ($\alpha = 2$)	0.079
NLP_PRG	0.98699 ($\alpha = 5$)	0.98706 ($\alpha = 4$)	0.336

TABLE VII. p-values for paired t-test on ROC curve AUC (reverse)

Data sets	Frequency-based	Proposed	p-value
ALG_PRG	0.94245 ($\alpha = 4$)	0.94256 ($\alpha = 3$)	0.689
DM_ALG	0.97828 ($\alpha = 5$)	0.97822 ($\alpha = 5$)	−0.384
DM_PRG	0.97235 ($\alpha = 3$)	0.97214 ($\alpha = 3$)	−0.195
NLP_ALG	0.98942 ($\alpha = 5$)	0.98941 ($\alpha = 5$)	−0.774
NLP_DM	0.95450 ($\alpha = 6$)	0.95422 ($\alpha = 6$)	**−0.024**
NLP_PRG	0.98563 ($\alpha = 3$)	0.98548 ($\alpha = 5$)	−0.054

cannot give these two types of trigrams separately, because all trigrams are mixed in the same ranking, as shown in the left columns of Tables IV and V.

Further, we can observe that the frequency-based ranking tends to provide trigrams having a general meaning like "lower bounds for", "the power of", "with applications to", "bounds on the", "lower bounds on", "a case study", "the case of", etc., where the original form is recovered from the root form of each word. In contrast, the proposed method tends to provide trigrams having a specific meaning, e.g. like "in polynomial time ", "in linear time", "in planar graphs", "term rewriting systems", "constraint satisfaction problems", "of complexity classes", "information retrieval system", "an empirical study", etc., with respect to the forward co-occurrence graph. Also with respect to the reverse co-occurrence graph, many trigrams given by the proposed method have at least as specific a meaning as the trigrams given by the frequency-based method. Therefore, it can be said that, at least with respect to the forward co-occurrence graph, the top-ranked trigrams obtained by the proposed method have a more specific meaning than those obtained by the frequency-based method.

However, it is possible that the proposed method may degrade the quality of the extracted trigrams by providing them in two separate rankings. Therefore, we compared the proposed method with the frequency-based method also in text classification task described in Section IV-A. We also used SVM (Support Vector Machine) for classification and checked if the trigrams extracted by the proposed method were as useful as the trigrams extracted by the frequency-based method.

To obtain the best classification accuracy in terms of Area Under the ROC curve (AUC), SVM was trained with two different kernels, namely linear kernel by setting $C = 1.0$ and rbf (Radial Basis Function) kernel by setting $C = 2.0$ and $gamma = 2.0$. We selected the term reweighting parameter α yielding the best case from each kernel and recorded the mean and standard deviation of AUC in the 10-fold cross validation.

Tables VI and VII summarize the p-values obtained by comparing the frequency-based method and the proposed method based on the TF modified by Eqs. (3) and (4) in terms of AUC. The *p*-values are obtained in a paired two-sided *t*-test. The *p*-value is assigned with a minus symbol if the classification accuracy of the proposed method is not as high as the frequency-based method. When the *p*-value is less than 0.05, it can be said that the improvement is statistically significant. The results of Table VI are given by using the trigrams obtained from the forward co-occurrence graph, where SVM is trained by using the rbf kernel. On the other hand, the results of Table VII are given by using the trigrams obtained from the reverse co-occurrence graph, where SVM is trained by using the linear kernel. For all but one case in Tables VI and VII, we could get as good an accuracy as the frequency-based method. We could not get a comparable accuracy only for the NLP_DM data set pair in Table VII. Consequently, the result showed that the proposed method at least could extract as effective trigrams as the frequency-based method. It can be said that the bidirectional nature of the proposed method is an extra gain, which cannot be achieved by the frequency-based method.

C. A Possible Application: Query Expansion

Based on the experimental results, it can be said that the trigrams extracted by the proposed method represent technical research topics well. We here discuss how such trigrams can be used in query expansion for information retrieval.

For example, as presented in Fig. 3, the query word "clustering" can be expanded to the right by the trigrams "in sensor networks", "for web search", "for text categorization", etc., which are obtained by the proposed method from the forward co-occurrence graph. These trigrams can be used for the *right* expansion in this manner, because their first word (i.e., "for","in","of", etc.) is a function word that mainly follows a noun. As we discussed in Section IV-B, the trigrams obtained by the proposed method tend to represent a specific meaning, especially with respect to the forward co-occurrence graph. Therefore, we may expect that the search results obtained by the queries expanded in this manner will relate to specific research topics. Fig. 4 gives another example. The query word "clustering" is expanded to the left by the trigrams "a framework for", "automatic extraction of", "efficient algorithm for", etc., which are obtained from the reverse co-occurrence graph. These trigrams can be used for the *left* expansion, because their last word (i.e., "for", "of", etc.) is a function word that is mainly followed by a noun.

It should be noted that a similar expansion cannot be straightforwardly achieved by the trigrams obtained by the frequency-based method, because the trigrams that can be used for the right expansion and those that can be used for the left expansion are mixed in the same ranking as shown in the left columns of Tables IV and V. However, the proposed method provides two types of trigrams in two different rankings, as shown in the middle and right columns of Tables IV and V.

We here verify how the search results obtained by the expanded queries can focus on more specific research topics. Fig. 5 shows the three types of search results obtained from Google Scholar. Fig. 5(a) gives the search results for the query "clustering". Fig. 5(b) gives the search results obtained by the proposed method from the forward co-occurrence graph. Fig. 5(c) gives the search results obtained by the proposed

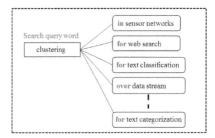

Fig. 3.   Example of possible right expansions of the query word 'clustering' by using the trigrams obtained from the forward co-occurrence graph for DM data set

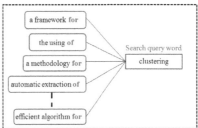

Fig. 4.   Example of possible left expansions of the query word 'clustering' by using the trigrams obtained from the reverse co-occurrence graph for DM data set

method from the reverse co-occurrence graph. As presented in Fig. 5(a), we can get the search result having a general meaning when we only input a single query word "clustering". For example, the topics like "Algorithms for clustering data" and "A comparison of document clustering techniques" tend to provide a general meaning consisting of the words like "algorithms" and "techniques". These words tend to represent a wide range of topics. Consequently, the single query word has not exploited users' needs and interests, but users can't get relevant topics when each user has a specific need.

In contrast, when a single query word "clustering" is expanded to the right by the phrase "in sensor networks", we can get the search results focusing on more specific topics as shown in Fig. 5(b). Most of the words or phrases appearing in the search results, e.g., "hybrid", "ad hoc", "hierarchical", and "wireless", have a specific meaning. On the other hand, when a single query word "clustering" is expanded to the left by the phrase "a framework for", we can also get the search results focusing on more specific topics as shown in Fig. 5(c). Some of the words or phrases occurring in the search results, e.g., "data streams", "high dimensional", and "Text and Categorical", represent a specific meaning.

Therefore, it is found that the query expansion, i.e., the expansion to the left and the expansion to the right, can give the search results relating to specific topics. Further, we can get two different types of search results due to the bidirectional nature of the proposed method. These results are more specific when we expand query words than when we only use a single query word. We can observe that the proposed method works as a new query expansion scheme more oriented toward actual user needs and interests for informational retrieval.

V. Conclusion

In this paper, we proposed a new method for query expansion based on bidirectional extraction of phrases. The proposed

Fig. 5.   Example of the search results for the query (a) 'clustering' (b) 'clustering in sensor networks' and (c) 'a framework for clustering'

method extracted important phrases as trigrams based on a procedure consisting of four processing steps. The trigrams extracted by the proposed method were evaluated as additional features in the paper title classification task using SVM. The experimental results showed that the accuracy was improved. We also compared the trigrams given by the proposed method with those given by the frequency-based method. According to the experimental results, the proposed method could provide as good trigrams as the frequency-based method. However, the proposed method has an extra gain, i.e., the bidirectional nature of trigrams extraction, which cannot be achieved by the frequency-based method. Further, we discussed how we could use such trigrams for query expansion. A search system using this type of query expansion can give search results relating to specific topics.

We have a future plan to perform a quantitative evaluation of the search results obtained by the query expansion based on the proposed method in information retrieval task.

ACKNOWLEDGMENT

This work has been supported by the Grant-in-Aid for enhancement of engineering higher education of the Japan International Cooperation Agency (JICA) from 2014 to 2017. We are grateful for their support.

REFERENCES

[1]   K.S. Hasan and V. Ng, "Conundrums in unsupervised keyphrase extraction: making sense of the start-of-the-art," in *Proc. of the 23rd International Conference on COLING 2010*, Beijing, pp. 365–373, August 2010.

[2]   C. Caragea, F. Bulgarov, A. Godea, S.D. Gollapalli, "Citation-enhanced keyphrase extraction from research papers: a supervised approach," in *Proc. of the 2014 Conference on Empirical Methods in Natural Language Processing (EMNLP)*, pp. 1435-1446, Doha, Qatar, October 2014.

[3]   R. Mihalcea, P. Tarau and E. Figa, "PageRank on semantic networks with application to word sense disambiguation," in *Proc. of the 20th International Conference on Computational Linguistics*, Stroudsburg, PA, USA, no. 1126, August 2004.

[4]   S.D. Gollapalli and C. Caragea, "Extracting keyphrases from research papers using citation networks," in *Proc. of the 28th AAAI Conference on Artificial Intelligence*, pp. 1629–1635, June 2014.

[5]   S. Choi, S. Cha, and C.C. Tappert, "A survery of binary similarity and distance measures," *J.Syst. Cybern, Inf.*, vol. 8, no. 1, pp. 43–48, 2010.

[6]   D. X. Wang, X. Gao, and P. Andreae, "Automatic keyword extraction from single-sentence natural language queries," in *Proc. of the 12th Pacific Rim International Conference on Artificial Intelligence*, Kuching, Malaysia, pp. 637–648, September 2012.

[7]   K. Zhang, H. Xu, J. Tang, and J. Li, "Keyword extraction using support vector machine," in *Proc. of the 7th International Conference on WAIM*, Hong Kong, China, pp. 85–96, June 2006.

[8]   T. Nomoto and Y. Matsumoto, "A new approach to unsupervised text summarization," in *Proc. of SIGIR2001*, pp. 26–34, 2001.

[9]   M. R. Amini and P. Gallinari, "The use of unlabeled data to improve supervised learning for text summarization," in *Proc. of SIGIR2002*, 105-112, 2002.

[10] S. N. Kim and M. Kan, "Re-examining automatic keyphrase extraction approaches in scientific articles," in *Proc. of 2009 Workshop on Multiword Expressions, ACL-IJCNLP 2009*, pp. 9-16, Suntec, Singapore, August 2009.

[11] Z. Liu, P. Li, Y. Zheng and M. Sun, "Clustering to find exemplar terms for keyphrase extraction," in *Proc. of 2009 Conference on Empirical Methods in Natural Language Processing*, pp. 257-266, Singapore, August 2009.

[12] T. Tomokiyo and M. Hurst, "A language model approach to keyphrase extraction," in *Proc. of ACL 2003 Workshop on Multiword Expressions: Analysis, Acquisition and Treatment*, vol. 18, pp. 33–40, 2003.

[13] K. Barker and N. Cornacchia, "Using noun phrase heads to extract document keyphrases," in *Proc. of 13th Biennial Conference of the Canadian Society for Computational Studies of Intelligence*, pp. 40–52, May 2000.

[14] S. N. Kim, T. Baldwin and M. Kan, "Evaluating n-gram based evaluation metrics for automatic keyphrase extraction," in *Proc. of 23rd international conference on COLING 2010*, pp. 572-580, Beijing, August 2010.

[15] R. Mihalcea, "Graph-based ranking algorithms for sentence extraction, applied to text Summarization," in *Proc. of ACL 2004 on Interactive Poster and Demonstration Sessions*, Article no. 20, Stroudsburg, PA, USA, July 2004.

[16] X. Wan and J. Xiao, "CollabRank: towards a collaborative approach to single-document keyphrase extraction," in *Proc. of 22nd International Conference on COLING 2008*, pp. 969–976, Manchester, August 2008.

[17] M. Litvak and M. Last, "Graph-based keyword extraction for single-document summarization," in *Proc. of 08 MMIES on COLING 2008*, pp. 17-24, Manchester, August 2008.

[18] Y. Win and T. Masada, "Exploring technical phrase frames from research paper titles," in *Proc. of 29th IEEE International Conference on WAINA-2015*, pp. 558–563, Gwangju, South Korea, 2015.

13

Comparative Analysis of Improved Cuckoo Search(ICS) Algorithm and Artificial Bee Colony (ABC) Algorithm on Continuous Optimization Problems

Shariba Islam Tusiy[1], Nasif Shawkat[2], Md. Arman Ahmed[3], Biswajit Panday[4], Nazmus Sakib[5]

Ahsanullah University of Science & Technology (AUST), Dhaka, Bangladesh

Abstract—**This work is related on two well-known algorithm, Improved Cuckoo Search and Artificial Bee Colony Algorithm which are inspired from nature. Improved Cuckoo Search (ICS) algorithm is based on Lévy flight and behavior of some birds and fruit flies and they have some assumptions and each assumption is highly observed to maintain their characteristics. Besides Artificial Bee Colony (ABC) algorithm is based on swarm intelligence, which is based on bee colony with the way the bees maintain their life in that colony. Bees' characteristics are the main part of this algorithm. This is a theoretical result of this topic and a quantitative research paper.**

Keywords—Artificial Bee Colony (ABC) algorithm; Bioinformatics; Improved Cuckoo Search (ICS) algorithm; Lévy flight; Meta heuristic; Nature Inspired Algorithms

I. INTRODUCTION

Beautiful nature is full of surprises and mystery. People have learnt a lot from the Mother nature. By analyzing symptoms people manage to reveal the mystery of nature. As time changes, humans also change their characteristics and their behavior to the nature. Now a day's people find solutions of their daily life problems with the help of nature and that is known as meta-heuristic solutions. The bee colony and the improved cuckoo search algorithm elevate the eco-life system in a new level. On the basis of key functions and iteration number, the comparison between Artificial Bee Colony and Improved Cuckoo Search algorithm is done. Artificial Bee Colony works on the optimization algorithm introduced by D. Karaboga[1]. And the Improved Cuckoo Search algorithm is extended to more complicated cases in which each nest has multiple eggs representing a set of solutions[2][3][4]. Within last few decades, dozens of meta-heuristic algorithms are published and still been publishing. Among them Bat [5][6], Firefly [7][8], Flower Pollination [9], Artificial Bee Colony [10], Improved Artificial Bee colony [11], Ant Colony [12], Cuckoo search [13] is highly recommended algorithms. The algorithms which have mentioned above are upgrading day by day. So, here it has been focused on the implementation and the operations of the iteration number, and the tested functions for both algorithms that mentioned above are same. For preparing this research, first of all, the data of mean and median for improved cuckoo search have been measured and the algorithm is obtained. Then the comparison makes them different from each other. By producing graphical outcome, it

is observed that improved cuckoo search is good enough. Improved cuckoo search (ICS) & its algorithm is being described in section II. Then in section IV the artificial bee colony (ABC) is being described with its algorithm. After that in section V the simulation & analysis part is being described and then the findings in section VI. Finally, in section VII the total work is being summarized in short in the conclusion.

II. CUCKOO SEARCH

A. Basic Ideas of Cuckoo Search

Cuckoo Search (CS) is used to solve optimization problems which are a meta–heuristic algorithm, developed by 'Xin-She Yang' that is based on the manner of the cuckoo species with the combination of Lévy flight behavior of some birds and fruit flies [14][15]. The inspiration behind developing Cuckoo Search Algorithm is the invasive reproductive strategy and the obligate brood parasitism of some cuckoo species by laying their eggs in the nest of host birds [16]. Some female cuckoo like Guira and Ani can copy the patterns and colors of few chosen host species. This imitates power is used to increase the hatching probability which bring their next generation. The cuckoo has an amazing timing of laying eggs. Parasitic cuckoos used to choose a nest where the host birds lay their own eggs and it takes less time to hatch cuckoo's egg than the host bird's eggs. After hatching the first egg, the first instinct, action is to throw out the host eggs or to propel the eggs out of the nest to ensure the food from the host bird.

B. Basic Points of Cuckoo Search

Each Cuckoo's egg in a nest illustrates a new solution. The aim of Improve Cuckoo Search is to serve the new and potentially better solutions to replace the previous solutions in the Cuckoo Search. The algorithm can be extended to more complicated cases in which each nest has multiple eggs that represent a set of solutions. The CS is based on three idealized rules that are given bellow:

1) Each cuckoo lays one egg at a certain time, and dumps it in a nest which is randomly chosen [17].

2) The best nests provide high quality of eggs (solutions) that will carry over to the next generations [17].

3) A host bird can discover an alien egg from his nest with probability of Pa ∈ [0, 1]. In this case, the host bird can either throw the egg away or abandon or can completely build a new nest in a new location [17].

C. Lévy Flights

Generally, the foraging path of an animal is successful a random walk as the next step is based on both the current location and the transition probability to the next location. The chosen direction implicitly depends on a probability, which can be modeled mathematically. The flight behavior of many animals and insects demonstrates the typical characteristics of Lévy flights. A Lévy flight is a random walk in which the step-lengths are distributed according to a heavy probability distribution. After a large number of steps, the distance from the origin of the random walk tends to a stable distribution [17][18].

III. IMPROVED CUCKOO SEARCH (ICS)

A. Characteristics of Improved Cuckoo Search

The parameters Pa, λ and α introduced in the CS, help the algorithm to find globally and locally improved solutions, respectively. The parameters Pa and α is very important parameters in fine-tuning of solution vectors, and can be potentially used to adjust the convergence rate of the algorithm. The traditional CS algorithm uses a fixed value for both Pa and α. The key difference between ICS and CS is the way of adjusting Pa and α. To improve the performance of CS algorithm and eliminate the drawbacks lies with fixed values of Pa and α, the ICS algorithm uses variables Pa and α. The values of Pa and α dynamically change with the number of generations and have been expressed in equations 1-3, where NI and gn are the number of total iterations and the current iteration respectively.

$$P_a(g_n) = P_{a\,max} - \frac{g_n}{NI}(P_{a\,max} - P_{a\,min}) \qquad (1)$$

$$\alpha(g_n) = \alpha_{max}\exp(c.g_n) \qquad (2)$$

$$c = \frac{1}{NI}L_n\left(\frac{\alpha_{min}}{\alpha_{max}}\right) \qquad (3)$$

B. Algorithm of ICS

Begin

Objective function f(x), x = (x₁, ..., x_d)ᵀ ;
Initial a population of *n* host nests xi (i = 1, 2, ..., n);
while (t<MaxGeneration) or (stop criterion)
  Get a cuckoo (say i) randomly by *Lévy* flights;
  Evaluate its quality/fitness F_i;
  Choose a nest among *n* (say j) randomly;
  if(F_i>F_j) Replace j by the new solution; *end*
A fraction (P_a) of worse nests are abandon and new once are built.
Keep the best solutions (or nests with quality solutions);
Rank the solutions and find the current best;
end while
Post-process results and visualization;

End

When generating new solutions X_i (t +1) for the iᵗʰ cuckoo, the following Lévy flight is performed

$$X_i(t+1) = X_i(t) + \alpha \oplus \text{Lévy}(\lambda) \qquad (4)$$

Where α>0 is the step size, which should be related to the scale of the problem of interest. The product $\oplus$ means an entry-wise multiplications. According to Yang's research work, it has considered that a *Lévy* flight in which the step-lengths are distributed according to the following probability distribution

$$\text{Lévy } u = t - \lambda, \ 1 < \lambda \le 3 \qquad (5)$$

This has an infinite variance. Here, the consecutive steps of a cuckoo essentially form a random walk process which obeys a power law step length distribution with a heavy tail.

It is worth pointing out that, in the real world, if a cuckoo's egg is very similar to a host's egg, then this cuckoo's egg is less likely to be discovered, thus the fittest should be related to the difference in solutions. Therefore, it is a good idea to do a random walk in a biased way with some random step sizes.

IV. ARTIFICIAL BEE COLONY (ABC)

A. Basic Ideas of Artificial Bee Colony (ABC)

The ABC algorithm is of wide range of insects that are dependent and meta-heuristic algorithm that is developed on the provision behavior of honey bee colonies [19]. The ABC is an algorithm which describes the intelligent provision behavior of honey bee swarms. It is simple, vigorous, strong and healthy and population dependent randomly determined optimization algorithm [20]. The ABC algorithm which may be used for explanation of multidimensional and multimodal optimization matters [21].

B. Some Common Mistakes

- In ABC, honey bees are classified into three groups that are named as employed bees, onlooker bees and scout bees.

- The employed bees are the bee which searches for the food source and gather the information about the quality of the food source.

- Onlooker bees stay in the hive and search the food sources on the basis of the information gathered by the employed bees.

- The scout bee, searches new food sources randomly in places of the abundant food sources.

C. Algorithm of ABC

1) Algorithm 1 Artificial Bee Colony Algorithm
Initialize the parameters;

While Termination criteria is not satisfied do

Step 1: Employed bee phase for computing new food sources.

Step 2: Onlooker bees phase for updating the location of food sources based on their amount of nectar.

Step 3: Scout bee phase for searching about new food sources in place of rejected food sources.

Step 4: Memorize the best food source identified so far.

End of while

Output The best solution obtained so far.

2) Algorithm 2 Solution update in Employed bee phase
Input: solution xi, probi and j ∈ (1, D);
for j ∈ {1 to D} do
    if U (0, 1) >probi then
     vij = x ij + φ ij (x ij – x kj) + ψ ij (x best j – x ij) ;
    else
     vij = x ij ;
    end if
end for

V. SIMULATION AND ANALYSIS

A. GRAPH with Parameter settings

In this paper, 70 independent runs on each algorithm to get the result from the test functions which are rowed in Table-1. The population for each function is set for 14. Maximum cycle has been used 70 for both algorithms. And the Dimension for each function for each algorithm are set for D=5, 10, 15, 25 respectively. Of ICS, we use Pa, α, λ for improved the result for ICS globally and locally. Here in ABC Pa=0.25, αmin=.05, αmax=0.5. Finding the best, worst, mean, median and Standard deviation value for both algorithms is the main goal. On the basis of the result of finding the best and new place as well as nest or colony the3D surface and mesh are simulated for Rosen rock function, Ackley functions have shown in two views. In X-axis the objective value and in Y-axis two variable values is plotted. For this MATLAB R2013a version is used for simulation with 4th generation Intel i5 processor 2.7GZ with 4GB RAM of PC.

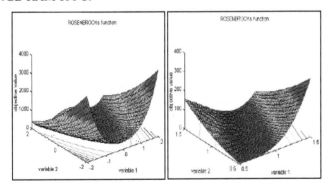

Fig. 1.   3D surface plots (2 view) of Rosenbrock function that best for ICS

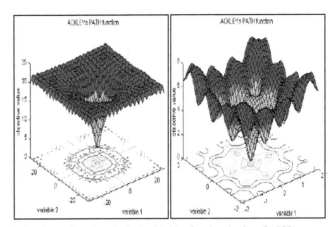

Fig. 2.   3D surface plots (2 view) of Ackley function that best for ICS.

VI. FINDINGS

In ABC algorithm the fitness and global min is compared with ICS algorithm. The work has mainly focused on the differences between these two algorithms and its basis on the mean value and on the basis of time. It is true that, if the exploitation is too high and the exploration is too low, then algorithms may trap into locally optimal points. So, some methods are followed and tried to avoid any kinds of trap that may cause trouble. Because this could affect find the global optimum. Or if the exploration is too high and exploration is too low than exploration then the convergence speed will decrease. In the ABC algorithm we use population for about 14 and tested in f1, f2, f3, f4, f5, f6 functions with runtime 70 for both ABC and ICS algorithm. When the dimension increases ABC gives poorer results than ICS but gives good result in lower dimensions. That means ICS gives the best result in high Dimension. So, it can be said that ABC works well in exploitation, but in the exploration it works poorly. But ICS works better in exploration. In this experiment ICS shows better results for dimension 10 and 25.For dimension 10, the ICS gives better result than ABC for f1, f3, f4, f5 functions on the basis of the mean value. And for dimension 25 ICS gives better result for f1, f3, f4, f5. And for other two dimensions, it works equally as ABC. Among these functions Rosenbrock gives the best result for ICS. Basically, in ICS cuckoo search his food within a wide range of area, not in a limited range of area. That means its food area is large. On the other hand In ABC a bee only finds its honey on its own place where the least and maximum capacity honey holder bees are present .If the bee fails to find honey from other sources that has the maximum capacity of honey than the bee turns back and looks for other bees that has the maximum amount of honey [22][23][24].

TABLE I.   Benchmark Functions used in the Experimental Studies. here, D: Dimensionality of the Function, S: Search Space, C Function Characteristics with Values — U: Unimodal and M: Multimodal

func	Name	D	C	S	Function Definition	f_{min}
$f1$	Sphere	5,10,15,25	U	[-5.12, 5.12]D	$f(x) = \sum_{i=1}^{d} x_i^2$	0.0
$f2$	Griewank	5,10,15,26	M	[-15, 15]D	$f(x) = \frac{1}{4000} \sum_{i=1}^{d} x_i^2 - \prod_{i=1}^{d} cos\frac{x_i}{\sqrt{i}} + 1$	0.0
$f3$	Rastrigin	5,10,15,27	M	[-15, 15]D	$f(x) = \sum_{i=1}^{d} [x_i^2 - 10\cos(2\pi x_i) + 10]$	0.0
$f4$	Rosenbrock	5,10,15,28	U	[-15, 15]D	$f(x) = \sum_{i=1}^{d-1} [100(x_{i+1} - x_i^2)^2 + (x_i - 1)^2]$	0.0
$f5$	Ackley	5,10,15,29	M	[-32, 32]D	$f(x) = -20\exp\left(-0.2\sqrt{\frac{1}{d}\sum_{i=1}^{d} x_i^2}\right) - \exp\left(\frac{1}{n}\sum_{i=1}^{n} cos\, 2\pi x_i\right) + 20 + e$	0.0
$f6$	Schwefel	5,10,15,25	M	[-500, 500]D	$f(x) = 418.9829 * d - \sum_{i=1}^{n} -x_i \sin\left(\sqrt{|x_i|}\right)$	0.0

TABLE II.   Comparison Between ABC & ICS on 6 Standard Benchmark Functions. All Algorithms are Run 24 Different Times on Each of the Functions. the Best Result for Each Function with Each Dimension is Marked Bold

func	Name	Algorithm	Dim	Best	Worst	Mean	Median	SD
$f1$	Sphere	ABC	5	1.42E-15	4.02E-12	**6.78E-13**	4.04E-14	2.10E-12
		ICS	5	6.90E-02	2.00E+00	1.00E+00	9.32E-01	7.92E-01
		ABC	10	0.0011	396.638	59.95378571	0.6839	137.54161
		ICS	10	0.0562095	6.667	**2.399571888**	0.475504	3.02238008
		ABC	15	8.15E-07	0.406	**0.077864353**	0.0018	0.21592723
		ICS	15	0.0515647	11.4665	3.957276326	0.353732	5.31127886
		ABC	25	2.8065	4808.4	1494.409743	882.56	2418.11741
		ICS	25	0.053437	9.13084	**3.194834213**	0.400225	4.19977765
$f2$	Griewank	ABC	5	0.0099	0.1052	**0.041857143**	0.0296	0.04682883
		ICS	5	0.0622457	3.70203	1.476183907	0.664273	1.59298316
		ABC	10	0.0197	0.2396	**0.0757**	0.0557	0.0693187
		ICS	10	0.0569547	6.15002	2.234685552	0.497087	2.77257803
		ABC	15	0.0322	0.1881	**0.073914286**	0.0375	0.09214291
		ICS	15	0.0596114	4.6883	1.775403562	0.578299	2.0705851
		ABC	25	0.9752	4.1287	**1.953914286**	1.1097	1.76112841
		ICS	25	0.0516801	11.3016	3.903275186	0.356502	5.23291644

f3	Rastrigin	ABC	5	8.79E-10	1.0087	**0.428446591**	0.00012567	0.75568014
		ICS		0.05697	6.13833	2.230957736	0.497574	2.76877797
		ABC	10	3.57E-11	7.44E+09	1298.930031	0.0777	3180.44622
		ICS		0.0669566	2.39661	**1.103110736**	0.845762	0.96833078
		ABC	15	2.0855	10.7955	6.661642857	7.5693	2.8464014
		ICS		0.0556383	7.09699	**2.537333517**	0.45937	3.22837601
		ABC	25	0.9989	9.4733	4.462328571	3.7386	4.27762977
		ICS		0.0553482	7.34006	**2.61550469**	0.451101	3.34467278
f4	Rosenbrock	ABC	5	4381.7	95705	19810.21429	8493.5	47395.5977
		ICS		0.0534698	9.09728	**3.183930407**	0.401039	4.18377811
		ABC	10	570.316	2.27E+07	4704766.516	767.163	8153588.31
		ICS		0.0547036	7.88175	**2.790039516**	0.433669	3.60370266
		ABC	15	25.9912	9.64E+09	1376914539	254.39	5151941350
		ICS		0.0518323	11.1223	**3.844661369**	0.359818	5.14762771
		ABC	25	0.2996	6459.9	962.3464143	68.6211	3428.69951
		ICS		0.051982	10.914	**3.77656329**	0.363682	5.04855246
f5	Ackley	ABC	5	1.67E+01	20.0102	19.53458571	20	1.74778989
		ICS		0.935065	12.2685	**5.024255357**	1.86915	5.13666397
		ABC	10	20	21.2061	20.31675714	20.0979	0.44159039
		ICS		0.0689939	2	**3.002425739**	0.933432	2.15179115
		ABC	15	20	20.4971	20.09742857	20.0045	0.26100873
		ICS		0.0563243	6.58119	**2.372110814**	0.478821	2.98126045
		ABC	25	20.0354	21.5963	20.4319	20.3623	0.77635362
		ICS		0.0530613	9.54667	**3.330121363**	0.39063	4.39792509
f6	Schwefel	ABC	5	-318.175	14.2288	**-238.4171**	-252.299	166.583262
		ICS		0.0615128	3.94352	1.548055762	0.639132	1.71018618
		ABC	10	-577.248	-599.325	**-690.0644574**	-607.041	88.8336281
		ICS		0.0602585	4.39303	1.684305824	0.599625	1.92797598
		ABC	15	-917.593	-864.837	**-931.2613**	-944.684	45.1666012
		ICS		0.0535746	8.98318	3.146848581	0.403787	4.12938848
		ABC	25	-1590.9	-693.199	**-1457.342729**	-1581.6	476.567607
		ICS		0.0515618	0.353507	3.968892143	0.353507	5.32786014

On the X-axis the number of generations and in Y-axis fitness is set and plot this graph (fig-3). From this graph the comparison is clarified clearly.

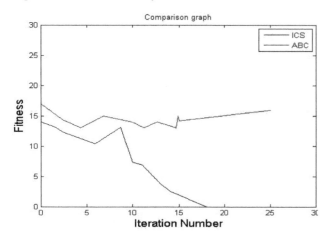

Fig. 3.   2D plot of ICS vs ABC algorithm

VII.   Conclusion

a) This paper represents the comparative study between swarm intelligence base and Lévy flight behavior base algorithms the Artificial Bee Colony (ABC) algorithm [1] and the Improved Cuckoo Search (ICS) algorithm [2][3][4].Optimization results in the standard benchmark problems for Artificial Bee Colony (ABC) algorithm and Improved Cuckoo Search (ICS) algorithm exhibit the effective results and competitive results of the algorithms. The main reason of the performance difference is basically in ICS where cuckoo search his food within a vast area rather than limited. On the contrary, in ABC a bee only finds its own light in its own place even though the light holder bees are present which have the minimum and maximum intensity of light. And if the bee fails to find from others that hold the highest capacity of light, then it turns back and search for other bees which have the larger intensity of light. And its place is limited not wide while searching for bees. So, that means cuckoo works on a wide range of area and it needs more dimension than ABC. On the other side, ABC needs fixed area to search its best. Last of all, it can be assumed that ICS and ABC can be improved more than before in the future.

References

[1]   D. Karaboga, "An Idea Based On Honey Bee Swarm For Numerical Optimization," Technical Report-TR06,Erciyes University, Engineering Faculty, Computer Engineering Department, 2005.

[2]   X. S. Yang, "Nature-Inspired Metaheuristic Algorithms," LuniverPress,UK, 2008.

[3]   M. Gendreau, "Handbook of Metaheuristics," in An introduction to tabu search, Kluwer Academic Publishers, 2003, p. 37–54.

[4]   X. Yang and S. Deb, "Engineering Optimisation by Cuckoo Search," in Int. J. Mathematical Modelling and Numerical Optimisation, 2010, p. 330–343.

[5]   X.-S. Yang, "Bat algorithm: literature review and applications," Int. J. Bio-Inspired Computation, vol. 5, p. 141–149, 2013.

[6]   I. J. Fister, D. Fister and X.-S. Yang, "A hybrid bat algorithm," in Elektrotehniski vestnik, press, 2013.

[7]   X.-S. Yang and s. H. Xing, "Firefly Algorithm: Recent Advances and Applications," Int. J. Swarm Intelligence, vol. 1, p. 36–50, 2013.

[8]   X.-S. Yang, "Firefly algorithms for multimodal optimization," Stochastic Algorithms: Foundations and Applications, SAGA 2009, Lecture Notes in Computer Sciences, vol. 5792, pp. 169-178, 2009.

[9]   X.-S. Yang, "Flower pollination algorithm for global optimiza-tion," Unconventional Computation and Natural Computation 2012,Lecture Notes in Computer Science, vol. 7445, pp. 240-249, 2012.

[10]   D. Karaboga, "An Idea Based on Honey Bee Swarm for Numerical Optimization.Technical Report-TR06," in Erciyes University, Computer Engineering Department, 2005.

[11]   M. Kiran and A. Babalik, "Improved Artificial Bee Colony Algorithm for Continuous Optimization Problems," Journal of Computer and Communications, vol. 2, pp. 108-116, 2014.

[12]   A. Farzindar and V. (. Keselj, Canadian AI 2010, LNAI 6085, 2010.

[13]   I. J. Fister, X. Yang, D. Fister and I. Fister, "Cuckoo search: A brief literature review," Cuckoo Search and Firefly Algorithm: Theory and Applications, Studies in Computational Intelligence, vol. 516, pp. 49-62, 2014.

[14]   X.-S. Yang, Optimization Problem, Department of Engineering,University of Cambridge, Trumpinton Street, Cambridge CB2 1PZ, UK.

[15]   M. Gendreau, "An introduction to tabu search," in Handbook of Metaheuristics, Kluwer Academic Publishers, 2003, p. 37–54.

[16]   E. Valian, S. Mohanna and S. Tavakoli, "Improved Cuckoo Search Algorithm for Global Optimization," University of Sistan and Baluchestan, Dec. 2011.

[17]   X.-S. Yang, Cuckoo Search via L´evy Flights, Department of Engineering,University of Cambridge, Trumpinton Street, Cambridge CB2 1PZ, UK.

[18]   Clifford T. Brown, "Lévy Flights in Dobe Ju/'hoansi Foraging Patterns," 777 Glades Road, Boca Raton, FL 33431, USA, Department of Anthropology, Florida Atlantic University , 6 December 2006, p. 129–138.

[19]   B. Kumar and D. Kumar, A review on Artificial Bee Colony algorithm, Dept. of Computer Science & Engineering, Guru Jambheswar University, Hisar, Haryana, India.

[20]   D. Karaboga and C. Ozturk, "A novel clustering approach: Artificial Bee Colony (ABC) algorithm," Kayseri, Turkey, Erciyes University, Intelligent Systems Research Group, Department of Computer Engineering.

[21]   G. YAN and Chuangqin, "Effective Refinement Artificial Bee Colony Optimization algorithm Based On Chaotic Search and Application for PID Control Tuning," Taiyuan, China, College of Information Engineering, Taiyuan University of Technology.

[22]   IJCIT, International Journal of Communications and Information Technology, vol. 1, Dec. 2011.

[23]   IJAIA, "International Journal of Artificial Intelligence & Applications," vol. 2, July 2011.

[24]   IJAIS, International Journal of Applied Information Systems, vol. 7, no. ISSN: 2249-0868, September 2014.

Vital Sign and Location/Attitude Monitoring with Sensor Networks for the Proposed Rescue System for Disabled and Elderly Persons Who Need a Help in Evacuation from Disaster Areas

Kohei Arai [1]

Graduate School of Science and Engineering

Saga University

Saga City, Japan

Abstract—Method and system for vital sign (Body temperature, blood pressure, bless, Heart beat pulse rate, and consciousness) and location/attitude monitoring with sensor network for the proposed rescue system for disabled and elderly persons who need a help in evacuation from disaster areas is proposed. Experimental results show that all of vital signs as well as location and attitude of the disabled and elderly persons are monitored with the proposed sensor networks.

Keyword—vital sign; heart beat puls ratee; body temperature; blood pressure; blesses; consciousnes; seonsor network

I. INTRODUCTION

Handicapped, disabled, diseased, elderly persons as well as peoples who need help in their ordinary life are facing too dangerous situation in event of evacuation when disaster occurs. In order to mitigate victims, evacuation system has to be created. Authors proposed such evacuation system as a prototype system already [1]-[4]. The system needs information of victims' locations, physical and psychological status as well as their attitudes. Authors proposed sensor network system which consist GPS receiver, attitude sensor, physical health monitoring sensors which allows wearable body temperature, heart beat pulse rates; bless monitoring together with blood pressure monitoring [5]-[7]. Also the number of steps, calorie consumptions is available to monitor. Because it is difficult to monitor the blood pressure with wearable sensors, it is done by using the number of steps and body temperature. In addition to these, psychological status is highly required for vital sign monitoring (consciousness monitoring). By using EEG sensors, it is possible to monitor psychological status in the wearable sensor. These are components of the proposed physical health and psychological monitoring system.

Method and system for vital sign (Body temperature, blood pressure, bless, Heart beat pulse rate, and consciousness) and location/attitude monitoring with sensor network for the proposed rescue system for disabled and elderly persons who need a help in evacuation from disaster areas is proposed. Experimental results show that all of vital signs as well as location and attitude of the disabled and elderly persons are monitored with the proposed sensor networks.

Section 2 describes the proposed acceleration sensor system followed by experiment method and results. Then conclusion is described together with some discussions..

II. PROPOSED SENSOR NETWORK SYSTEM

A. System Configuration

Figure 1 shows the entire system configuration of the proposed physical and psychological health monitoring system. Patients have physical and psychological health sensors and send the acquired data through Bluetooth and Internet to the Health Data Collection Center: HDCC server. On the other hand, volunteers receive health data of the previously designated several patients together with traffic flow information and appropriate route information. When something wrong occurs on the designated patients, HDCC provides information which required for rescue to the designated volunteers then the volunteers rescue patients in an efficient and an effective manner.

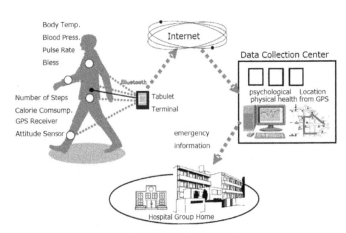

Fig. 1. Entire system configuration of the proposed wearable physical and psychological health monitoring system

B. Sensor and Communication Ssystem

In order for evacuation and rescue, victims' location and attitude is important. Therefore, GPS receiver and accelerometer are added to the aforementioned measuring sensors for body temperature pulse rate, blood pressure, bless, and eeg, emg. All sensors should be wearable and can be attached to ones' tall forehead. Acquired data can be transmitted to mobile devices in ones' pockets. Through WiFi network or wireless LAN connection, acquired data can be collected in the designated information collection center. Then acquired data can be refereed from the designated volunteers who are responsible to help victims for evacuation and rescue.

III.   EXPERIMENTS

A.   Experimental Method

Four patients are participated to the experiments. The difference due to gender can be discussed through a comparison between patients A and C while the difference due to age can be discussed through a comparison between patients B and C. Meanwhile, the difference due to the degree of Alzheimer can be discussed through a comparison between patients B and D as shown in Table 1.

TABLE I.   FOUR PATIENTS

Patient	Male/Female	Age	Remarks
1	Male	37	Good in Health
2	Female	47	Good in Health
3	Female	39	Good in Health
4	Female	91	Weak Alzheimer
5	Male	36	Good in Health
6	Male	39	Good in Health
7	Male	49	Good in Health
8	Female	29	Good in Health
9	Female	53	Good in Health
10	Female	56	Good in Health
11	Female	58	Good in Health

Experiments are conducted for eight hours a day for almost every working day (Monday to Friday) for six months starting from May 2012. Measuring time intervals are different by the measuring items. GPS location can be measured every two seconds while accelerometer data can be obtained every 10 seconds.

Meanwhile, body temperature, pulse rate can be measured every one minutes while blood pressure is measured every one hour together with EEG and EMG signals. The number of steps is measured when the walking event happened. At the end of day, four patients evaluate their physical and psychological conditions which are listed in Table 2.

The 20 items listed in the Table 2 are questionnaires for four patients. In the Table, Ai is questionnaire for physical health while Bi is questionnaire for psychological health. The patients respond to the questionnaire above with five levels range from 0 to 4 grades. Total Score is defined as sum of the aforementioned self evaluation of 20 items including physical and psychological health items.

TABLE II.   SELF EVALUATION ITEMS

A1	Feel fever
B1	Loosing thinking capability
A2	Feel tiredness
B2	Could not sleep well
A3	Get tired after exercise
B3	Feel bad
A4	Muscle hurt
B4	Unconfident about health
A5	Feel depression
B5	Do not want to work
A6	Limper hurt
B6	Cannot remember something
A7	Head ach
B7	Loosing balance
A8	Cannot recover after sleep
B8	Cannot think deeply
A9	Throat hurt
B9	Loosing concentration
A10	Joint hurt
B10	Sleep for too long time

B.   Experimental Results

Figure 2 shows physical and psychological stress for the patients. Physical and psychological stress is different each other participants as shown in Figure 2. There are patients who are sensitive to their stress such as Patient No.8. There are also patients who are not so sensitive to their stress such as patient No. 5.

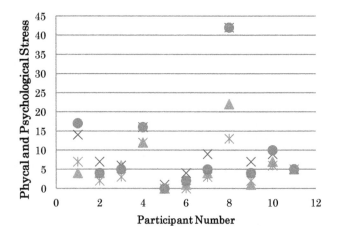

Fig. 2.   Physical and psychological stress for the patients

Physical and psychological stress does not depend on male / female difference as shown in Figure 3. Also it is found that physical and psychological stress does not depend on age as shown in Figure 4 (a). Also Figure 4 (b) shows age dependency on physical conditions of body temperature, blood pressure, pulse rate, and the number of steps.

An example of relation between total score of stress (sum of physical and psychological stress) and measured physical conditions of body temperature, blood pressure, pulse rate, and the number of steps (for the patient with weak Alzheimer) is shown in Figure 5.

As shown in Figure 5, there is no relation between total score of stress (sum of physical and psychological stress) and measured physical conditions of body temperature, blood pressure, pulse rate, and the number of steps (for the patient with weak Alzheimer). This is same thing for the other patients. During the experiments, patients have to repeat the following cycle of "walk for 10 minutes and then take a rest for 10 minutes" for 10 times. Therefore, only the number of steps is proportional to the total score.

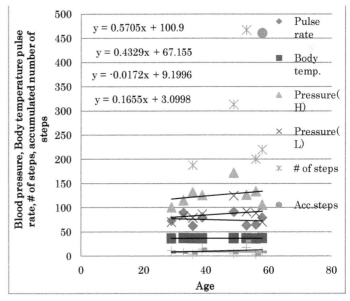

(b) Relation between physical conditions and age

Fig. 4. Age independency on physical and psychological stress as well as age dependency on physical conditions

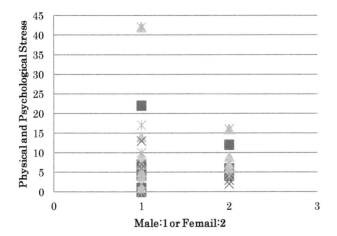

Fig. 3. Sex independency on phycal and psychological stress

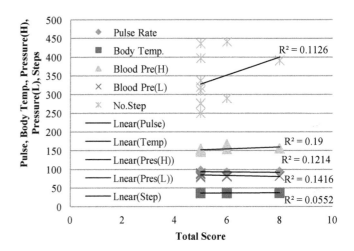

Fig. 5. An example of relation between total score of stress (sum of physical and psychological stress) and measured physical conditions of body temperature, blood pressure, pulse rate, and the number of steps (for the patient with weak Alzheimer)

In accordance with increasing of time duration, the number of steps is increased obviously. In accordance with increasing of the number of steps, blood pressure (High) and blood pressure (Low) is increased usually. It, however, is not always that pulse rate is proportional to the number of steps as shown in Figure 6.

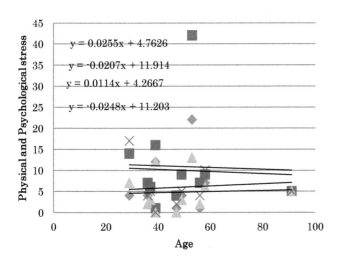

(a) Age independency on physical and psychological stress

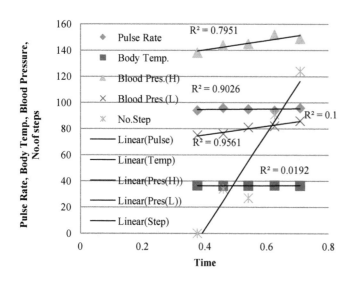

(a)Patient with weak Alzheimer (Minimum total score)

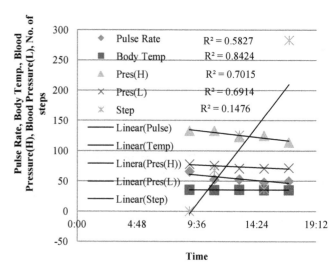

(d) Male patient whose age is 37 (Maximum total score)

Fig. 6. Examples of relation between the number of steps (Time) and measured physical conditions, blood pressure, pulse rate, body temperature

In more detail, relation between physical stress as well as psychological stress and measured physical conditions of blood pressure, pulse rate, body temperature and the number of steps are shown in Figure 7 together with relation between standard deviation of measured physical conditions of blood pressure, pulse rate, body temperature and the number of steps and physical and psychological stress. Figure 8-13 shows relations between physical stress as well as psychological stress and measured physical conditions of blood pressure, pulse rate, body temperature and the number of steps together with relation between standard deviation of measured physical conditions of blood pressure, pulse rate, body temperature and the number of steps and physical and psychological stress for the male whose age is 49 years old, the female whose age is 58 years old, the female whose age is 56 years old, the male whose age is 39 years old, the female whose age is 53 years old, the female whose age is 29 years old, respectively.

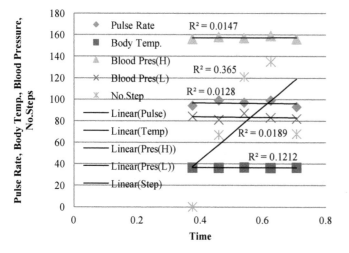

(b)Patient with weak Alzheimer (Maximum total score)

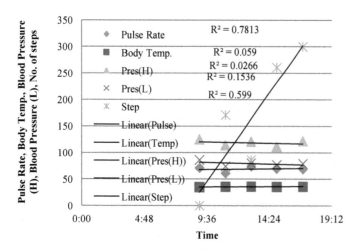

(c)Male patient whose age is 37 (Minimum total score)

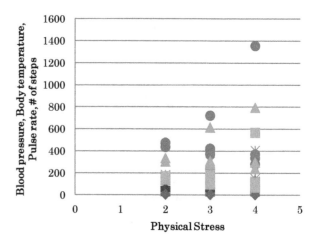

(a)Mean of physical condition (Physical stress)

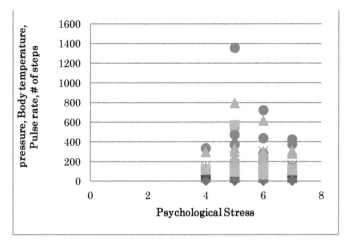

(b) Mean of physical condition (Psychological stress)

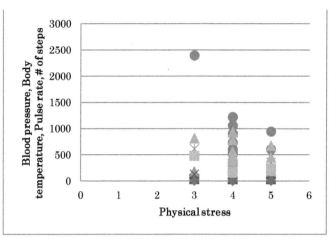

(a)Mean of physical condition (Physical stress)

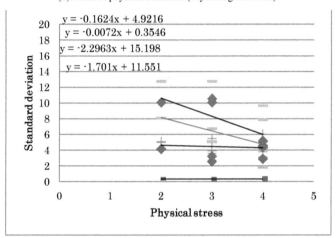

(c) Standard deviation of physical condition (Physical stress)

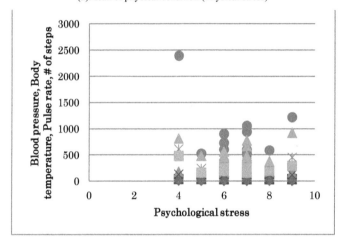

(b) Mean of physical condition (Psychological stress)

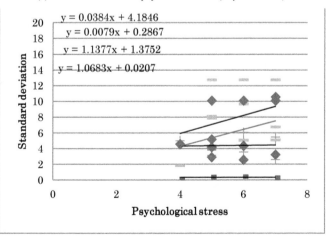

(d)Standard deviation of physical condition (Psychological stress)

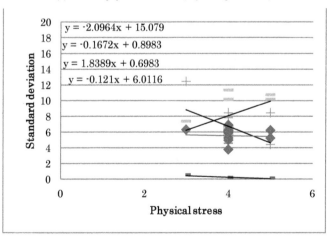

(c) Standard deviation of physical condition (Physical stress)

Fig. 7. Relation between physical stress as well as psychological stress and measured physical conditions of blood pressure, pulse rate, body temperature and the number of steps together with relation between standard deviation of measured physical conditions of blood pressure, pulse rate, body temperature and the number of steps and physical and psychological stress for the male whose age is 36 years old

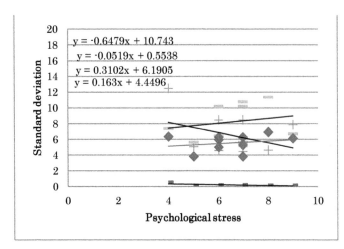

(d)Standard deviation of physical condition (Psychological stress)

Fig. 8.   Relation between physical stress as well as psychological stress and measured physical conditions of blood pressure, pulse rate, body temperature and the number of steps together with relation between standard deviation of measured physical conditions of blood pressure, pulse rate, body temperature and the number of steps and physical and psychological stress for the male whose age is 49 years old

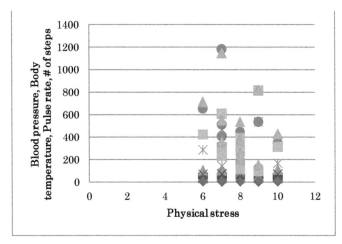

(a)Mean of physical condition (Physical stress)

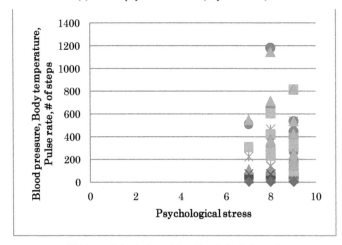

(b) Mean of physical condition (Psychological stress)

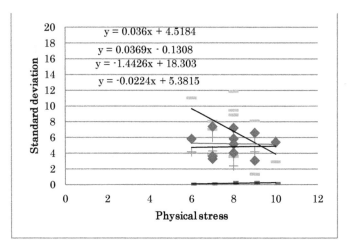

(c) Standard deviation of physical condition (Physical stress)

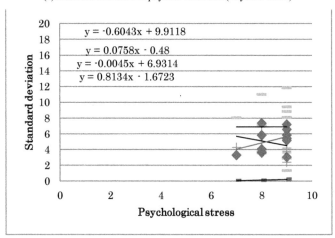

(d)Standard deviation of physical condition (Psychological stress)

Fig. 9.   Relation between physical stress as well as psychological stress and measured physical conditions of blood pressure, pulse rate, body temperature and the number of steps together with relation between standard deviation of measured physical conditions of blood pressure, pulse rate, body temperature and the number of steps and physical and psychological stress for the female whose age is 58 years old

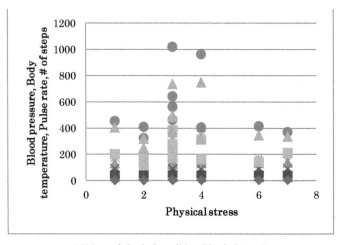

(a)Mean of physical condition (Physical stress)

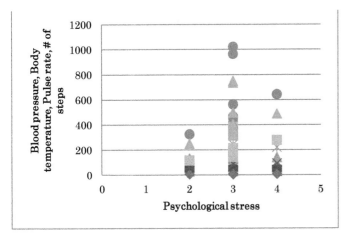

(b) Mean of physical condition (Psychological stress)

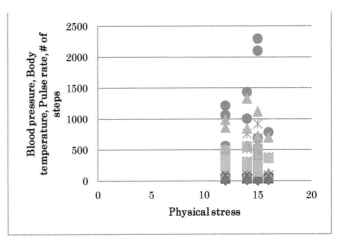

(a)Mean of physical condition (Physical stress)

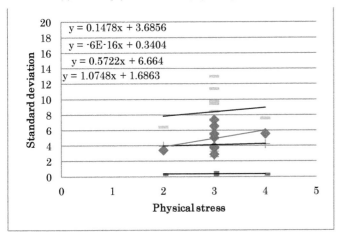

(c) Standard deviation of physical condition (Physical stress)

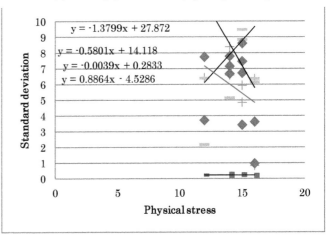

(b) Mean of physical condition (Psychological stress)

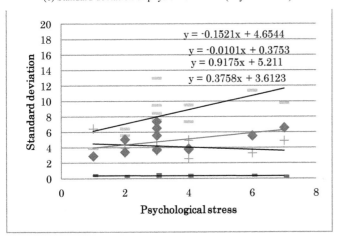

(d)Standard deviation of physical condition (Psychological stress)

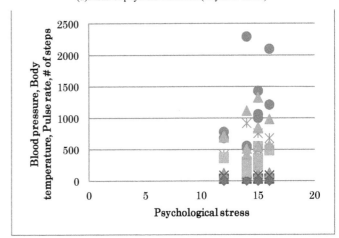

(c) Standard deviation of physical condition (Physical stress)

Fig. 10. Relation between physical stress as well as psychological stress and measured physical conditions of blood pressure, pulse rate, body temperature and the number of steps together with relation between standard deviation of measured physical conditions of blood pressure, pulse rate, body temperature and the number of steps and physical and psychological stress for the female whose age is 56 years old

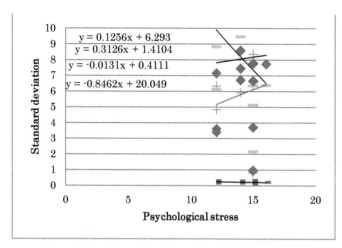

(d)Standard deviation of physical condition (Psychological stress)

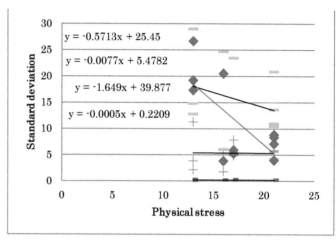

(c) Standard deviation of physical condition (Physical stress)

Fig. 11. Relation between physical stress as well as psychological stress and measured physical conditions of blood pressure, pulse rate, body temperature and the number of steps together with relation between standard deviation of measured physical conditions of blood pressure, pulse rate, body temperature and the number of steps and physical and psychological stress for the male whose age is 39 years old

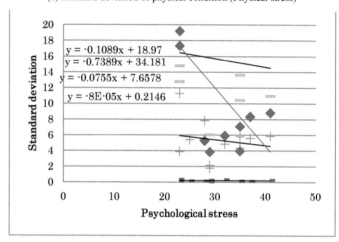

(d)Standard deviation of physical condition (Psychological stress)

Fig. 12. Relation between physical stress as well as psychological stress and measured physical conditions of blood pressure, pulse rate, body temperature and the number of steps together with relation between standard deviation of measured physical conditions of blood pressure, pulse rate, body temperature and the number of steps and physical and psychological stress for the female whose age is 53 years old

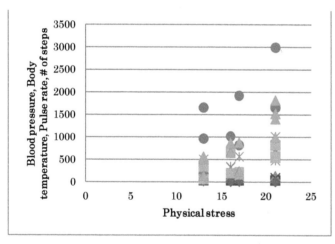

(a)Mean of physical condition (Physical stress)

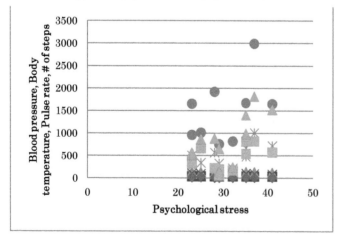

(b) Mean of physical condition (Psychological stress)

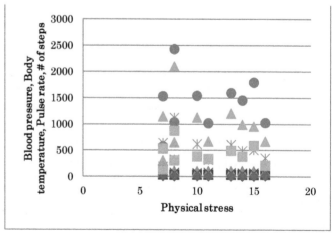

(a)Mean of physical condition (Physical stress)

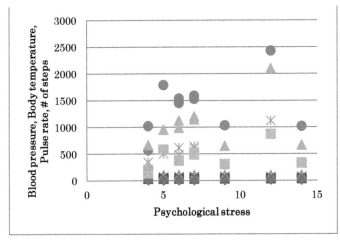

(b) Mean of physical condition (Psychological stress)

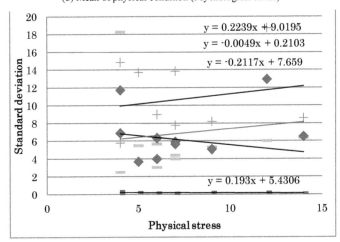

(c) Standard deviation of physical condition (Physical stress)

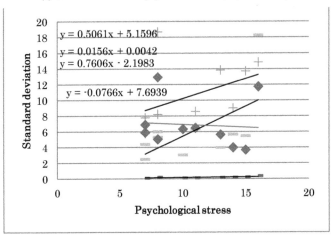

(d)Standard deviation of physical condition (Psychological stress)

Fig. 13. Relation between physical stress as well as psychological stress and measured physical conditions of blood pressure, pulse rate, body temperature and the number of steps together with relation between standard deviation of measured physical conditions of blood pressure, pulse rate, body temperature and the number of steps and physical and psychological stress for the female whose age is 29 years old

Through these experiments, it is found that the followings,

- There is no difference between male and female on physical and psychological stress

- There is difference between the person in healthy condition and the patient with weak Alzheimer

- There are age dependencies on physical and psychological stress as well as blood pressure

- In accordance with increasing of the number of steps, physical stress is increased while psychological stress is decreased. This trend is observed from the relations between standard deviation of physical conditions, blood pressure, body temperature, pulse rate and physical and psychological stress. Also this trend is remarkable for young generation of patients.

Consciousness is measured with EEG sensor together with eye movement observation. Quick eye movements (Succored movements) are highly related to EEG sensor signals. This fact is verified with the following experiments.

By using EEG analyzer tools, we analyze the fatigue effect between the condition when user is looking at one point and condition when user is looking at four points. In order to analyze fatigue effect, we use Peak Alpha Frequency: PAF [8]-[11]. It is possible to measure psychological status by using PAF derived from EEG signal. Psychological health condition is measured with Bio Switch MCTOS of Brain Wave Measuring instrument (BM-Set1) manufactured by Technos Japan Co. Ltd. every one hour.

Figure 14 shows alpha and beta frequency components of EEG signals measured with the male patient whose age is 49. At the begging of the experiment, he surprised so much that beta signal raised remarkably. At that time, his eye moved as quickly as shown in Figure 15. This situation is same for the other patients. Therefore, it is said that there is high relation between eye movement and psychological status.

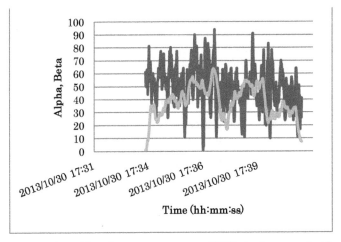

Fig. 14. alpha and beta frequency components of EEG signals measured with the male patient whose age is 49

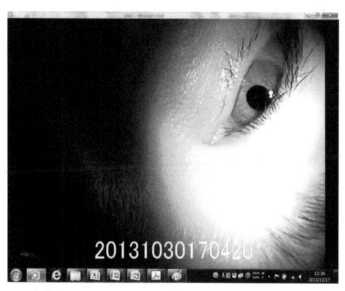

Fig. 15. Quick eye movements observed for the male patient whose age is 49

IV. CONCLUSION

Method and system for vital sign (Body temperature, blood pressure, bless, Heart beat pulse rate, and consciousness) and location/attitude monitoring with sensor network for the proposed rescue system for disabled and elderly persons who need a help in evacuation from disaster areas is proposed. Experimental results show that all of vital signs as well as location and attitude of the disabled and elderly persons are monitored with the proposed sensor networks.

Through the experiments with 11 patients, it is found the followings,

- There is no relation between pulse rate and age

- Body temperature is stable for time duration through the experiments

- There is no age dependency on the number of steps (calorie consumption)

- Psychological status can be estimated with eye movements→There is relation between EEG signal and eye movements (psychological status)

- There is no difference between male and female on physical and psychological stress

- There is difference between the person in healthy condition and the patient with weak Alzheimer

- There are age dependencies on physical and psychological stress as well as blood pressure

- In accordance with increasing of the number of steps, physical stress is increased while psychological stress is decreased. This trend is observed from the relations between standard deviation of physical conditions, blood pressure, body temperature, pulse rate and physical and psychological stress. Also this trend is remarkable for young generation of patients.

ACKNOWLEDGMENT

The author would like to thank all the patients who are contributed to the experiments conducted. The author also would like to thank Professor Dr. Takao Hotokebuchi, President of Saga University for his support this research works.

REFERENCES

[1] Kohei Arai, Tran Xuan Sang, Decision making and emergency communication system in rescue simulation for people with disabilities, International Journal of Advanced Research in Artificial Intelligence, 2, 3, 77-85, 2013.

[2] K.Arai, T.X.Sang, N.T.Uyen, Task allocation model for rescue disable persons in disaster area with help of volunteers, International Journal of Advanced Computer Science and Applications, 3, 7, 96-101, 2012.

[3] K.Arai, T.X.Sang, Emergency rescue simulation for disabled persons with help from volunteers, International Journal of Research and Review on Computer Science, 3, 2, 1543-1547, 2012.

[4] K. Arai, and T. X. Sang, "Fuzzy Genetic Algorithm for Prioritization Determination with Technique for Order Preference by Similarity to Ideal Solution", International Journal of Computer Science and Network Security, vol.11, no.5, 229-235, May 2011.

[5] Arai K., R. Mardiyanto, Evaluation of Students' Impact for Using the Proposed Eye Based HCI with Moving and Fixed Keyboard by Using EEG Signals, International Journal of Review and Research on Computer Science(IJRRCS), 2, 6, 1228-1234, 2011

[6] K.Arai, Wearable healthy monitoring sensor network and its application to evacuation and rescue information server system for disabled and elderly person, International Journal of Research and Review on Computer Science, 3, 3, 1633-1639, 2012.

[7] K.Arai, Wearable Physical and Psychological Health Monitoring System, Proceedings of the *Science and Information Conference 2013 October 7-9, 2013 | London, UK*

15

Analysis and Prediction of Crimes by Clustering and Classification

Rasoul Kiani

Department of Computer
Engineering, Fars Science and
Research Branch, Islamic Azad
University, Marvdasht, Iran

Siamak Mahdavi

Department of Computer
Engineering, Fars Science and
Research Branch, Islamic Azad
University, Marvdasht, Iran

Amin Keshavarzi

Department of Computer
Engineering, Marvdasht Branch,
Islamic Azad University, Marvdasht,
Iran

Abstract—Crimes will somehow influence organizations and institutions when occurred frequently in a society. Thus, it seems necessary to study reasons, factors and relations between occurrence of different crimes and finding the most appropriate ways to control and avoid more crimes. The main objective of this paper is to classify clustered crimes based on occurrence frequency during different years. Data mining is used extensively in terms of analysis, investigation and discovery of patterns for occurrence of different crimes. We applied a theoretical model based on data mining techniques such as clustering and classification to real crime dataset recorded by police in England and Wales within 1990 to 2011. We assigned weights to the features in order to improve the quality of the model and remove low value of them. The Genetic Algorithm (GA) is used for optimizing of Outlier Detection operator parameters using RapidMiner tool.

Keywords—crime; clustering; classification; genetic algorithm; weighting; rapidminer

I. INTRODUCTION

A. Crime Analysis

Today, collection and analysis of crime-related data are imperative to security agencies. The use of a coherent method to classify these data based on the rate and location of occurrence, detection of the hidden pattern among the committed crimes at different times, and prediction of their future relationship are the most important aspects that have to be addressed.

In this regard, the use of real datasets and presentation of a suitable framework that does not be affected by outliers should be considered. Preprocessing is an important phase in data mining in which the results are significantly affected by outliers. Thus, the outlier data should be detected and eliminated though a suitable method. Optimization of Outlier Detection operator parameters through the GA and definition of a Fitness function are both based on Accuracy and Classification error. The weighting method was used to eliminate low-value features because such data reduce the quality of data clustering and classification and, consequently, reduce the prediction accuracy and increase the classification error.

The main purposes of crime analysis are mentioned below [1]:

- Extraction of crime patterns by crime analysis and based on available criminal information,

- Prediction of crimes based on spatial distribution of existing data and prediction of crime frequency using various data mining techniques,

- Crime recognition.

B. Clustering

Division of a set of data or objects to a number of clusters is called clustering. Thereby, a cluster is composed of a set of similar data which behave same as a group. It can be said that the clustering is equal to the classification, with only difference that the classes are not defined and determined in advance, and grouping of the data is done without supervision [2].

C. Clustering by K-means Algorithm

K-means is the simplest and most commonly used partitioning algorithm among the clustering algorithms in scientific and industrial software [3] [4] [5]. Acceptance of the K-means is mainly due to its being simple. This algorithm is also suitable for clustering of the large datasets since it has much less computational complexity, though this complexity grows linearly by increasing of the data points [5]. Beside simplicity of this technique, it however suffers from some disadvantages such as determination of the number of clusters by user, affectability from outlier data, high-dimensional data, and sensitivity toward centers for initial clusters and thus possibility of being trapped into local minimum may reduce efficiency of the K-means algorithm [6].

D. Classification

Classification is one of the important features of data mining as a technique for modeling of forecasts. In other words, classification is the process of dividing the data to some groups that can act either dependently or independently [7]. Classification is used to make some examples of hidden and future decisions on the basis of the previous decision makings [8]. Decision tree learning, neural network, nearest neighborhood, Nave Bayes method and support vector machine are different algorithms which are used for the purpose of classification [9].

E. Genetic Algorithm

In [10] S. Sindhiya and S. Gunasundari, have discussed about Genetic Algorithm (GA) as an evolutionary algorithm. "GA starts with an initial population called a candidate

solution. After a sequence of iterations it achieves the optimal solution. The fitness is used to estimate the quality of each candidate solution. The chromosome, which has the highest fitness is to be kept in the next iteration. The crossover and mutation are the two basic operators of GA. The crossover is the procedure of taking above one parent solutions and generating a child solution. The mutation is used to preserve the genetic diversity from one iteration to the next iteration. And again the fitness function and the genetic operators are used to generate successive generations of individuals and are repeated several times until a suitable solution is found. The performance of GA depends on a number of issues such as crossover, mutation, fitness function and the various user determined parameters such as population size, probability of genetic operators."

The rest of the paper is organized as follows. Section 2 describes the existing systems for analyzing crimes. The New framework and experimental results are presented in section 3. Section 4 contains the conclusion. Finally, section 5 discusses the future scope of this paper.

II.   LETRATURE REVIEW

J. Agarwal, R. Nagpal and R. Sehgal in [1] have analyzed crime and considered homicide crime taking into account the corresponding year and that the trend is descending from 1990 to 2011. They have used the k-means clustering technique for extracting useful information from the crime dataset using RapidMiner tool because it is solid and complete package with flexible support options. Figure1 shows the proposed system architecture.

Priyanka Gera and Dr. Rajan Vohra in [11] have used a linear regression for prediction the occurrence of crimes in Delhi (India). They review a dataset of the last 59 years to predict occurrence of some crimes including murder, burglary, robbery and etc. Their work will be helpful for the local police stations in decision making and crime supervision.

"After training systems will predict data values for next coming fifteen years. The system is trained by applying linear regression over previous year data. This will produce a formula and squared correlation(r^2).

The formula is used to predict values for comong future years. The coefficent of determination, r^2, is useful because is gives the proportion of variance of one variable that is predictable from other variable." Figure 2 shows the proposed system architecture.

In [12] an integrated system called PrepSearch have proposed by L. Ding et al. It has been combined using two separate categories of visualization tools: providing the geographic view of crimes and visualization ability for social networks. "It will take a given description of a crime,

including its location, type, and the physical description of suspects (personal characteristics) as input.

To detect suspects, the system will process these inputs through four integrated components: geographic profiling, social network analysis, crime patterns and physical matching." Figure 3 shows the system design and process of PrepSearch.

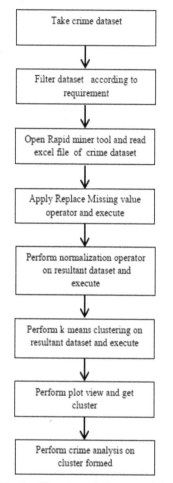

Fig. 1.   Flow chart of crime analysis [1]

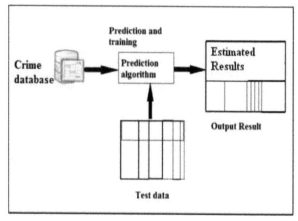

Fig. 2.   Predicting future crime trends [11]

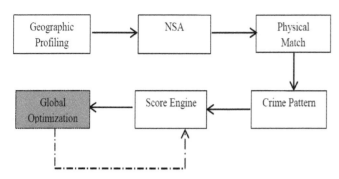

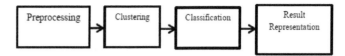

Fig. 4.   New framework

Fig. 3.   System design and process of PrepSearch [12]

In [13] researches have introduced intelligent criminal identification system called ICIS which can potentially distinguish a criminal in accordance with the observations collected from the crime location for a certain class of crimes.

The system uses existing evidences in situations for identifying a criminal by clustering mechanism to segment crime data in to subsets, and the Nave Bayesian classification has used for identifying possible suspect of crime incidents. ICIS has been used the communication power of multi agent system for increasing the efficiency in identifying possible suspects. In order to describe the system ICIS is divided to user interface, managed bean, multi agent system and database. Oracle Database is used for implementing of database, and identification of crime patterns has been implemented using Java platform.

In [14] an improved method of classification algorithms for crime prediction has proposed by A. Babakura, N. Sulaiman and M. Yusuf. They have compared Naïve Bayesian and Back Propagation (BP) classification algorithms for predicting crime category for distinctive state in USA. In the first step phase, the model is built on the training and in the second phase the model is applied. The performance measurements such as Accuracy, Precision and Recall are used for comparing of the classification algorithms. The precision and recall remain the same when BP is used as a classifier.

In [15] researches have introduced crime analysis and prediction using data mining. They have proposed an approach between computer science and criminal justice to develop a data mining procedure that can help solve crimes faster. Also they have focused on causes of crime occurrence like criminal background of offender, political, enmity and crime factors of each day. Their method steps are data collection, classification, pattern identification, prediction and visualization.

III.   NEW FRAMEWORK

In this section a new framework is introduced for clustering and prediction of cluster members to analyze crimes.   A dataset of crimes recorded by police in England and Wales[1] within 1990 to 2011 has been used, and RapidMiner will be used for the purpose of implementation.

A.   Preprocessing Phase

1) Read the dataset of crime using Read Excel operator: The dataset selectd by the operator is read in RapidMiner tool.

2) Filter dataset according to requirement: Since there may be data in the read dataset that would not be used according to our method, the unnecessary data have to be filtered.

3) Apply Replace Missing Value operator: This operator replaces the dataset missing values with a new value, and adds it to our previous dataset. This can be done by one of the "Minimum", "Maximum", "Average" and "None" functions which is determined by the Default parameter. If "None" were selected, it will be leaded to no replacement. To achieve this purpose there is an accessible wizard using Column parameter.

Each one of these features chooses one of the functions through Column parameter. If one feature name were be shown as a key in the list, function name will be used as the key value. If the feature name does not exist in the list, a selected function with default parameters can be used for it. For nominal features the Mod function, nominal value that has been occurred the most in the dataset, can be replaced with the Average function as an example. For the nominal features which their replacement parameter has been assigned Zero for them, the first nominal defined value for the feature will be replaced with the missing values. We can also use the Replenishment Value parameter to specify the replacement values.

4) Outlier detection using Outlier Detection(Distance) operator: Outlier detection goals:

- Improving the quality of clusters in clustering phase,

- Increasing the accuracy of Decision tree in classification phase,

- Decreasing the classification error in classification phase.

This operator tries to detect outliers in the dataset according to their distance with their neighbors. This operator discovers outliers by a kind of search that can be known as a statistical search. This method starts the search according to the distance of K-th nearest neighbors, and then it tries to sort the search result by their local position. The ones which are farther than their K-th nearest neighbor will be specified as the outlier in the dataset more probably. Theory says dispersion and distance of outliers is more than the average of dataset. Then according to the distance of each data with its K-th nearest neighbor, all data will be ranked and then we can say the higher ranking shows the outlier of the dataset.

[1] www.gov.uk/government/publications/offences

Parameters:

- Number of neighbors: Specifies the value for the k-th nearest neighbors,

- Number of outliers: The number of top-n outliers need to be looked for.

5) Outlier Distance operator parameters optimization using GA: At this point, as shown in Figure 5, given that the process of optimizing the Outlier Detection operator parameters must improve the results of the predicted cluster members in the clusters, parameters of Accuracy and Classification error are used to define the fitness function
Optimize Parameters (Evolutionary) operator values:

- Max generation= 50

- Population size= 5

- Crossover prob= 0.9

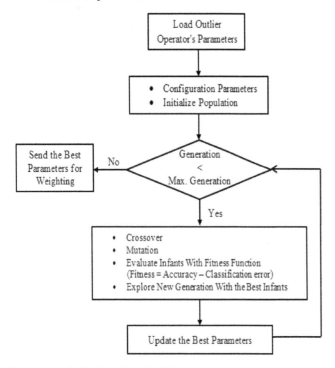

Fig. 5.   Genetic algorithm for optimizing

6) Store a new dataset: After applying the changes, the dataset is stored for further use.

B. Clustering Phase

1) Apply Weight by Deviation operator: One of the clustering algorithm challenges is high-dimensional data, so to deal with this challenge is using weighted features and removing low-value featuers. A suitable operator to apply our idea is Weight by Deviation. It creates weights from the standard deviations of all featuers. The values can be normalized by the "Average", "Minimum", or "Maximum" of the featuers.

Parameters:

- Normalize weights: Activates the normalization of all weights,

- Normalize: Indicates that the standard deviation should be divided by the minimum, maximum, or average of the featuers.

2) Thershold selection: There is not specific critria for selecting the threshold. It is selected based on the Trial and error method for removing low-value featuers. The threshold is determined, and all the featuers, that their values are equal or less than it, will be removed from the dataset.

3) Store a new dataset: After applying the changes, the dataset is stored for further use.

4) Performe K-means algorithm on result dataset: At this step, the clustering process is carried out on the dataset using K-means operator.

5) Enable K-means Operator Parameters: The classification process is performed on the data after data clustering; therefore, the target class is defined in this step. The goal is that the cluster obtained in the previous step be defined as the target class. The k-means operator parameters are used for this purpose.
Parameters:

- Add cluster attribute: If enabled, a cluster id is generated as new special attribute directly in this operator. Othrwise this operator does not add an id attribute,

- Add as label: If true, the cluster id is stored in an attribute with the special role "label" instead of "cluster".

C. Classification Phase

1) Training and Testing Data: In this phase production of training and testing data is done using Sample (Stratified) and Set Minus operators for increasing confidence in the response without replacement.

- Sample operator is used to reserve 10% of data,

- Set Minus operator is used to reduce training data from the dataset.

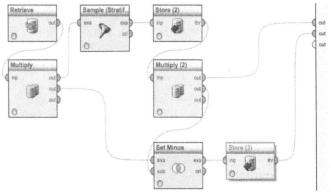

Fig. 6.   Production of training and testing data

2) Decision Tree: We use Decision Tree operator to learn decision tree model, and the value of Criterion parameter is selected "gini_index".

3) Apply the model and test data: The model and test data have been produced in the previous steps are applied on Apply Model operator inputs to predict the cluster members.

D. Result Presentation Phase

At this phase, the following operators are used to show the results obtained from the presented framework.

- The model accuracy and classification error are calculated by Performance operators,

- Log operator is used to record and save performance report,

- A comparison of accuracy and classification error are used to evaluate the effect of optimization Outlier Detection operator parameters,

- Analysis of crimes based on the new framework.

Figure 7 shows the new framework scheme with details. Comparison between results in Table 1 shows that when the number of clusters is similar, after optimizing the parameters of the Outlier Detection operator, the classification accuracy increased and classification error decreased and, consequently, the obtained fitness function was optimized. Figure 8 shows the predicted occurrence rate for the crimes of buggery, homicide, and robbery. Also Figure 9 shows the implementation of the model by Rapid Miner tool.

TABLE I. RESULTS

Mode	Number of Cluster	Accuracy of Prediction	Classification Error	Fitness Function
Optimized	6	91.64%	8.36%	83.28
Non-Optimized	6	85.74%	13.26%	72.48

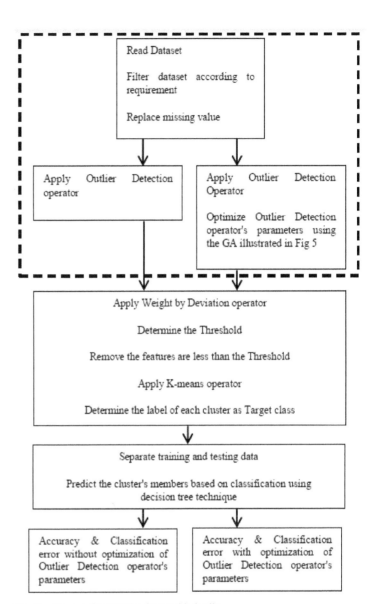

Fig. 7. The new framework scheme with details

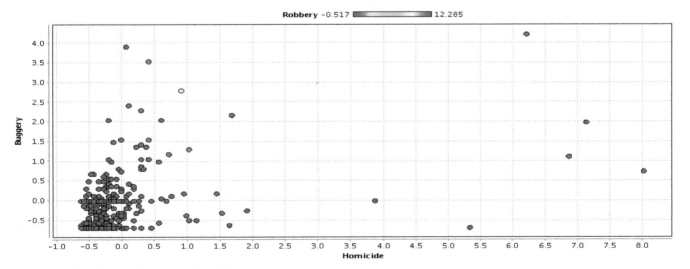

Fig. 8. Prediction of robbery, buggery and homicide

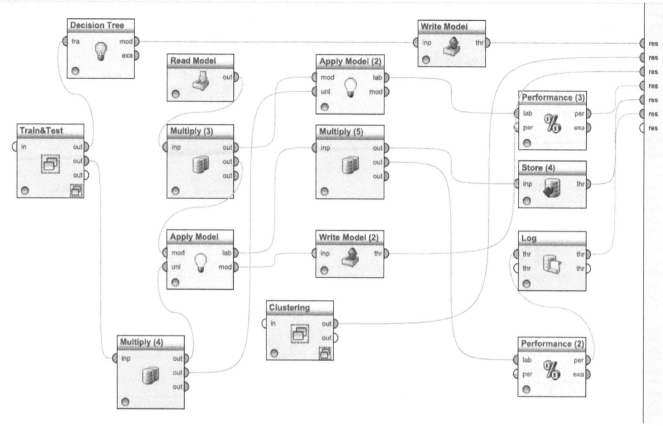

Fig. 9. Model for prediction of criems using decision tree

IV. CONCLUSION

This paper presents a new framework for clustering and predicting crimes based on real data. Examining the methods proposed for crime prediction shows that the parameters such as the effect of outliers in the data mining preprocessing, quality of the training and testing data, and the value of features have not been addressed before. In this framework, the GA was used to improve outlier detection in the preprocessing phase, and the fitness function was defined based on accuracy and classification error parameters. In order to improve the clustering process, the features were weighted, and the low-value features were deleted through selecting a suitable threshold. The proposed method was implemented, and the results of the optimized and non-optimized parameters were compared to determine their quality and effectiveness.

The main purposes of the new framework for clustering and classification of crimes are mentioned below:

- Generation of training and testing data,

- Removing low-value attributes using weighting technique to deal with high-dimensional data challenge,

- Optimization of Outlier operator parameters using GA.

V. FUTURE SCOPE

One of the most important issues that should be addressed in the model presented in this paper to improve the clustering process and crime detection is the optimization of the number of clusters in the clustering process and the optimization of the technique used in the prediction phase of model development.

ACKNOWLEDGMENT

Hereby, we would like to kindly acknowledge sincere cooperation and valuable instructions of Mr. Mojtaba Montazeri and Mr. Mohammad Sarabi.

REFERENCES

[1] J. Agarwal, R. Nagpal, and R. Sehgal, "Crime analysis using k-means clustering," International Journal of Computer Applications, Vol. 83 – No4, December 2013.

[2] J. Han, and M. Kamber, "Data mining: concepts and techniques," Jim Gray, Series Editor Morgan Kaufmann Publishers, August 2000.

[3] P. Berkhin, "Survey of clustering data mining techniques," In: Accrue Software, 2003.

[4] W. Li, "Modified k-means clustering algorithm," IEEE Congress on Image and Signal Processing, pp. 616- 621, 2006.

[5] D.T Pham, S. Otri, A. Afifty, M. Mahmuddin, and H. Al-Jabbouli, "Data clustering using the Bees algorithm," proceedings of 40th CRIP International Manufacturing Systems Seminar, 2006.

[6] J. Han, and M. Kamber, "Data mining: concepts and techniques," 2nd Edition, Morgan Kaufmann Publisher, 2001.

[7] S. Joshi, and B. Nigam, "Categorizing the document using multi class classification in data mining," International Conference on Computational Intelligence and Communication Systems, 2011.

[8] T. Phyu, "Survey of classification techniques in data mining," Proceedings of the International Multi Conference of Engineers and Computer Scientists Vol. IIMECS 2009, March 18 - 20, 2009, Hong Kong.

[9] S.B. Kim, H.C. Rim, D.S. Yook, and H.S. Lim, "Effective Methods for Improving Naïve Bayes Text Classifiers," In Proceeding of the 7th Pacific Rim International Conference on Artificial Intelligence, Vol.2417, 2002.

[10] S. Sindhiya, and S. Gunasundari, "A survey on Genetic algorithm based feature selection for disease diagnosis system," IEEE International Conference on Computer Communication and Systems(ICCCS), Feb 20-21, 2014, Chermai, INDIA.

[11] P. Gera, and R. Vohra, "Predicting Future Trends in City Crime Using Linear Regression," IJCSMS (International Journal of Computer Science & Management Studies) Vol. 14, Issue 07Publishing Month: July 2014.

[12] L. Ding et al., "PerpSearch: an integrated crime detection system," 2009 IEEE 161-163 ISI 2009, June 8-11, 2009, Richardson, TX, USA.

[13] K. Bogahawatte, and S. Adikari, "Intelligent criminal identification system," IEEE 2013 The 8th International Conference on Computer Science & Education (ICCSE 2013) April 26-28, 2013. Colombo, Sri Lanka.

[14] A. Babakura, N. Sulaiman, and M. Yusuf, "Improved method of calssification algorithms for crime prediction," International Symposium on Biometrics and Security Technologies (ISBAST) IEEE 2014.

[15] S. Sathyadevan, and S. Gangadharan, "Crime analysis and prediction using data mining," IEEE 2014.

16

Zernike Moment Feature Extraction for Handwritten Devanagari (Marathi) Compound Character Recognition

Karbhari V. Kale, *Senior Member, IEEE* *, **Prapti D. Deshmukh** †, **Shriniwas V. Chavan,** *Student Member, IEEE* ‡,
Majharoddin M. Kazi, *Student Member, IEEE*§, **Yogesh S. Rode,** *Student Member, IEEE* ¶

*‡§Department of Computer Science and IT, Dr. B. A. M. University, Aurangabad, Maharashtra, India - 431004
†Dr. G. Y. Pathrikar College of Computer Science, MGM, Maharashtra, India - 431005
‡Department of Computer Science, MSS's ACS College, Ambad, Jalna, Maharashtra, India - 431203
**Email: kvkale91@gmail.com,* †*prapti.research@gmail.com,*‡*shripc@gmail.com,*§*mazhar940@gmail.com,*¶*ys.rode@gmail.com*

Abstract—Compound character recognition of Devanagari script is one of the challenging tasks since the characters are complex in structure and can be modified by writing combination of two or more characters. These compound characters occurs 12 to 15% in the Devanagari Script. The moment based techniques are being successfully applied to several image processing problems and represents a fundamental tool to generate feature descriptors where the Zernike moment technique has a rotation invariance property which found to be desirable for handwritten character recognition. This paper discusses extraction of features from handwritten compound characters using Zernike moment feature descriptor and proposes SVM and k-NN based classification system. The proposed classification system preprocess and normalize the 27000 handwritten character images into 30x30 pixels images and divides them into zones. The pre-classification produces three classes depending on presence or absence of vertical bar. Further Zernike moment feature extraction is performed on each zone. The overall recognition rate of proposed system using SVM and k-NN classifier is upto 98.37%, and 95.82% respectively.

Keywords—Handwritten Character, Devanagari Compound, Zernike, SVM, k-NN.

I. Introduction

Handwritten character recognition is gaining popularity for many years and attracting researchers for the purpose of potential application development. These potential applications reduce the cost of human efforts and save the time. Some of its potential application areas are like bank automation, postal automation [1]–[3] etc. Similarly the biometric and criminal identification system uses scanned handwritten script for forensic and Historic Document Analysis (HDA) and represents an excellent study area within the research field of biometrics and forensic science.

The technical challenge in handwritten character recognition comes from three sources: *Symbol*: an ideal shape that occurs in hierarchy and symbol are arranged in complex form at different level in organization. *Deformation*: shape variation in each symbol to undergoes geometric transformation (translation, rotation, scaling, stretching) and complex representation. *Defect* flaw in image owing to print, scan, quantized, binary etc. Handwritten and Printing character demands diverse approach, handwritten consist of extended stroke and printed consist of normal shaped blobs.

Research in handwritten character recognition focuses on two main approaches i.e. on-line and off-line. In on-line character recognition system captures data by the sensors during writing process, which makes the information dynamically available according to the strokes. While, off-line character recognition takes place in static form where images are captured or scanned after completion of the writing process on paper/sheets. Both the tasks are challenging for automatic character recognition, specifically in off-line character recognition requires more efforts due to various reasons viz. large variations in shape of characters due to pen ink, pen width, and accuracy of devices, stroke size and location, effect of physical and mental situation of the writer on writing style, in turns effect the recognition accuracy.

Character recognition problem becomes more challenging even in on-line and off-line in Indian Language Scripts due to several reasons [4]. The Indian scripts have character set with large number of characters. The shape of the characters in Indian scripts is more complex and may have modifiers. These modifiers may found at above, below or in-line with the character. The modifiers are the vowel that changes their shapes when they get connected with the consonants. The scripts may have some character pairs that are looks alike and cause difficulty in classification. Some of Indian languages like Devanagari, Bangla are having the specific problem in compound characters where two or more consonants join with each other to form a special character [5], [6].

The research work on character recognition of Devanagari script was started in 1970, where Sinha and Mahabala [7] were presented a syntactic pattern analysis system for the recognition of Devanagari characters (DC). First research report on handwritten Devanagari Characters (HDC) was published in 1977 by Sethi and Chatterjee [8], very few work were reported on OCR in the literature and later on in the next decade S. Kumar and et. al. contributed more in this domain [9]. An extensive research work on printed Devanagari Characters and Handwritten Characters was carried out by Bansal [10]–[12] and Reena et.al, [13], [14] respectively. Recognition of characters in different languages using Zernike Moments was reported in [9], [15]–[22]. Researchers have proposed Chain Code Histogram and directional information gradient based feature extraction in [22]–[24]. A significant contribution by Arora and et. al., proposed feature extraction techniques namely, intersection, shadow feature, chain code histogram and straight line fitting features in [25]–[28]. Deshpande and et. al. [29] has proposed fine classification and recognition of Devanagari characters. S. Kumar in [30] also extracted various features and performed comparison using SVM and MLP. Pal and et al. proposed SVM and MQDF based scheme for recognition of Devanagari Characters [31]. U. Pal and T. Wakabayashi [32] given a comparative study of different Devanagari Character recognizers which extracts features based on curvature and gradient information. Sushama Shelke and et. al. [33] presented a novel approach for recognition of unconstrained handwritten Marathi characters. Baheti M.J. and et. al. [34] proposed a method based on Affine Invariant Moment (AIM) for Gujarati numerals using k-NN and PCA classifiers. Elastic matching (EM) technique based on an Eigen Deformation (ED) for recognition of handwritten Devanagari characters is proposed by V. Mane and et.al, [35]. Recognition of handwritten Bangla compound characters was attempted by U. Pal and et al. [36] using gradient features. S. Shelke and S. Apte have reported work on handwritten Marathi compound characters using multi-stage multi-feature classifier [5], [6].

The literature evidence shows that moment can be considered as potential features for recognition of characters and numerals, which motivate us to enrich the several orthogonal and discrete moment features and test the efficacy of the system for compound characters. While significant advances have been achieved in recognizing Roman-based scripts like English, ideographic characters Chinese, Japanese, Korean, and Arabic, only few works on some of the major Indian scripts like Devanagari, Bangla, Gurumukhi, Tamil, Telugu, are available in the literature [37]–[41].

This paper proposes a novel Zernike moment based feature descriptor followed by SVM and k-NN neural network approach for recognition of Marathi Script Basic and Compound Characters derived from Devanagari. The organization of the paper is as follows: Section 2 deals with properties of Devanagari derived Marathi script. Database designing and Proposed System has been discussed in Section 3. Section 4 deals with Zernike Moment based feature extraction technique. Details about the SVM and k-NN approach used for character recognition system are elaborated in Section 5. The experimental results are discussed in Section 6. Conclusion of the paper is given in Section 7.

II. PROPERTIES OF DEVANAGARI BASIC AND COMPOUND CHARACTER

The basic set of symbols of Devanagari script consists of 12 vowels (or swar), 36 consonants (or vyanjan). The alphabet of modern Devanagari script consists of 14 vowels and 33 consonants also called as basic characters. Writing style of the Devanagari script is from left to right and the concept of upper and lower case is absent in the script. In this script vowel following by a consonant takes a modified shape, these modified shapes are called modified characters. A consonant or vowel following a consonant sometime takes a compound orthographic shape, which we called as a compound character. Compound characters can be combination of two consonants as well as a consonant and a vowel. The compound characters are joined in various ways, by removing vertical line of the character and then to the other characters from the

left side like म्य, in another way it is joined side by side or one above the other like इ. The example of compound characters is shown in Fig (1). The split character is half of the basic character which gets connected to other characters. The example of split component of compound character is shown in Fig (2). Compounding of three or four characters also exists in the script. There are about 280 compound characters in the Devanagari script [4], [31].

Figure 1.   Sample images of Compound Character

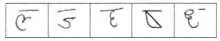

Figure 2.   Sample images of Split Component of Compound Character

Marathi script is one of the derived script from Devanagari, and it is an official language of Maharashtra. Marathi script consists of 16 vowels and 36 consonants making 52 alphabets. Marathi script is written from left to right, which does not have upper and lower case characters. Similar to Devanagari it has nearly the similar type of compound characters property. However, the occurrence of compound characters in Marathi is found to be about 11 to 12%, whereas in other scripts of Devanagari, it is about 5 to 7% [42].

III.   DATABASE DESIGNING AND PROPOSED SYSTEM

A. Database

At present no dataset of handwritten compound characters is available for Marathi script derived from Devanagari and hence we have created handwritten compound characters dataset for this work and it has been tested with our proposed system for its recognition, this adds a new contribution in the literature. Details of this database are provided in Table I.

The database of Handwritten Characters of Marathi from Devanagari script is created for the purpose of this work, which contains basic, compound and split components of compound characters. These data characters were recorded in written form on special paper sheet from 250 different volunteers of different age group. (in between 20-40 year old). The recorded character is then scanned with Flatbed Scanner at 300 dpi. The size of the image of each character is considered 90x90 pixels and it is stored in TIFF image format.

Table I.   DATASET OF HANDWRITTEN DEVANAGARI BASIC AND COMPOUND CHARACTERS

Property	Descriptions
Number of subjects	250
Number of basic character	48
Number of compound characters	45
Number of split compound characters	15
Number of images given by each subject	48+45+15=108
Gray/Color	Color
Resolution	90*90 pixels
DPI	300
Format	TIFF
Total Number of images	27000

B. Proposed System

In the proposed system, we aim at recognizing handwritten Marathi Devanagari compound characters. This is done by employing Zernike moment feature extraction using SVM and k-NN neural network approach. Fig. (3) shows the basic block diagram of the proposed recognition system, which consists of different phases begin with input character images, preprocessing, pre-classification of the characters, Zernike moment based feature extraction and character recognition. The brief phase wise explanation of the recognition system as follows:

Fig. (3) shows the basic block diagram of the recognition system. It shows that the handwritten Devanagari character are scanned and a digitized document is obtained. From it a particular character is selected, the image character is cropped and resized into fix row and columns. Each block of the recognition system is elaborated in following sections.

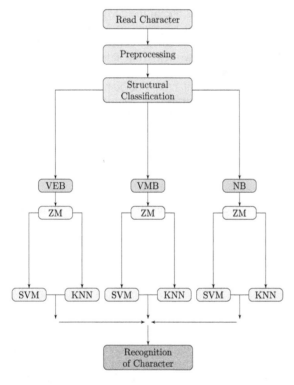

Figure 3. Block Diagram of Proposed Character Recognition System

C. Preprocessing

Pre-processing step is performed on the character image to remove the noise from it and also to minimize the variations in character styles. Occasionally, the document while scanning was not clean and so it has produced small dots in the images. Noise generated by the shaded areas and dots must be filtered during preprocessing step. Moreover, the characters in scanned images may found to be skewed, slant and varied in sizes due to cropping. This has been processed in this step. The typical flow of preprocessing step is shown in following Fig (4).

1) RGB to Gray Image: The database contains color character images. In preprocessing the character images are converted to binary images using rgb2gray utility in MATLAB.

2) Thresholding: This preprocessing step also termed as binarization process and converts the pixels that are above the threshold to white and those which are below the threshold to black. We have set the threshold value Th= 190 to produce good quality binarized images.

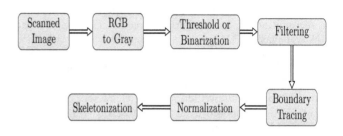

Figure 4. Steps in Preprocessing of Image

3) Filtering: To remove the noise present in the binarized image filtering has been done. We have used Median filter to remove small black spot in the image and the black shade appearing at the edges. Further, the documents were cropped from the edges. Thresholding and filtering steps often resulted in some broken characters. To rejoin the broken characters, image dilation operation on the filtered images has performed.

4) Boundary Tracing: Tracing of the boundary identifies the connected components of the characters in the filtered images and stores it in array. To find the connected components, the algorithm starts by traversing the rows of filtered image. It searches for a foreground pixel, and then it marks that pixel and picks it. Similarly, marking of all the neighbors of found pixel in all search directions completed till all the pixels of the possible character have been traversed and marked. Otherwise, it will continue the search in the next row. If the size of any picked connected component is too small than the actual required size, then the algorithm treats that component as noise and neglects that component.

5) Normalization: During normalization step, slant in characters is removed and resized to a window. Slant is the average divergence of the vertical strokes of the character from the right side of the character. To remove the slant, we used imrotate with angle θ. At each angle the sum of vertical projection of the transformed characteris calculated. The angle with maximum sum of vertical projection is used to finally perform shear transformation on the character and estimated the slant angle.

6) Skeletonization: In skeletonization, the thickness of the character is reduced to one-pixel character bound. We have applied the thinning operation on the character and taken the precaution, do not to break the character. These operations were used not only to find the vertical bar and position of vertical bar in the character, but also to extract endpoints, junction in the character. This features helps in the pre-classification of the characters. A sample output of the preprocessed character क is shown in Fig. (5).

D. Pre-classification

Character Images after preprocessing stage consists of some global and local features. The global feature consists of presence of vertical line, position of vertical bar in the

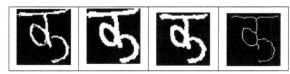

Figure 5.　Character Images after Preprocessing

character and enclosed region in the character. The local features consists of the end points and junction position in the character. On the basis of global feature, the character is classified into three major categories based on the presence of vertical bar i.e. a) character with vertical bar at right (**VEB**: *Vertical End Bar*), b) character with vertical bar at middle (**VMB**: *Vertical Mid Bar*), and c) the character with absence of vertical bar (**NB**: *No Bar*). Vertical bar at right are further classified into two categories based on whether the vertical bar and rest of the character are connected or not to the bar. These pre-classification of characters are shown in the following Table II and III.

Table II.　CLASSIFICATION OF DEVANAGARI BASIC CHARACTER

Sr. No.	Pre Classification	Character
1	Character connected vertical bar at right side	ख, घ, च, ज, झ, ञ, त, थ, द, ध, न, प, ब, भ, म, य, र, ल, व, स, ष, क्ष, ज्ञ
2	Character not connected with vertical bar at right side	ग, ण, श
3	Character with absence of vertical bar	ङ, छ, ट, ठ, ड, ढ, द, र, ह, ळ

Table III.　CLASSIFICATION OF DEVANAGARI COMPOUND CHARACTER

Sr. No.	Pre Classification	Character
1	Character connected vertical bar at right side	ण्य, ज्ग, ग्य, त्य, ध्य, भ्य, ल्य, व्य, घ्य, म्य, ब्य, न्य, च्य, स्य, स्म, न्म, ज्म, त्म, स्प, ल्प, क्य, क्ल, फ्य, ग्य, ण्य, श्व, ज्व, स्व, भ्व, म्न, ब्ध, प्त, ब्ज, ख्ख, त्त, न्न
2	Character not connected with vertical bar at right side	स्क, त्क, स्फ
3	Character with absence of vertical bar	ध्द, च्छ, म्ह, न्ह, ष्ट, ब्द, ब्द

E. Local Structural Classification

The local features are detected on the basis of the end points of the character. We have firstly partitioned the character into 3x3 image i.e. 9 quadrants and extracted the end points and junctions in each individual block as shown in Fig. (6).

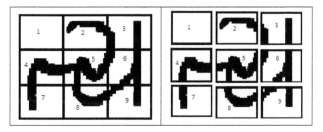

Figure 6.　Presence of End Points in partition block of Character

Thus pre-classification of character is done and put the character in proper class like VEB, VMB and NB and then Zernike moment features are extracted for its final classification under SVM and k-NN.

IV.　ZERNIKE MOMENT BASED FEATURE EXTRACTION

Zernike moments are complex number by which an image is mapped on to a set of two-dimensional complex Zernike polynomials. The magnitude of Zernike moments is used as a rotation invariant feature to represent a character image patterns [43]. Zernike moments are a class of orthogonal moments and have been shown effective in terms of image representation. The orthogonal property of Zernike polynomials enables the contribution of each moment to be unique and independent of information in an image. A Zernike moment does the mapping of an image onto a set of complex Zernike polynomials. These Zernike polynomials are orthogonal to each other and have characteristics to represent data with no redundancy and able to handle overlapping of information between the moments [26]. Due to these characteristics, Zernike moments have been utilized as feature sets in applications such as pattern recognition [27] and content-based image retrieval [28]. These specific aspects and properties of Zernike moment are supposed to found to extract the features of compound handwritten characters. Teague [16] has introduced the use of Zernike moments to overcome the shortcomings of information redundancy due to geometric moments.

The Zernike moment were first proposed in 1934 by Zernike [44]. Their moment formulation appears to be one of the most popular, outperforming the alternatives [45] (in terms of noise resilience, information redundancy and reconstruction capability). Complex Zernike moments [46] are constructed using a set of complex polynomials which form a complete orthogonal basis set defined on the unit disc $(x^2+y^2) \leq 1$. They are expressed as A_{pq}. Two dimensional Zernike moments:

$$A_{mn} = \frac{m+1}{\pi} \int_x \int_y f(x,y)[V_{mn}(x,y)]^* dx\, dy$$
$$where\ x^2 + y^2 \leq 1 \bigtriangledown_y (x_i + h, y_j + k) \tag{1}$$

where $m = 0, 1, 2, ..., \infty$ and defines the order, $f(x,y)$ is the function being described and * denotes the complex conjugate. While n is an integer (that can be positive or negative) depicting the angular dependence, or rotation, subject to the conditions:

$$m - |n| = even, |n| \leq m \tag{2}$$

and $A_{mn}^* = A_{m,-n}$ is true. The Zernike polynomials [20] $V_{mn}(x,y)V_{mn}(x,y)$ Zernike polynomial expressed in polar coordinates are:

$$V_{mn}(r, \theta) = R_{mn}(r)exp(jn\theta) \tag{3}$$

where (r, θ) are defined over the unit disc, $j = \sqrt{-1}$ and $R_{mn}(r)$ and is the orthogonal radial polynomial, defined as $R_{mn}(r)$ Orthogonal radial polynomial:

$$R_{mn}(r) = \sum_{s=0}^{\frac{m-|n|}{2}} (-1)^s\ F(m, n, s, r) \tag{4}$$

where:

$$F(m,n,s,r) = \frac{(m-s)!}{s!\left(\frac{m+|n|}{2}-s\right)!\left(\frac{m-|n|}{2}-s\right)!}r^{m-2s} \tag{5}$$

where $R_{mn}(r) = R_{m,-n}(r)$ and it must be noted that if the conditions in Eq. 2 are not met, then $R_{mn}(r) = 0$. The first six orthogonal radial polynomials are:

$$R_{00}(r) = 1 \quad R_{11}(r) = r$$
$$R_{20}(r) = 2r^2 - 1 \quad R_{22}(r) = r^2$$
$$R_{31}(r) = 3r^3 - 2r \quad R_{33}(r) = r^3 \tag{6}$$

So for a discrete image, if P_{xy} is the current pixel then Eq. (1) becomes:

$$A_{mn} = \frac{(m+1)}{\pi}\sum_x\sum_y P_{xy}[V_{mn}(x,y)]^*$$
$$where \; x^2 + y^2 \leq 1 \tag{7}$$

To calculate the Zernike moments, the image (or region of interest) is first mapped to the unit disc using polar coordinates, where the centre of the image is the origin of the unit disc. Those pixels falling outside the unit disc are not used in the calculation. The coordinates are then described by the length of the vector from the origin to the coordinate point, r, and the angle from the x axis to the vector $r.r$ Polar co-ordinate radius, $\theta.\theta$ Polar co-ordinate angle, by convention measured from the positive x axis in a counter clockwise direction. The mapping from Cartesian to polar coordinates is:

$$x = r\cos\theta \quad y = r\sin\theta \tag{8}$$

where,

$$r = \sqrt{x^2 + y^2} \quad \theta = \tan^{-1}\left(\frac{y}{x}\right) \tag{9}$$

However, $\tan^{-1}$ in practice is often defined over the interval $\frac{x}{2} \leq \theta \leq \frac{x}{2}$, so care must be taken as to which quadrant the Cartesian coordinates appear in. Translation and scale invariance can be achieved by normalising the image using the Cartesian moments prior to calculation of the Zernike moments [47]. Translation invariance is achieved by moving the origin to the image's COM, causing $m_{01} = m_{10} = 0$. Following this, scale invariance is produced by altering each object so that its area (or pixel count for a binary image) is $m_{00} = \beta$, where β is a predetermined value. Both invariance properties (for a binary image) can be achieved using :

$$h(x,y) = f\left(\frac{x}{a}+\bar{x}, \frac{y}{a}+\bar{y}\right)$$
$$where \; a = \sqrt{\frac{\beta}{m_{00}}} \tag{10}$$

and $h(x,y)$ is the new translated and scaled function. The error involved in the discrete implementation can be reduced by interpolation. If the coordinate calculated by Equation 58 does not coincide with an actual grid location, the pixel value associated with it is interpolated from the four surrounding pixels. As a result of the normalization, the Zernike moments $|A_{00}|$ and $|A_{11}|$ are set to known values. $|A_{11}|$ is set to zero, due to the translation of the shape to the center of the coordinate system. This however will be affected by a discrete

implementation where the error in the mapping will decrease as the shape (being mapped) size (or pixel-resolution) increases. $|A_{00}|$ is dependent on m_{00}, and thus on β

$$|A_{00}| = \frac{\beta}{\pi} \tag{11}$$

Further, the absolute value of a Zernike moment is rotation invariant as reflected in the mapping of the image to the unit disc. The rotation of the shape around the unit disc is expressed as a phase change, if ϕ is the angle of rotation, A_{mn}^R is the Zernike moment of the rotated image and A_{mn} is the Zernike moment of the original image then:

$$A_{mn}^R = A_{mn}exp(-jn\phi) \tag{12}$$

Moment based features are extracted from the each zone of the scaled character bitmapped image. The image is partitioned into zone and features are extracted from each zone. In this paper Zernike moments based feature extraction is proposed for off-line Devnagari Handwritten Basic and Compound Character. To get the feature set, at first, the image is segmented to 30 x 30 blocks, and partitioned as feature set as follows and the List of the first 8 order Zernike moments is given in Table IV.

Feature set 1: Fig. 7 (a) is considered as a whole character image.

Feature set 2: Fig. 7 (b) shows the image divided into four equal zones.

Feature set 3: Fig. 7 (c) shows the image divided into three vertical equal zones.

Feature set 4: Figure 7 (d) shows the image divided into three horizontal equal zones.

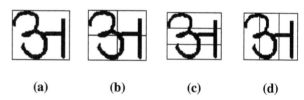

(a)   (b)   (c)   (d)

Figure 7. Partition of Devanagari Character into feature set

Table IV. THE FIRST 8 ORDER ZERNIKE MOMENTS

Order	Dimensionality	Zernike Moments
0	1	$A_{0,0}$
1	2	$A_{1,1}$
2	4	$A_{2,0}, A_{2,2}$
3	6	$A_{3,1}, A_{3,3}$
4	9	$A_{4,0}, A_{4,2}, A_{4,4}$
5	12	$A_{5,1}, A_{5,3}, A_{5,5}$
6	16	$A_{6,0}, A_{6,2}, A_{6,4}, A_{6,6}$
7	20	$A_{7,1}, A_{7,3}, A_{7,5}, A_{7,7}$

V. CLASSIFICATION AND RECOGNITION

The classification stage is the decision making part of a recognition system and it uses the features extracted in the previous stage. We have used Support Vector Machine (SVM) and k-NN for the purpose of Classification and recognition.

A. Support Vector Machine (SVM)

The support vector machine (SVM) is capable of learning and to achieve good generalization performance. If SVM is given a finite amount of training data, it is striking a balance between the goodness of fit on a given training and testing datasets. The SVM shows high ability to achieve error-free recognition. With this concept as the basis, support vector machines have proved to achieve good generalization performance with no prior knowledge of the data. The SVM is nonlinearly map the input data onto a higher dimensional feature space and determines a separating hyper plane with maximum margin between the two classes. A support vector machine is a maximal margin hyper plane in feature space built by using a kernel function. This results a nonlinear boundary in the input space. The optimal separating hyper plane can be determined without any computations in the higher dimensional feature space by using kernel functions in the input space [48].

The SVM produces a model (based on the training data) which predicts the target values of the test data features. Given a training set of instance-label pairs $(x_i, y_i), i = 1, 2, ...l$ where $x_i \in R^n$ and $y \in \{1, -1\}$, the SVM require the solution of the following optimization problem:

$$\min_{w,b,\xi} = \frac{1}{2} w^T w + C \sum_{i=1}^{l} \xi_i$$

$$Subject\ to\ y_i(w^t \phi(x_i) + b) \geq 1 - \xi_i, \xi_i > 0 \tag{13}$$

Here the training vectors x_i is mapped into a higher dimensional space by the function ϕ. SVM finds the optimal hyperplane which maximizes the distance, or more specifically the margin, between the nearest examples of both the classes. These nearest examples are called as support vectors (SVs). Where, $C > 0$ is the penalty parameter of the error term. Furthermore, $K(x_i\ x_j) \equiv \phi(x_i)^T \phi(x_j)$ is called the kernel function. We have used the radial basis function (RBF) kernel in our work given by

$$K(x_i, x_j) \equiv e^{(-\gamma\|x_i - x_j\|)}, \gamma > 0 \tag{14}$$

A search is applied to find the value of γ which is parameter of RBF. The value of both variance parameter are selected in the range of (0, 1) for gamma γ and (0, 1000) for cost (c) for support vectors and examines the recognition rate.

B. k-Nearest Neighbor (k-NN) Classifier

In the k-NN based classification similar observations belongs to similar classes. The test numeral feature vector is classified to a class, depending upon nearest neighbor distance. The nearest factor is based on minimum Euclidean Distance. Prior features are used to decide the k-nearest neighbor of the given feature vector. The most common similarity measure for k-NN classification is the Euclidian distance metric, defined between feature vectors as:

$$euc(x^\rho, y^\rho) = \sqrt{\sum_{i=1}^{f} (x_i - y_i)^2} \tag{15}$$

Where, f represents the number of features. The less distance values represent greater similarity [18], [19].

VI. EXPERIMENTS AND RESULTS

The performance evaluation using SVM and k-NN based classification has be performed on the database of Handwritten Devanagari Marathi Characters. The training dataset consists of 9600 basic character, 9000 Compound and 3000 split component of compound characters. The testing images are preprocessed and pre-classified as discussed in Section III (A). This gives 30x30 blocks segmented images of each character. Depending on the zones decided in the preprocessing step we have classified the feature sets as discussed in Section V. Then, moment based features are extracted from the each zone of the scaled character bitmapped image. Table IV shows first eight ordered Zernike moments extracted from each character using equation Zernike Moment. The Zernike moments are further divided into five folded cross validation parameters for each Devanagari Marathi Basic and Compound Characters shown in Table II and III. We experiment with different 2 values of the gamma (γ) and cost function c. The value of gamma (γ) = 0.5 and cost (c) = 1000. For the value of k-NN with K=3 is selected. The results are promising for both basic and compound character. The overall recognition accuracy is 98.37% for SVM, and 95.82% for k-NN for basic character and 98.32% for SVM and 95.42% for k-NN towards Compound character. The results on some sample are placed in Table V and VI.

The performance of the proposed method in terms of the recognition rate is compared with the other reported work and is given in Table VII. On the basis of the Table VII our proposed method shows the enhancement in the recognition rate i.e. 98.37% for basic characters and 98.32% for compound characters.

Table V.   RECOGNITION RESULT FOR BASIC CHARACTER

Basic Character	Classifier		Basic Character	Classifier	
	SVM	k-NN		SVM	k-NN
क	99.26	95.14	ध	98.49	96.12
ख	98.45	95.39	न	98.82	96.57
ग	99.12	96.00	प	98.78	96.62
घ	98.40	95.30	फ	99.23	96.21
ङ	97.66	94.73	ब	98.58	95.82
च	98.73	94.50	भ	98.61	97.26
छ	97.84	95.19	म	98.59	96.36
ज	98.62	95.87	य	98.50	95.45
झ	98.53	95.09	र	98.53	96.33
ञ	97.43	95.94	ल	98.62	96.10
ट	98.43	95.43	व	98.63	96.06
ठ	98.52	95.02	श	99.27	97.25
ड	97.40	95.29	स	98.76	96.00
द	98.29	95.30	ह	97.74	94.49
ण	99.23	96.36	ळ	98.13	95.51
त	98.82	95.95	क्ष	98.20	95.36
थ	97.79	96.37	ज्ञ	98.23	96.94
द	98.09	95.90	ज्ञ	98.48	96.26
Average Recognition Rate SVM: 98.37 and k-NN: 95.82					

VII. CONCLUSION

This paper presents a system for offline handwritten simple and compound character recognition for Marathi derived Devanagari script. Huge compound a basic and compound character dataset is collected from various age groups of writers and which has been used for database creation and named as KVKPR2013. This database further utilize for classification and recognition purpose specifically for compound characters.

Table VI.   RECOGNITION RESULT FOR COMPOUND CHARACTER

Compound Character	Classifier		Compound Character	Classifier	
	SVM	k-NN		SVM	k-NN
क्य	98.86	96.13	व्य	98.80	96.82
त्त	98.50	96.48	स्व	98.04	96.56
प्य	98.75	96.90	त्म	98.63	96.36
क्ल	98.56	96.30	न्ह	97.71	93.45
ज्य	97.76	96.78	ह	97.06	93.28
स्म	97.99	94.55	घ्य	98.54	96.09
त्न	98.09	94.25	म्य	98.70	94.66
ध्व	97.85	94.86	न्य	98.86	94.31
ज्व	97.33	94.92	व्द	98.52	94.42
ग्य	98.64	96.14	म्न	98.28	94.49
च्छ	97.24	95.28	स्फ	98.48	94.62
फ्य	98.92	95.02	ब्य	98.59	95.36
व्य	98.62	94.28	प्त	98.79	96.54
त्य	98.76	95.26	ब्ज	98.37	96.46
स्क	98.36	95.76	च्य	98.47	95.67
त्क	98.18	95.45	ख्ख	97.64	94.48
म्ह	97.88	94.72	ण्य	98.56	96.57
न्म	98.49	94.86	स्य	98.54	96.74
ध्य	98.36	95.42	स्प	98.60	95.64
ज्म	98.38	96.57	ल्य	98.51	95.76
भ्य	98.72	95.78	ब्द	97.89	94.29
त्य	98.89	95.54	श्व	98.47	95.47
			भ्व	97.25	94.45
Average Recognition Rate SVM: 98.32 and k-NN: 95.42					

Table VII.   COMPARISON OF RESULTS OF PROPOSED METHOD WITH OTHER METHODS IN LITERATURE

Method	Features	Classifier	Dataset (size)	Accuracy (%)
Sharma [22]	Chain Code	Quadratic	11270	80.36
Deshpande [29]	Chain Code	RE and MED	5000	82.00
Arora [26]	Structural	FFNN	50000	89.12
Hanmandlu [23]	Vector Distance	Fuzzy set	4750	90.65
Arora [28]	Shadow and CH	NKO and MED	7154	90.74
Kumar [9]	Gradient	SVM	25000	94.10
Pal et. al [24]	Gradient and Gaussian Filter	Quadratic	36172	94.24
Mane [35]	Eigen Deformation (ED)	Elastic Matching	3600	94.91
Pal [26]	Gradient	SVM and MQDF	36172	95.13
Pal [32]	Gradient	MIL	36172	95.19
Arora [27]	Chain code, Shadow	MLP	1500	98.16
S. Shelke and S. Apte [49]	NPD, ED, Wavelet	MLP	37000	97.95
Proposed (Basic)	Zernike Moment	SVM	12000	98.37
Proposed (Compound)	Zernike Moment	SVM	15000	98.32

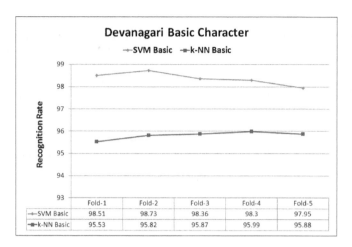

Figure 8.   Recognition Rate of Basic Characters through SVM and k-NN Classifiers

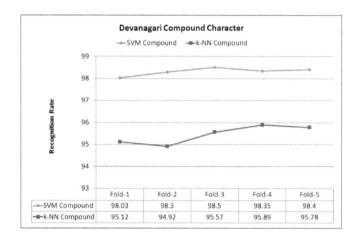

Figure 9.   Recognition Rate of Basic Characters through SVM and k-NN Classifiers

only 45 compound characters that extracts Zernike moment features and classified using SVM and k-NN approach. In future the system can be extended to handle more compound characters with other features like orthogonal moments and can classified through advanced patterns classification and neural network approaches.

ACKNOWLEDGMENT

Authors would like to acknowledge and thanks to University Grants Commission (UGC), India for granting UGC SAP (II) DRS Phase-I F. No.-3-42/2009 Biometrics: Multimodal System Development Laboratory facility and One Time Research Grant F. No. 4-10/2010 (BSR) supporting to this work.

REFERENCES

[1] K. Roy, S. Vaidya, U. Pal, B. B. Chaudhuri, and A. Belaid, "A system for indian postal automation," in *Proc. 8th Int. Conf. Document Analysis and Recognition, Seoul, Korea, Aug. 31-Sep. 1*, 2005, pp. 1060–1064.

[2] U. Pal, R. K. Roy, and F. Kimura, "Indian multi script full pin-code string recognition for postal automation," in *Proc. 10th Int. Conf. Document Analysis and Recognition, Barcelona, Spain, Jul. 26-29*, 2009, pp. 456–460.

Prior to feature extraction the character is pre-classified into three categories using structural features. Various complex features of compound characters from the database has been created through Zernike moment approach and implemented successfully for its classification and recognition under SVM and k-NN approach. Zernike moment feature for Devnagari has given better result for compound character. The proposed system gives improved recognition rate of 0.37% than other handwritten character recognition system. The system has been evaluated on a huge amount of Handwritten Character Database i.e. 12000 basic and 15000 compound character dataset created in our laboratory. Since, no work has been reported on Devanagari Compound Character recognition in the last decade, the system handles the problem with structural and statistical features of compound character. The system handles

[3] B. B. Chaudhari, "Digital document processing - major directions and recent advances," *London:Springer*, 2007.

[4] U. Pal and B. B. Chaudhari, "Indian script character recognition: a survey," *Pattern Recognition*, vol. 37, pp. 1887–1899, 2004.

[5] S. Shelke, S. Apte, and et. al., "A novel multistage classification andwavelet based kernel generation for handwritten marathi compound character recognition," in *Proc. Int. Conf. Communications and Signal Processing, Kerala, India*, 2011, pp. 193–197.

[6] S. Shelke and S. Apte, "A multistage handwritten marathi compound character recognition scheme using neural networks and wavelet features," *International Journal of Signal Processing, Image Processing and Pattern Recognition*, vol. 4, pp. 81–94, 2011.

[7] R. K. Sinha and Mahabala, "Machine recognition of devnagari script," *IEEE Trans. System Man Cyber*, pp. 435–441, 1979.

[8] I. K. Sethi and B. Chatterjee, "Machine recognition of constrained hand printed devnagari," *Pattern Recognition*, vol. 9, pp. 69–75, 1977.

[9] S. Kumar and C. Singh, "A study of zernike moments and its use in devnagari handwritten character recognition," in *Proc. Intl. Conf. Cognition and Recognition, Mandya (India)*, 2005, pp. 514–520.

[10] V. Bansal and et. al., *Integrating Knowledge Sources in Devanagari Text Recognition*. IIT, Kharagpur: Ph.D. Thesis, 1999.

[11] V. Bansal and R. M. K. Sinha, "Partitioning and searching dictionary for correction of optically read devanagari character strings," in *Proc. 5th Int. Conf. Document Analysis and Recognition, Bangalore, India, Sept. 20-22*, 1999, pp. 53–656.

[12] V. Bansal. and R. M. K. Sinha., "On how to describe shapes of devanagari characters and use them for recognition," in *Proc. 5th Int. Conf. Document Analysis and Recognition, Bangalore, India, Sept. 20-22*, 1999, pp. 410–413.

[13] R. Bajaj, L. Dey, and S. Chaudhury, "Devnagari numeral recognition by combining decision of multiple connectionist classifiers," *Sadhana*, vol. 27, pp. 59–72, 2002.

[14] P. M. Patil and T. R. Sontakke, "Rotation, scale and translation invariant handwritten devanagari numeral character recognition using general fuzzy neural network," *Pattern Recognition*, vol. 40, pp. 2110–2117, 2007.

[15] L. C. Barczak, M. J. Johnson, and C. H. Messom, "Revisiting moment invariant: Rapid feature extraction and classification for handwritten digit," in *Proceeding of Image and Vision Computing, Hamilton, New Zealand, December*, 2007, pp. 137–142.

[16] R. O. Duda, P. E. Hart, and D. G. Stork, *Pattern Classification, Second ed.* John Wiley and Sons, Inc. 14, 2001.

[17] H. R. Boveiri and et. al., "Persian printed numeral character recognition using geometrical central moments and fuzzy min max neural network," *International Journal of Signal Processing*, pp. 226–232, 2009.

[18] H. R. Boveiri., "Persian printed numeral classification using extended moment invariants," *World Academy of Science, Engineering and Technology 63*, pp. 167–174, 2010.

[19] S. Arora, D. Bhattacharjee, M. Nasipuri, D. K. Basu, and M. Kundu, "Application of statistical features in handwritten devanagari character recognition," *International Journal of Recent Trends in Engineering*, vol. 2, pp. 40–42, 2009.

[20] R. S. Kunte and R. D. S. Samuel, "A simple and efficient optical character recognition system for basic symbols in printed kannada text," *Sadhana*, vol. 32, pp. 21–533, 2007.

[21] S. N. Nawaz and et. al., "An approach to offline arabic character recognition using neural network," in *Proceeding of IEEE ICECS*, 2003, pp. 1325–1331.

[22] N. Sharma, U. Pal, F. Kimura, and S. Pal, "Recognition of offline handwritten devnagari characters using quadratic classifier," in *Proc. Indian Conf. Computer Vision Graphics and Image Processing, Madurai (India)*, 2006, pp. 805–816.

[23] M. Hanmandlu, O. V. R. Murthy, and V. K. Madasu, "Fuzzy model based recognition of handwritten hindi characters," in *Proc. Ninth Biennial Conf. Australian Pattern Recognition Society on Digital Image Computing Techniques and Applications, Glenelg (Australia)*, 2007, pp. 454–461.

[24] U. Pal, N. Sharma, T. Wakabayashi, and F. Kimura, "Off-line handwritten character recognition of devnagari script," in *Proc. Ninth Intl. Conf. Document Analysis and Recognition, Curitiba (Brazil)*, 2007, pp. 496–500.

[25] S. Arora, D. Bhattacharjee, M. Nasipuri, D. K. Basu, and M. Kundu, "Combining multiple feature extraction techniques for handwritten devnagari character recognition," in *Proc. IEEE Region 10 Colloquium and Third Intl. Conf. Industrial and Information Systems, Kharagpur (India)*, 2008.

[26] S. Arora, D. Bhatcharjee, M. Nasipuri, and L. Malik, "A two stage classification approach for handwritten devanagari characters," in *Proc.Int. Conf. Comput. Intell. Multimedia Appl.*, 2007, pp. 399–403.

[27] S. Arora, D. Bhattacharjee, M. Nasipuri, D. K. Basu, M. Kundu, and L. Malik, "Study of different features on handwritten devnagari character," in *Proc. 2nd Emerging Trends Eng. Technol.*, 2009, pp. 929–933.

[28] S. Arora, D. Bhattacharjee, M. Nasipuri, D. K. Basu, and M. Kundu, "Recognition of non-compound handwritten devnagari characters using a combination of mlp and minimum edit distance," *Int. J. Comput. Sci. Security*, vol. 4, pp. 1–14, 2010.

[29] P. S. Deshpande, L. Malik, and S. Arora, "Fine classification and recognition of hand written devnagari characters with regular expressions and minimum edit distance method," *Journal of Computers*, vol. 3, pp. 11–17, 2008.

[30] S. Kumar, "Performance comparison of features on devanagari handprinted dataset," *Int. J. Recent Trends*, vol. 1, pp. 33–37, 2009.

[31] U. Pal, S. Chanda, T. Wakabayashi, and F. Kimura, "Accuracy improvement of devnagari character recognition combining svm and mqdf," in *Proc. Eleventh Intl. Conf. Frontiers in Handwriting Recognition, Montreal (Canada)*, 2008, pp. 367–372.

[32] U. Pal, T. Wakabayashi, and F. Kimura, "Comparative study of devnagari handwritten character recognition using different feature and classifiers," in *Proc. Tenth Intl. Conf. Document Analysis and Recognition, Barcelona (Spain)*, 2009, pp. 1111–1115.

[33] S. Shelke and S. Apte, "A novel multi-feature multi-classifier scheme for unconstrained handwritten devanagari character recognition," in *12th International Conference on Frontiers in Handwriting Recognition*, 2010.

[34] M. J. Baheti, K. V. Kale, and M. E. Jadhav, "Comparison of classifiers for gujarati numeral recognition," *International Journal of Machine Intelligence (IJMI)*, vol. 3, pp. 160–163, 2011.

[35] V. Mane and L. Ragha, "Handwritten character recognition using elastic matching and pca," in *Proc. Int. Conf. Adv. Comput., Commun. Control*, 2009, pp. 410–415.

[36] U. Pal, T. Wakabayashi, and F. Kimura, "Handwritten bangla compound character recognition using gradient feature," in *Proc. 10th Int. Conf. Information Technology, Orissa, India*, 2007, pp. 208–213.

[37] B. B. Chaudhuri and U. Pal, "A complete printed bangla ocr system," *Pattern Recognition*, vol. 31, pp. 531–549, 1998.

[38] K. Roy, T. Pal, U. Pal, and F. Kimura, "Oriya handwritten numeral recognition system," in *Proc. 8th Int. Conf. Document Analysis and Recognition, vol. 2, Seoul, Korea, Aug. 31-Sep. 1*, 2005, pp. 770–774.

[39] A. Pujari, C. D. Naidu, M. S. Rao, and B. C. Jinaga, "An intelligent character recognizer for telugu scripts using multi resolution analysis and associative memory," *Image and Vision Computing*, vol. 22, pp. 1221–1227, 2004.

[40] U. Bhattacharya, S. K. Ghosh, and S. K. Parui, "A two stage recognition scheme for handwritten tamil characters," in *Proc. 9th Int. Conf. Document Analysis and Recognition, Parana, Sept. 23-26*, 2007, pp. 511–515.

[41] M. Hasnat, S. M. Habib, and M. Khan, "A high performance domain specific ocr for bangla script," in *Novel Algorithms and Techniques In Telecommunications, Automation and Industrial Electronics*. Springer Netherlands, 2008, pp. 174–178.

[42] U. ., G. ., and B. B. Chaudhari, "Segmentation of touching character in printed devanagari and bangla script using fuzzy multifactorial analysis," *IEEE Trans. Systems, Man and Cybernetics-Part C, Applications and Reviews*, vol. 32, pp. 449–459, 2002.

[43] H. J. Kim and W.-Y. Kim, "Eye detection in facial images using zernike moments with svm," *ETRI Journal*, vol. 32, pp. 335–337, 2008.

[44] F. Zernike, "Beugungstheorie des schneidenverfahrens und seiner verbesserten form, der phasenkontrastmethode (diffraction theory of

the cut procedure and its improved form, the phase contrast method),"
Physica, vol. 1, pp. 689–704, 1934.

[45] C. Teh and R. T. Chin, "On image analysis by the method of moments,"
IEEE Trans. on Pattern Analysis and Machine Intelligence, vol. 10, pp.
496–513, 1988.

[46] M. R. Teague, "Image analysis via the general theory of moments,"
Journal of the Optical Society of America, vol. 70, pp. 920–930, 1979.

[47] A. Khotanzad and Y. H. Hongs, "Invariant image recognition by zernike
moments," *IEEE Trans. on Pattern Analysis and Machine Intelligence*,
vol. 12, pp. 489–497, 1990.

[48] V. Vapnik., "The nature of statistical learning theory," *Springer, N.Y.
ISBN 0-387-94559-8*, 1995.

[49] S. Shelke and S. Apte, "Multistage handwritten marathi compound character recognition using neural networks," *Journal of Pattern Recognition
Research 2 (JPPR)*, pp. 253–268, 2011.

Method and System for Human Action Detections with Acceleration Sensors for the Proposed Rescue System for Disabled and Elderly Persons Who Need a Help in Evacuation from Disaster Area

Kohei Arai [1]
Graduate School of Science and Engineering
Saga University
Saga City, Japan

Abstract—**Method and system for human action detections with acceleration sensors for the proposed rescue system for disabled and elderly persons who need a help in evacuation from disaster areas is proposed. Not only vital signs, blood pressure, heart beat pulse rate, body temperature, bless and consciousness, but also, the location and attitude of the persons have to be monitored for the proposed rescue system. The attitude can be measured with acceleration sensors. In particular, it is better to discriminate the attitudes, sitting, standing up, and lying down. Also, action speed has to be detected. Experimental results show that these attitude monitoring can be done with acceleration sensors.**

Keywords—*vital sign; heart beat puls ratee; body temperature; blood pressure; blesses, consciousnes; seonsor network*

I. INTRODUCTION

Handicapped, disabled, diseased, elderly persons as well as peoples who need help in their ordinary life are facing too dangerous situation in event of evacuation when disaster occurs. In order to mitigate victims, evacuation system has to be created. Authors proposed such evacuation system as a prototype system already [1]-[4].

The system needs information of victims' locations, physical and psychological status as well as their attitudes. Authors proposed sensor network system which consist GPS receiver, attitude sensor, physical health monitoring sensors which allows wearable body temperature, heart beat pulse rates; bless monitoring together with blood pressure monitoring [5]-[7]. Also the number of steps, calorie consumptions is available to monitor. Because it is difficult to monitor the blood pressure with wearable sensors, it is done by using the number of steps and body temperature. In addition to these, psychological status is highly required for vital sign monitoring (consciousness monitoring). By using EEG sensors, it is possible to monitor psychological status in the wearable sensor. These are components of the proposed physical health and psychological monitoring system.

Method and system for human action detections with acceleration sensors for the proposed rescue system for disabled and elderly persons who need a help in evacuation from disaster area is proposed. Experimental results show that human actions can be estimated with acceleration sensors.

Section 2 describes the proposed acceleration sensor system followed by experiment method and results. Then conclusion is described together with some discussions..

II. PROPOSED SENSOR NETWORK SYSTEM

A. Acceleration Sensor Used

Figure 1 shows outlook of the acceleration sensor used in the proposed rescue system. It is the Small Sized Wireless Hybrid Sensor WAA-010

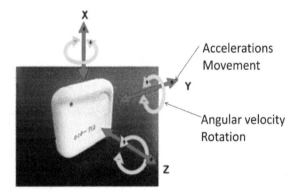

Fig. 1. Outlook of the Small Sized Wireless Hybrid Sensor WAA-010.

It allows measurements of movements in x, y, and z directions and roll, pitch and yaw rotations. WAA-010 is a multi sensor which allows measurements of attitude, angular velocity (gyro), Earth magnet in three axes. Also WAA-010 has communication capability to PC through Bluetooth communication links.

B. Acceleration Data Acquisition

Figure 2 (a) shows the toggle of the Bluetooth while Figure 2 (b) shows installation window of Bluetooth on PC display.

(a)Toggle of the Bluetooth

(b) Installation window of Bluetooth on PC display.

Fig. 2. Toggle of the Bluetooth and installation window of Bluetooth on PC display.

After the installation of acceleration sensor driver, communication pot assignments, batteries of the three axis of acceleration, angular velocity as well as Earth magnet-meter can be monitored through PC screen which is shown in Figure 3.

(a)Port asignment

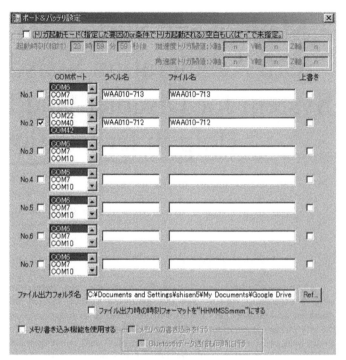

(b)Battery monitor

Fig. 3. Communication port assignment and battery monitor

III. EXPERIMENTS

A. Examples of Acceleration and Angular Velocity Data

Figure 4 (a) shows an example of acceleration sensor data with the maximum range of 5000mG while Figure 4 (b) shows and example of measured acceleration sensor data with the maximum range of 2000mG. As shown in Figure 4, the maximum range has to be adjusted for the acceleration of the target objects. Meanwhile, Figure 5 (a) and (b) shows an example of measured angular velocity with maximum of 500 dps and 200 dps, respectively. These examples are obtained through the experiments of which the acceleration sensor are held by hands. Therefore, these performances are just for the acceleration sensor only.

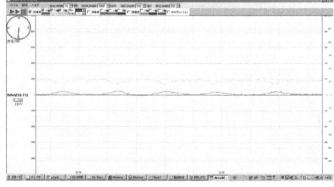

(a)5000mG

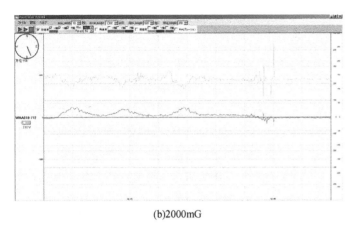

(b)2000mG

Fig. 4. Examples of acceleration sensor data

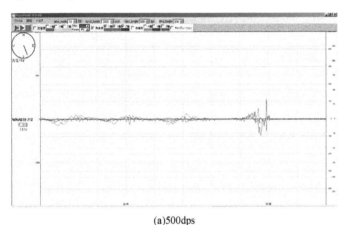

(a)500dps

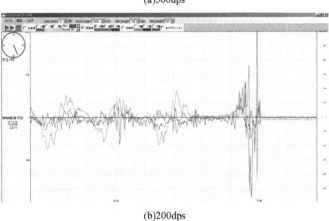

(b)200dps

Fig. 5. Examples of angular velocity measured

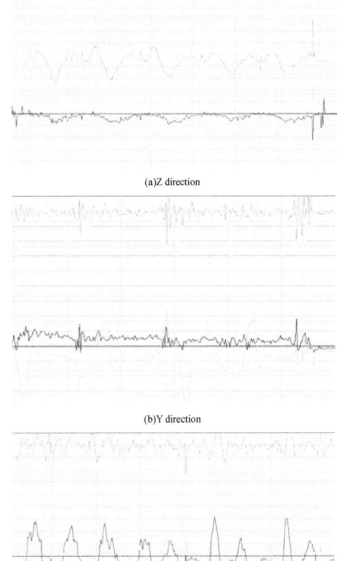

(a)Z direction

(b)Y direction

(c)X direction

Fig. 6. Examples of acceleration sensor data for the motions in X, Y, and Z directions

Meanwhile, example of acceleration sensor data (200mG of maximum range) for the motion in Z direction is shown in Figure 6 (a).

On the other hand, Figure 6 (b) and (c) shows examples of acceleration sensor data for the motion in Y and X directions, respectively.

Figure 7 (a), (b), and (c) shows examples of angular velocity data (500 dps of maximum range) for the rotations in Z, Y, X directions, roll, pitch and yaw angles, respectively.

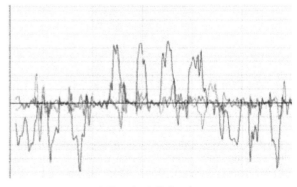

(a)Rotation in Z direction

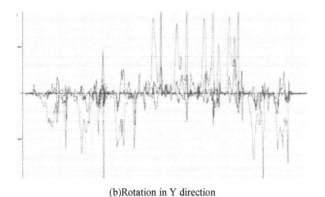

(b)Rotation in Y direction

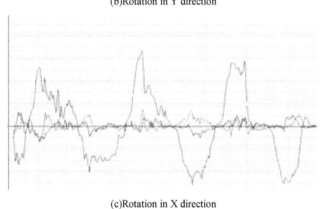

(c)Rotation in X direction

Fig. 7.   Examples of angular velocity data

B. *Examples of Acceleration and Angular Velocity Data Attached to the Wearing Glass End*

Acceleration sensor is then attached to the wearing glass end as is shown in Figure 8. Not only the acceleration sensor but also body temperature, heart beat pulse rate sensors as well as EEG sensor head are attached to the glass end together with battery and Bluetooth communicator. At the forehead, EEG sensor head is attached with spring wire extended from the center of the glass. Therefore most of vital signs can be measured with the glass.

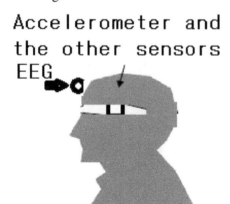

Fig. 8.   Location of the acceleration sensor (at the glass end)

Figure 9 shows examples of acceleration sensor and angular velocity data when the user lie down and stand up slowly. Meanwhile, Figure 10 shows examples of those when the user sit down on a chair slowly.

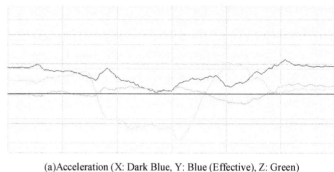

(a)Acceleration (X: Dark Blue, Y: Blue (Effective), Z: Green)

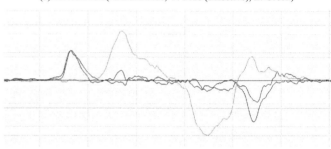

(b)Angular velocity (X: Black, Y: Green, Z: Red (Effective))

Fig. 9.   Examples of acceleration sensor and angular velocity data when the user laid down and stand up slowly.

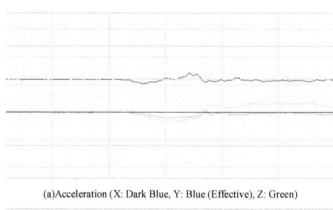

(a)Acceleration (X: Dark Blue, Y: Blue (Effective), Z: Green)

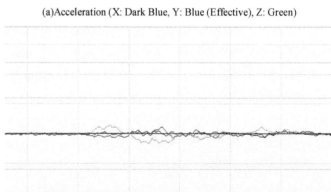

(b)Angular velocity (X: Black, Y: Green, Z: Red (Effective))

Fig. 10. Examples of acceleration sensor and angular velocity data when the user sit down on chair slowly.

Figure 11 shows examples of acceleration sensor and angular velocity data when the user lie down and stand up quickly. Meanwhile, Figure 12 shows examples of those when the user sit down on a chair quickly. On the other hand, Figure 13 (a) shows Y axis data when the user lies down and stand up slowly while Figure 13 (b) shows X axis data when the user sit down on the chair slowly .

Furthermore, Figure 13 (c) shows Y axis data when the user lies down and the stand up quickly while Figure 13 (d) shows X axis data when the user sit on the chair quickly.

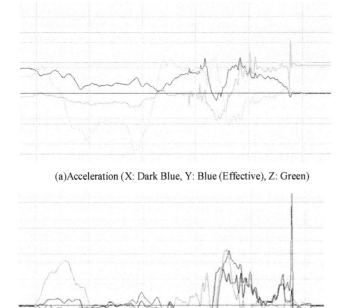

(a)Acceleration (X: Dark Blue, Y: Blue (Effective), Z: Green)

(b)Angular velocity (X: Black, Y: Green, Z: Red (Effective))

Fig. 11. Examples of acceleration sensor and angular velocity data when the user lie down and the stand up quickly.

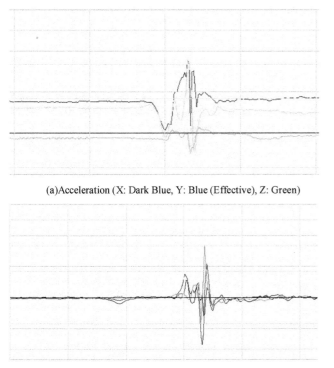

(a)Acceleration (X: Dark Blue, Y: Blue (Effective), Z: Green)

(b)Angular velocity (X: Black, Y: Green, Z: Red (Effective))

Fig. 12. Examples of acceleration sensor and angular velocity data when the user sit down on chair quickly.

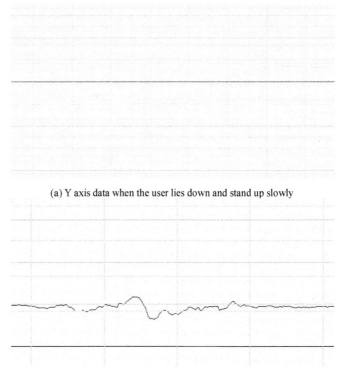

(a) Y axis data when the user lies down and stand up slowly

(b) X axis data when the user sit down on the chair slowly

(c) Y axis data when the user lies down and the stand up quickly

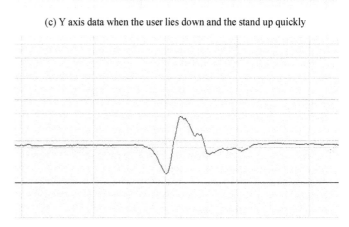

(d) X axis data when the user sit on the chair quickly

Fig. 13. Specific axis data for the action of lie down and then stand up as well as sit down on the chair, slowly and quickly

C. Validation of Acceleration and Angular Velocity Data

Three axes of acceleration sensor and angular velocity sensor data are validated for the actions "Lie down and then stand up slowly", "Sit down on the chair slowly", "Lie down and then stand up quickly", "Sit down on the chair quickly". The results are shown in Figure 14. Through these validations, it is found that the most effective signal for detection of lie down and then stand up is acceleration in Y axis while that for detecting of sit down action is angular velocity sensor data in Z axis as are shown in Table 1.

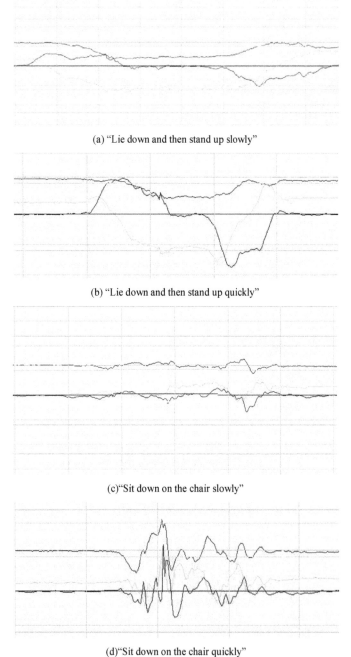

(a) "Lie down and then stand up slowly"

(b) "Lie down and then stand up quickly"

(c) "Sit down on the chair slowly"

(d) "Sit down on the chair quickly"

Fig. 14. Validation of three axes of acceleration sensor and angular velocity sensor data (X: Dark Blue, Y: Blue, Z: Green for acceleration sensor, and X: Black, Y: Green, Z: Red for angular velocity sensor)

TABLE I.   SUMMARY OF VALIDATION RESULTS FOR DETCTION OF USERS' ACTIONS OF "LIE DOWN AND THEN STAND UP" AND "SIT DOWN ON THE CHAIR"

	Action	Slow action		Quick action	
		Lie down and stand up	Sit on chair	Lie down and stand up	Sit on chair
On	Accelerations X	0	2	2	3
On	Accelerations Y	3	1	3	2
Off	Accelerations Z	0	1	2	1
Off	Angular Velocity X	2	0	2	1
Off	Angular Velocity Y	1	0	2	1
On	Angular Velocity Z	3	1	3	2

IV. CONCLUSION

Method and system for human action detections with acceleration sensors for the proposed rescue system for disabled and elderly persons who need a help in evacuation from disaster areas is proposed. Not only vital signs, blood pressure, heart beat pulse rate, body temperature, bless and consciousness, but also, the location and attitude of the persons have to be monitored for the proposed rescue system. The attitude can be measured with acceleration sensors.

In particular, it is better to discriminate the attitudes, sitting, standing up, and lying down. Also, action speed has to be detected. Experimental results show that these attitude monitoring can be done with acceleration sensors.

Through these validations, it is found that the most effective signal for detection of lie down and then stand up is acceleration in Y axis while that for detecting of sit down action is angular velocity sensor data in Z axis. Namely, validation results for detection of users' actions of "Lie down and then stand up" and "Sit down on the chair"

ACKNOWLEDGMENT

The author would like to thank all the patients who are contributed to the experiments conducted. The author also would like to thank Professor Dr. Takao Hotokebuchi, President of Saga University for his support this research works.

REFERENCES

[1] Kohei Arai, Tran Xuan Sang, Decision making and emergency communication system in rescue simulation for people with disabilities, International Journal of Advanced Research in Artificial Intelligence, 2, 3, 77-85, 2013.

[2] K.Arai, T.X.Sang, N.T.Uyen, Task allocation model for rescue disable persons in disaster area with help of volunteers, International Journal of Advanced Computer Science and Applications, 3, 7, 96-101, 2012.

[3] K.Arai, T.X.Sang, Emergency rescue simulation for disabled persons with help from volunteers, International Journal of Research and Review on Computer Science, 3, 2, 1543-1547, 2012.

[4] K. Arai, and T. X. Sang, "Fuzzy Genetic Algorithm for Prioritization Determination with Technique for Order Preference by Similarity to Ideal Solution", International Journal of Computer Science and Network Security, vol.11, no.5, 229-235, May 2011.

[5] Arai K., R. Mardiyanto, Evaluation of Students' Impact for Using the Proposed Eye Based HCI with Moving and Fixed Keyboard by Using EEG Signals, International Journal of Review and Research on Computer Science(IJRRCS), 2, 6, 1228-1234, 2011

[6] K.Arai, Wearable healthy monitoring sensor network and its application to evacuation and rescue information server system for disabled and elderly person, International Journal of Research and Review on Computer Science, 3, 3, 1633-1639, 2012.

[7] K.Arai, Wearable Physical and Psychological Health Monitoring System, Proceedings of the Science and Information Conference 2013 October 7-9, 2013 | London, UK

18

Application of distributed lighting control architecture in dementia-friendly smart homes

Atousa Zaeim
School of CSE
University of Salford Manchester
United Kingdom

Samia Nefti-Meziani
School of CSE
University of Salford Manchester
United Kingdom

Adham Atyabi
School of CSE
University of Salford Manchester
United Kingdom
School of CSEM
Flinders University of South Australia

Abstract—**Dementia is a growing problem in societies with aging populations, not only for patients, but also for family members and for the society in terms of the associated costs of providing health care. Helping patients to maintain a degree of independence in their home environment while ensuring their safeties is considered as a positive step forward for addressing individual needs of dementia patients. A common symptom for dementia patients including those with Alzheimer's Disease and Related Dementia (ADRD) is sleep disturbance, patients being awake at night and asleep during the day. One of the problems with night time sleep disturbance in dementia patients is the possible accidental falls of patients in the dark environment. An issue associated with un-hourly sleeping behavior in these patients is the lighting condition of their surroundings. Clinical studies indicate that appropriate level of lighting can help to restore the rest-activity cycles of ADRD patients. This study tackles this problem by generating machine learning solutions for controlling the lighting conditions of multiple rooms in the house in different hours based on patterns of behaviors generated for the patient. Several neural network oriented classification methods are investigated and their feasibilities are assessed with a collection of synthetic data capturing two conditions of balanced and unbalanced inter-class samples. The classifiers are utilized within two centric and distributed lighting control architectures. The results indicate the feasibility of the distributed architecture in achieving a high level of classification performance resulting in adequate control over lighting conditions of the house in various time periods.**

Keywords—*Smart Home; Ambient intelligence; Machine Learning; Distributed Learning*

I. INTRODUCTION

Intelligent health-care technologies often include utilizing smart devices or systems tuned to operate optimally under some specific conditions, fuse and interpret information, and make decisions that benefit the patients' care. In the context of smart home, such smart devices interact with each other through "plug and play" mechanisms, intelligent software agents and inter-connection messaging protocols. Smart home is an attractive concept with many potentials and applications in intelligent health-care due to its ability to provide cognitive assistance whenever required[1],[2]. In this study, we investigate feasibility of smart lighting control for improving living conditions of dementia patients in the context of smart home.

The effectiveness of traditional verbal commands/prompts may be questioned especially for individuals with advanced Alzheimer's disease and Related Dementia (ADRD). Sleep disturbance is a common problem in ADRD patients that can cause possible accidental falls of patients in the dark environment [7]. It is reported that 40% of patients suffering from dementia also suffer from sleep disorder and being frightened from room darkness while being incapable of turning on the lights[3]. Smart control over lighting conditions of patients' surroundings can effectively improve their quality of life due to minimizing chances of such accidents while allowing the patients to be less dependent to constant presence of caregivers.

This study focuses on investigating the feasibility of machine learning solutions for controlling the lighting conditions of multiple rooms in the house in different hours. This is to be facilitated by creating patterns of behaviors for the patients and carers and tuning the lighting conditions to the patients needs. In a small scale (single room or a small house), fuzzy logic systems are ideal candidates for developing a smart lighting condition monitoring and controlling system due to their design simplicity and their performance accuracy. However, extending such designs to larger scales such as special hospitals with multiple patients and multiple rooms might be problematic due to the required tedious design adaptation stage.

In this study a distributed lighting control architecture is proposed and its feasibility is assessed against a commonly used centric architecture. The study considers a set of synthetic data and investigates the potential of several classification methods. In order to provide clear indication on overall performance of the system both balanced and unbalanced data sample conditions are considered.

The structure of this paper is as follows: related works are presented in Section II. The architectural design is reported in Section III. The procedures considered for generating the synthetic data that is utilized in this study are discussed in Section IV. The results are reported in Section V and the final conclusions and the future works are discussed in Sections VI and VII.

---

[1]National Sleep Foundation,
http://sleepfoundation.org/ask-the-expert/sleep-and-alzheimers-disease/page/0/2

[2]Alzheimer's Society,
http://www.alzheimers.org.uk/site/scripts/documents_info.php?documentID=145

[3]Alzheimer's Association,
http://www.alz.org/alzheimers_disease_10429.asp

II. RELATED WORK

In the context of smart lighting control, Nagy et al [6] proposed an occupant centric lighting control strategy that utilizes occupant specific set-points with fixed minimum and maximum illuminance threshold set to the satisfaction of each occupant. The results indicated that within a six weeks duration in an environment with 10 offices total of 37% energy savings is achieved using the proposed strategy.

Gopalakrishna et al [8] investigated the use of prediction methods (decision trees) for intelligent lighting control in an office environment. The study is conducted on the basis of synthetic data within a large environment with 2 different areas for resting and having informal meetings. Various sample sizes in the range of 100 to 50k are considered with each sample reflecting mixtures of six sensory inputs. Although systematic experimentation is carried out to identify the maximum number of samples required to make an informed decision, the results are questionable due to possible repetition of samples caused by limitations imposed from using non-numeric sensory values.

In the context of smart home and elderly care, Mahmoud et al [1] investigated implications of various soft computing approaches for generating patterns of occupancy behaviour and predicting upcoming abnormal behaviours. The study considered both synthetic and physical data and identified non-linear autoregressive network with exogenous inputs as an ideal type of recurrent neural network for the prediction and forecasting task.

Lotfi et al. [2] investigated the use of simple network of sensors to monitor the behavior elderly people suffering from dementia with a focus on improving their ability to live independently. Standard home automation sensors such as motion and door entry sensors are utilized. Abnormal behaviors are identified using the acquired data and recurrent neural network is employed to forecast the upcoming events and possible sensory readings. Utilizing such predictions, the system transit certain messages to the caregivers informing them about any possible abnormal behavior in the near future.

In the context of smart home, Dawadi et al [3] and [4] utilized a machine learning based approach for monitoring the well being of individuals and assessing their cognitive health. combinations of principle component analysis, support vector machine, and logistic regression methods are utilized to assess the quality of an activity performed by participants. The study gathered sensory data from 263 participants and using the hybrid machine learning mechanism achieved a meaningful classification performance distinguishing two classes of dementia and cognitive healthy. The mixture of door, item, power usage, and motion sensors combined with activity time log are utilized to represent the activities conducted by participants. Similar study is conducted by Dawadi et al [5]. In the study sets of activities are defined and after their completions with participating elderly people, features representing each of the activities performed are assessed separately with machine learning based predictive models that are pre-trained only with features of such activity. The study considered Naive Bayse, J48, SMO, and Neural Network approaches among which Naive Bayse and Neural Network methods performed better than others.

Fig. 1: A sample layout of a house with four rooms being considered in this study.

III. SYSTEM ARCHITECTURE FOR DEMENTIA FRIENDLY SMART HOUSE

As mentioned earlier, in this study, a machine learning paradigm capable of providing intelligence control over lighting condition of a smart house is to be designed. Figure 1 depicts an schematic layout of the environment being consider in the study. In this study, two different designs are considered. In the first architecture, a single learning mechanism, e.g. a variation of Artificial Neural Network (ANN), with sufficient number of output nodes that is capable of handling the lighting condition of all rooms in the house is considered. The layout of this architecture is depicted in figure 2. Although this architecture is commonly employed in literature, such an architecture suffers from lack of flexibility. That is, similar to a fuzzy-based controller, adaptation of such system to a larger scale problem, e.g. controlling the lighting condition of a hospital for example, is likely to be tedious and problematic.

In order to provide a flexible and reconfigurable learning mechanism capable of being adapted to larger scale problems with minimum adaptation effort, an alternative architecture is proposed. In this architecture, the lighting condition of each room in the house is to be controlled by a separate classifier. That is, in this architecture, the lighting control is distributed across the rooms in the house. The structural design of this architecture is illustrated in figure 3. In here, the learning is distributed across classifiers that are controlling the rooms' lighting conditions. As a result, although the samples are to be representative of the status of the entire house, however, the expected reactions in terms of changes in the lighting conditions within each room are to be varied. This is facilitated by providing separated labels for each room with each sample (training and/or testing). The resulting distributed architecture

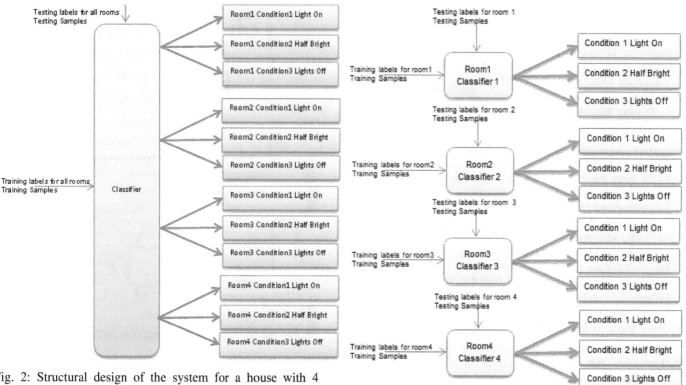

Fig. 2: Structural design of the system for a house with 4 rooms. The centric architecture.

Fig. 3: Structural design of the system for a house with 4 rooms. The distributed architecture.

is highly flexible and re-scalable.

IV. SYNTHETIC DATA

This section presents the procedures followed for generating the synthetic data. In this study, following sensory inputs are considered:

- Light Detection Range (LDR) sensor: The range of sensory outputs for this sensor is assumed to be varied by time. That is, the 24 hour (day) is divided to 6 equal length periods (4 hours length each). The sensory reading within each period follows normal distribution within period-specific ranges. These periods, their associated ranges and values are presented in table I.

- Thermal camera: The camera detects whether the patient or the caregivers are present in a room. Considering a house with 4 rooms, a binary representation is utilized in which bits with 0 and 1 values respectively represent absence and presence of patient and/or caregivers.

- Displacement sensor: the sensor detects if the patient is on bed. Similar to the LDR sensor, 6 periods of 4 hours long are considered and the sensory readings within each period follow normal distribution within a certain range. These periods and their associated ranges are presented in table II. Although all the ranges are selected arbitrary however, they are deliberately set to overlap each other.

In this study, each sample represents sensory readings originating from LDR, displacement and thermal sensors. The study considers sets of 100 datasets, each containing 100 random training samples and 10 random testing samples. Figure 4 depicts the rules utilized for generating room specific labels for each sample. Two sets of datasets capturing balanced and unbalanced condition within inter classes of 'Lights Off', 'Half Bright' and 'Lights On' are considered. The choice of considering both balanced and unbalanced conditions is made due to difficulties associated with conducting complete recording sessions with patients suffering from dementia. That is, in such patients, it often happens that the data collection procedures are left uncompleted with only a sub-set of samples recorded due to inability or unwillingness of the patients to finish the tasks or recording sessions. As a consequence, it is considered advantageous for a system to be able to maintain some degree of efficiency with both balanced and unbalanced sampling conditions.

V. EXPERIMENTS & RESULTS

As mentioned in previous sections, two neural network oriented architectures of centric and distributed are considered in this study. Two sets of experiments are considered in this section to help investigating the feasibility of these two architectures with unbalanced and balanced synthetic datasets. Four well-known variations of ANN (Perceptron, Single hidden layer feed-forward neural network (SLNN) with 80 hidden nodes, Probabilistic neural network (PNN) and Multilayer feed-forward back-propagation neural network (MLNN) with three hidden layers with 40, 20 and 20 nodes in each hidden layer respectively) are considered.

TABLE I: The time periods and the associated sensory value ranges for LDR sensor

Time	Range	Value
12.00-4.00	[0,1.5]	
4.00-8.00	[1.5,4]	
8.00-12.00	[2.5,4]	Dark: LDR < 1.5
12.00-16.00	[4,1.5]	Half Bright: $1.5 \leq LDR > 2.5$
16.00-20.00	[2,4]	Bright: $LDR \geq 2.5$
20.00-24.00	[2.5,1.5]	

TABLE II: The time periods and the associated sensory value ranges for displacement sensor

Time	Range	Value
12.00-4.00	[0,2]	
4.00-8.00	[0,3]	
8.00-12.00	[2,3]	Sleeping on the bed≤ 2
12.00-16.00	[1,3]	Not sleeping on the bed > 2
16.00-20.00	[2,3]	
20.00-24.00	[1,2]	

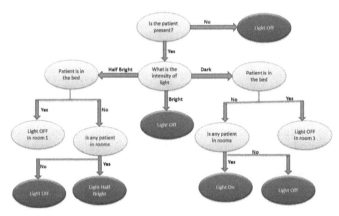

Fig. 4: Description of rules utilized for generating room specific labels for every given sample.

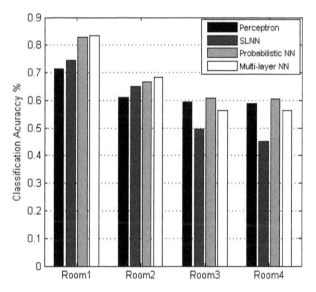

Fig. 5: Average classification accuracy achieved with the centric architecture on unbalanced datasets across four rooms in the house using variations of ANN.

A. Experiment 1: Unbalanced Data, Centric Architecture

The first experiment investigates the feasibility of the proposed architectures under unbalanced sampling condition. To do so, sets of 100 datasets utilized in this experiment are deliberately altered in a way to lack inter class balance within samples. That is, in all datasets, high percentage of samples in training and testing sets represent *'Lights Off'* class. In the training set, the percentage of samples reflecting *'Lights Off'* class in rooms one to four are set to 90%, 70%, 68% and 68% respectively. This distribution is in the range of 75%, 65%, 60% and 60% across rooms one to four respectively in the testing sets.

The overall results achieved with the centric architecture is reported in figure 5. In order to provide consistency with the distributed architecture in terms of general performance visualization, the performances are broken-down between rooms. The results indicate overall superiority of PNN across rooms closely followed by MLNN. A clear performance difference is observed between the first room (the room with a bed for the dementia suffering patient) and the other three rooms in the house.

Applying N-way ANOVA to the results revealed statistical significance across classifiers (p=$1.61728e^{-019}$) and rooms

(p=$8.0147e^{-117}$) and their interactions (p=$1.43541e^{-12}$). Between classifiers, Perceptron and SLNN are significantly different from each other and PNN and MLNN while the former two lack statistical significance from each other. Between rooms, rooms 1 and 2 are significantly different from each other and rooms 3 and 4 while the former two only lack significant difference from each other. Considering unbalance/unequal distribution of training and testing samples, as observed in previous section, it is important to further investigate the results in order to gain better understanding of performances achieved by different classifiers. This section utilizes confusion matrix to illustrate the differences on true positive (TP), true negative (TN), false positive (FP) and false negative (FN) conditions with each classifier for conditions of lights off, half bright and lights on. The results are depicted in figures 6 and 7. The results highlight inefficiency of the classifiers due to their inability to avoid False Positive (FP) and False Negative (FN) conditions across all classes and all rooms.

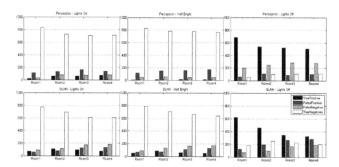

Fig. 6: Inter-class performances achieved with Perceptron and Single-Layer NN unbalanced datasets and centric architecture.

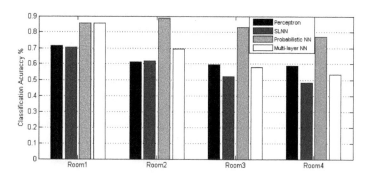

Fig. 8: Average classification accuracy achieved with the distributed architecture on unbalanced datasets across four rooms in the house using variations of ANN.

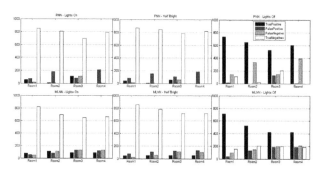

Fig. 7: Inter-class performances achieved with Probabilistic and Multi-Layer NNs unbalanced datasets and centric architecture.

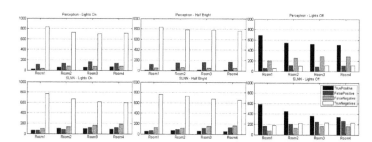

Fig. 9: Inter-class performance achieved with Perceptron and Single-Layer NN unbalanced datasets and distributed architecture.

B. Experiment 1: Unbalanced Data, Distributed Architecture

In this section the classification performance achieved with the introduced distributed architecture using the generated unbalanced synthetic data are reported. The results are illustrated on the basis of average accuracy achieved and it is categorized on the basis of the rooms in the house (see figure 8). As mentioned in previous sections, the classification approaches considered in the study includes Perceptron, Single-Layer NN, Probabilistic NN, and Multi-Layer NN. First, the differences between the performances achieved with various classifiers in each room are discussed and later the inter-class performance of the classifiers are reported.

The results indicate overall superiority of PNN across rooms followed by MLNN. Considering the best performing classification method in experiment 1, e.g. PNN, comparison between results presented in figures 5 and 8 indicate a clear advantage over using the distributed architecture. It is noteworthy that unlike the clear performance difference of PNN observed across rooms in centric architecture, such performance degradation is less obvious when the distributed architecture is utilized.

Applying N-way ANOVA to the results revealed statistical significance across classifiers ($p=3.81547e^{-132}$) and rooms ($p=5.44377e^{-8}$) and their interactions ($p=2.03979e^{-23}$). The results indicate statistical significance among all classifiers and all rooms. Figures 9 and 10 present inter-class confidence of the classifiers using a representation inspired by table of confusion. It is noteworthy that Perceptron, SLNN, and MLNN are performing similarly on both centric and distributed

architectures. However, PNN improved its performance under distributed architecture by reducing number of FPs and increasing the number of TPs across rooms under *'Lights On'* and *'Half Bright'* classes in addition to reducing FNs and increasing TNs under *'Lights Off'* class.

The low number of TP and high number of TN instances on both *'Lights On'* and *'Half Bright'* classes are justifiable with the training and testing samples inter-class distributions. As mentioned earlier, the class distributions on both training and testing sets are highly biased towards having a high number of *'Lights Off'* (over 70% and 60% in training and testing sets respectively) samples. That is, the training samples used to train the classifiers to predict the lighting condition of rooms are heavily unbalanced with lights off condition having con-

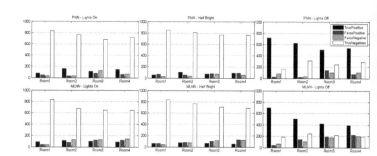

Fig. 10: Inter-class performance achieved with Probabilistic and Multi-Layer NNs unbalanced datasets and distributed architecture.

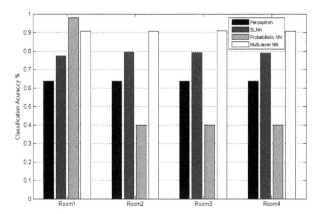

Fig. 11: Average classification accuracy achieved with the centric architecture on balanced datasets across four rooms in the house using variations of ANN.

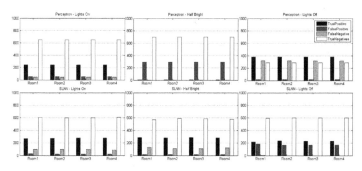

Fig. 12: Inter-class performances achieved with Perceptron and Single-Layer NN balanced datasets and centric architecture.

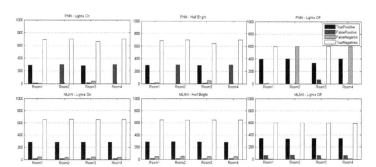

Fig. 13: Inter-class performances achieved with Probabilistic and Multi-Layer NNs balanced datasets and centric architecture.

siderably higher share of training samples compared with the other 2 conditions. This results in developing learning models that are incapable of recognizing and predicting *'Lights On'* and *'Half Bright'* conditions since their training is highly biased towards the *'Lights Off'* condition. Similar effect reported with testing sets indicate that the classification performance achieved from various classifiers studied in previous sections are likely to represent how well they can identify the *'Lights Off'* condition rather than their ability to adequately distinguish the three conditions from each other.

C. Experiment 2: Balanced Data, Centric Architecture

Considering that the results of TP, TN, FP, and FN conditions in experiment 1 indicated the variation in the performance across classifiers due to having unbalance/unequal number of training samples from different classes, the conclusion of PNN being the best performing classifier is at best restricted to the condition of lacking adequate balance between training samples. In order to better understand the feasibility of these methods for the dementia-friendly smart home scenario identified in the study, in here, the first experiment is repeated with new sets of synthetic datasets with balanced training and testing samples.

Figure 11 reports average classification performances (across 100 datasets) achieved by the classifiers under centric architecture with balanced datasets. The results indicate consistent average classification performance across classifiers in all rooms with the exception of PNN that reached to the best average classification performance in room 1 while it consistently performed poorly in all other rooms. From comparison of results presented in figures 5 and 11 it is noteworthy that with the exception of PNN, all other classifiers have considerable improvement in their average classification performances across all rooms with MLNN showing the highest performance improvement. It is also noteworthy that with the exception of room 1 in which PNN outperformed all other methods, a considerable performance degradation is observed in PNN in all other 3 rooms.

Closer look at the differences between reported inter-class performances in figures 12,13 and 6,7 reveals a noticeable

in FPs across rooms 2 and 4 in *'Lights On'*, *'Half Bright'* and *'Lights Off'* conditions with PNN. This phenomenon is followed by considerable increased TPs in rooms 1 and 3 under *'Lights On'* and *'Half Bright'* conditions. Finally, it is noticeable that although PNN reported increment TN in rooms 1 and 3 under *'Lights Off'* condition, the decrement in TNs across all rooms under that condition is another reason behind the declined classification performance. Unlike PNN, MLNN showed consistent decrease of FNs and increased TNs in most cases. Similarly, SLNN reported declines in FPs and FNs and increases in TPs and TNs (only under *'Lights Off'* condition).

Applying N-way ANOVA to the results revealed statistical significance across classifiers (p=0) and rooms (p=0) and their interactions (p=0). The results indicate statistical significance among all classifiers. Between rooms, with exception of rooms 1 that is significantly different from all other rooms, no other statistical significance is observed.

D. Experiment 2: Balanced Data, Distributed Architecture

This section replicates the previous experiment by utilizing distributed architecture. Figure14 depicts the classification performance achieved by various classification methods in each room (averaged across all synthetic datasets). Similar to previous experiment, the datasets have inter-class balance in their training and testing sets.

Comparison between results achieved with the centric (figure 11) and the distributed (figure 14) architectures indicate clear performance improvement when the distributed

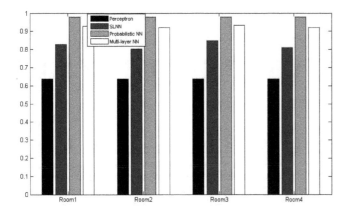

Fig. 14: Average classification accuracy achieved with the distributed architecture on balanced datasets across four rooms in the house using variations of ANN.

architecture is utilized (under inter-class balanced datasets). In addition, the consistencies across rooms and classifiers are noteworthy. Comparison between results achieved with balanced and unbalanced synthetic datasets using the distributed architectures (figures 14 and 8 respectively) indicate clear performance improvements across all classifiers in all rooms with the exception of Perceptron which demonstrated decreased performance in the first room and neglect-able performance increase in all other rooms under balanced dataset. Considering the differences between reported inter-class performances in figures 12,13 (centric architecture with inter-class balance datasets) and figures 15 and 16 indicate slight decrement of FN and slight increment of TP across all rooms in SLNN under 'Lights On' and 'Half Bright' classes. PNN illustrated considerable increase in TN and noticeable decrease in FN in rooms 2 and 4 under 'Lights Off' class in addition to considerable decrease of FP and noteworthy increase of TP in rooms 2 and 4 under 'Lights On' and 'Half Bright' classes. This phenomenon resulted in major improvement in classification performance of PNN when distributed architecture is utilized. Perceptron and MLNN did not illustrated any major difference in their inter-class performances across the two architectures.

Considering the inter-class performance differences of classifiers reported in figures 9 and 10 (distributed architecture with inter-class unbalanced datasets) and figures 15 and 16 (distributed architecture with inter-class balanced datasets) a clear decrement of FPs and FNs in addition to a clear increment of TNs and TPs are observed across all rooms with all classifiers across all three classes of 'Lights On', 'Half Bright' and 'Lights Off'. This is with the exception of Perceptron which reported increased FNs under 'Lights Off' class, increased FPs under 'Half Bright' class and increased TPs under 'Lights On' class.

Applying N-way ANOVA to the results revealed statistical significance only within classifiers (p=0) while no such significant difference is observed across rooms (p=0.1518) or the interactions of rooms and classifiers (p=0.4434). The results indicate statistical significance among all classifiers.

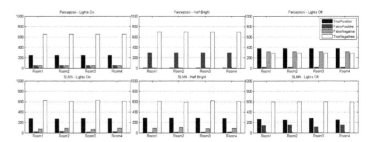

Fig. 15: Inter-class performances achieved with Perceptron and Single-Layer NN balanced datasets and distributed architecture.

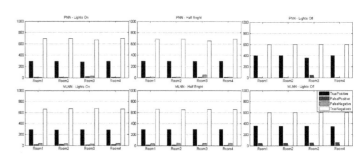

Fig. 16: Inter-class performances achieved with Probabilistic and Multi-Layer NNs balanced datasets and distributed architecture.

VI. CONCLUSION

Dementia is a growing problem in societies with aging populations due to intense level of care required and the associated costs in order to facilitate such care. Providing a safe environment in which the dementia patients (who are in the earlier stages of the disease) can live independently can improve the patients quality of life. Sleep disturbance is a common symptom in dementia patients. Patients inability to turn on the light by themselves, being frightened by room darkness, and possible accidental falls are some of the issues associated to such sleep disturbance. Clinical studies indicate that appropriate level of lighting can help to restore the rest-activity cycles of these patients. This study proposed a distributed learning mechanism on the backbone of machine learning solutions to address the required lighting control. The proposed architecture considered a collection of learning methods (classifiers) each being responsible to handle the lighting condition of a single room in the house. This architecture is scalable and provides efficient and smart control over lighting condition of multiple rooms in the house. An alternative lighting control architecture, called centric architecture, that employs a single learning mechanism, e.g. a classifier, to control the lighting condition of all rooms in the house is considered to assess feasibility of the proposed distributed architecture.

Four classification methods are considered and their feasibilities are assessed using a collection of synthetic data. Two sets of experiments reflecting balanced and unbalanced data sampling scenarios for a smart house with four rooms are designed. The samples reflected multiple scenarios with possibility of having a dementia patient and caregiver in different rooms of the house. The results indicated the feasibility of such distributed control approach for controlling the lighting conditions of multiple rooms irrespective to each other. Probabilistic neural network is identified as the most feasible classification method for such an architecture.

VII. FUTURE WORK

Several future directions can be considered for the current study including assessing the robustness of the distributed architecture proposed in this study in higher scale environments such as special dementia facilities and hospitals with multiple patients with higher variety of patients' behavioural patterns. Another possible future direction is to introduce more complex scenarios capturing more than three lighting conditions (e.g., *Lights on*, *Lights off*, and *Half Bright*) as well as extending the current architecture to cover multi-objective scenarios in which in addition to addressing the lighting needs of the patients and caregivers the overall energy consumption is reduced. The stated future directions are in addition to obvious necessity of assessing the feasibility of the architecture proposed in the study against real-world data to be captured from activities of real patients interacting with the system.

REFERENCES

[1] S. M. Mahmoud and A. Lotfi and C. Langensiepen, *Behavioural pattern identification and prediction in intelligent environments*, Applied soft computing, Elsevier, 13, 1813–1822, 2013.

[2] A. Lotfi and C. Langensiepen and S. M. Mahmoud and M. J. Akhlaghinia, *Smart homes for elderly dementia sufferers: Identification and prediction of abnormal behaviour*, Journal of Ambient Intelligence and Humanized Computing, 3(3), 205–218, 2012.

[3] P. N. Dawadi and D. J. Cook and M. Schmitter-Edgecombe and C. Parsey, *Automated assessment of cognitive health using smart home technologies*, Technol Health Care 21(4), 323–343, 2013.

[4] P. N. Dawadi and D. J. Cook and M. Schmitter-Edgecombe, *Automated cognitive health assessment using smart home monitoring of complex tasks*, IEEE transactions on systems, man, and cybernetics:systems 43(6) 1302–1313, 2013.

[5] P. N. Dawadi and C. Parsey and M. Schmitter-Edgecombe, *An approach to cognitive assessment in smart home*, Proceedings of the 2011 workshop on Data mining for medicine and healthcare, pp. 56–59, 2011.

[6] Z. Nagy and F. Y. Yong and M. Frei and A. Schlueter, *Occupant centered lighting control for comfort and energy efficient building operation*, J. Energy and buildings, Springer. doi.org/10.1016/j.enbuild.2015.02.053, 2015.

[7] K. Appold, *Lighting Affects Dementia Patients' Sleep*, Today's Geriatric Medicine 7(5) 10–10, 2014.

[8] A. K. Gopalakrishna and T. Ozcelebiy and A. Liotta and J. J. Lukkien, *Exploiting Machine Learning for Intelligent Room Lighting Applications*, 6th IEEE International Conference on Intelligent Systems (IS), 406–411, 2012.

19

Speech emotion recognition in emotional feedback for Human-Robot Interaction

Javier G. Rázuri*, David Sundgren*, Rahim Rahmani*, Aron Larsson*, Antonio Moran Cardenas[‡] and Isis Bonet[§]

*Dept. of Computer and Systems Sciences (DSV)
Stockholm University, Stockholm, Sweden

[‡]Pontifical Catholic University of Peru´ (PUCP)
Lima, Peru

[§]Antioquia School of Engineering (EIA)
Antioquia, Colombia

Abstract—**For robots to plan their actions autonomously and interact with people, recognizing human emotions is crucial. For most humans nonverbal cues such as pitch, loudness, spectrum, speech rate are efficient carriers of emotions. The features of the sound of a spoken voice probably contains crucial information on the emotional state of the speaker, within this framework, a machine might use such properties of sound to recognize emotions. This work evaluated six different kinds of classifiers to predict six basic universal emotions from non-verbal features of human speech. The classification techniques used information from six audio files extracted from the eNTERFACE05 audio-visual emotion database. The information gain from a decision tree was also used in order to choose the most significant speech features, from a set of acoustic features commonly extracted in emotion analysis. The classifiers were evaluated with the proposed features and the features selected by the decision tree. With this feature selection could be observed that each one of compared classifiers increased the global accuracy and the recall. The best performance was obtained with Support Vector Machine and bayesNet.**

Keywords—*Affective Computing; Detection of Emotional Information; Machine Learning; Speech Emotion Recognition.*

I. INTRODUCTION

Traditionally, emotions in machines have been presented as dissociated from any type of rationality having virtually no role in their internal decision systems. However, recent discoveries in neurosciences, together with the extension of notions like emotional intelligence and multilevel intelligence, has led to the emergence of the new framework "Affective Computing" [1], according to which, the main aim is to build machines that recognize, express, model, communicate and respond to users emotion indicators. In the new framework, emotions hold a key role in machines which could impact positively their future decisions, bringing closer to taking part in a more sociable loop of human-machine interaction. As main field of application, the research shall implement the connection between robots and humans that will involve an emotional feedback framework, in which robots can understand emotions from some cues from human speech. The idea is to use robots which may understand emotions, and take part in the society cooperatively, according to

the emotional state received from humans. Improving the communicative behavior of robots is urgent if people are to accept and integrate them in their world representation [2]. Robots have to be spontaneous, polite and must learn how to react according to the human being emotional charge, providing a friendly environment. Without the emotional feedback from humans, it will be very difficult for robots to interact with humans in a natural way [3], [4]. Within the context of human natural language, automatic emotional speech recognition by machines will expand the possibilities of interaction, since human speech provides a natural and intuitive interface for interaction with machines.

Emotions are visualized through various indicators in humans, many of these indicators have been previously analyzed to provide affective knowledge to machines, focusing on facial expressions [5], [6], vocal features [7], [8], [9], body movements and postures [10], [11], [12], [13] and the integration of all of them in emotion analysis systems [14], [15], [16]. But human beings cannot always hope that robots may be able to react in a timely and sensible manner, especially if they haven't be able to recover all the affective information through their sensors. Not always are the emotional features that the robot must capture provided by different sources from the human body at the same time. Maybe, all the information collected lacks robustness or, because the robot lacks the specific sensor to extract the emotional feature. Along the way to this goal, this research is based on the possible effects of some crucial speech features on the inference of emotions in communication with humans. It is known that emotions cause mental and physiological changes which are also reflected in uttered speech [17], [18], [19]. It is possible to find connections between emotional cues in speech and they can be utilized to learn about human emotions. Once such links are learned, theoretically, one can calculate the features and then automatically recognize the emotions present in human speech utterances, taking into account that the emotional content of speech does not depend on the speaker or the lexical content. Decrypting emotions in speech through several features has been a challenging research issue and one that has been of growing importance in robotics, because of the emotional factors that the robot can handle and learn in social situations. In emotional classification from speech a multitude of different features have been used and a rule to follow is not yet established.

The fields of psychology and psycholinguistics provided interesting results about how prosodic cues, fundamental frequencies and the intensity of the voice can show variability levels across different speakers [20]. Short-term spectral features and sound quality can reveal emotional indicators [21], [22]. To delimit the scope of features selection, the research focus on the most useful group of them. Prosodic features, like pitch, loudness, speaking rate, durations, pause and rhythm show have strong correlations between them, providing valuate emotional information. In the case of the analysis of entire segment of voice, statistical functions like mean, median, minimum, maximum, standard deviation, or more seldom third or fourth standardized moment are applied to the fundamental frequency (F0) base contour [23], [24], [25]. The speech signal contains other frequency related characteristics that are spectral features. Mel Frequency Cepstral Coefficients (MFCCs) are generally used in speech recognition with great accuracy in emotion detection [26]. Predictive Cepstral Coefficients (LPCC) or Mel Filter Bank (MFB) features have a more common use [27]. The same performance displayed by MFCCs, is showed by RASTA-PLP (Relative Spectral Transform - Perceptual Linear Prediction) [28]. Through the analysis of voice quality [29], [30] and linguistic features, it can clearly be seen that there is a strong correlation between voice, pronounced words and emotions [31], [32], [33]. Different levels of voice could be depicted by neutral, whispery, breathy, creaky, and harsh or falsetto voice. In the case of features extracted from chains of words, the relation is depicted by the affective states associated with specific words; many of them are related to the probability of one emotion giving a certain sequence of words.

The machine learning framework shows several classifiers used in several tasks related to emotion recognition. Each classifier has advantages and disadvantages in order to deal with the speech emotion recognition problem. The more common group used are composed of Hidden Markov Model (HMM) [34], [35] regarded as the simplest dynamic Bayesian networks, Gaussian Mixture Models (GMM) [36], Nearest-Neighbour classifiers [37], artificial neural networks (ANN) [38], support vector machine (SVM) [39], k-NN [40], Decision Trees [41] and many others. The vast majority of emotion recognition systems over speech have employed a highdimensional speech grouped in a big vector of features, so the main goal will be to handle the dimensionality in order to improve the emotion recognition performance.

In this paper, the most commonly used features in several researches for capturing emotional speech characteristics in time and frequency were selected. The performance of different well known classifiers was compared in order to select the best result to predict the emotion, based on speech emotional data. To effectively reduce the size of speech features and improve the results obtained by the classifiers, the output from a decision tree classifier like feature selection

method was used.

This paper is organized as follows. Section II describes the data set used in the research, the features extracted to represent the emotions from human speech, the machine learning techniques to perform the emotion classification experiments and the measures to evaluate the performance of classifiers. Section III describes the experimental results of all the several classification tests. Some conclusions are presented in Section IV.

II. METHODS

1) Dataset: The emotional speech characteristics were extracted from the eNTERFACE05 audio-visual emotion database [42]. The data base is based on six universal emotions [43] like anger, disgust, fear, joy, sadness, and surprise. The voice data are provided by 44 non-native English speakers from 14 nations. The individuals expressed six basic emotions through five different sentences portrayed in 1320 videos, with a duration ranging from 1.2 to 6.7 seconds. For this research only one sentence per each emotion was used, which leads to a total of 264 videos. Each video is subsequently converted to a Waveform Audio File Format, for this task the MultimediaFileReader object from the DSP System Toolbox Library of MATLAB [44] was used to read the group of audio frames from each multimedia file. Fig. 1 shows the process applied to each video to build the emotional data set.

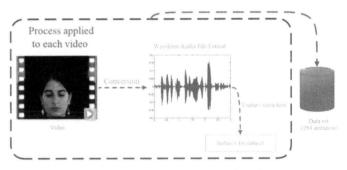

Fig. 1: Flowchart of the construction of dataset

2) Features: The data were acquired directly from the group of Waveform Audio files and they were transformed in 264 vectors of features. A wide range of possibilities exist for parametrically representing a speech signal and its content in a vector, with the intention of the extraction of relevant information from it. A variety of choices for this task can be applied to represent the speakers speech in a large number of parameters, in which the changes in these parameters will result in corresponding change in emotions. Taking into account that the system could be useful for a companion robot, the efforts should be focused on a system that contributing to a gradual gain of controllability and robustness that might save a substantial cost in computational efficiency. Not all the features that the robot can capture could be helpful and essential for its emotional feedback loop. Using all the features is not a guarantee to arrive the best performance, it could be better that the robot localizes the own best features and discard useless features from the data base.

The kind of extracted features used in the research have been commonly used in music information retrieval (MIR), much of the research is based on the extraction mechanism from musical pieces, retrieval methodologies covered in various tasks related to different music representation media. It is attainable that the variability of emotions can be explained by a small set of acoustic features, for this task in order to identify objective acoustic features MATLAB was used, most of them developed in [45]. The spectral change of a signal is measured by the Spectral Flux (SF) feature [46]; the value is calculated through differences between each magnitude spectrum bin in the current frame to the corresponding value related to the magnitude spectrum of the previous frame. The result is the sum of the squares of the differences. Spectral Centroid (SC) [47] measures the center of mass of the power spectrum, it weighs the mean of the frequencies present in the speech signal. The SC uses the highest concentration point of energy in the spectrum and is correlated with the dominant frequency over the signal. Spectral Roll off Point [48] is often used as an indicator of the slant of the frequencies depicted in a frame. It is represented by a measure of the right-skewedness of the power spectrum. It increases with the bandwidth of a signal. Root Mean Square (RMS) [49] measures the power of a signal over a frame; the squares of each sample are summed and divided by the number of samples contained in frames. The value is square root of the total sum. Spectral Centroid Variability (SCV) [50] is the standard deviation of the magnitude spectrum, it measures the variability of the speech signal. Zero Crossing rate (ZCR) [50] provides an approximate estimation of dominant frequency and the spectral centroid and is described as the number of zero crossings during one second in the temporal domain. Compactness [51] is an indicator of the levels of noise in a signal; it is calculated by comparisons of components in a magnitude spectrum of a frame and the magnitude spectrum of its neighboring frames. Mel-Frequency Cepstral Coefficients (MFCCs) [52] is used to describe a spectrum frame, its first and second derivative in time are used to reflect dynamic changes. The first 8-13 MFCC coefficients are commonly used to describe the shape of the spectrum. They represent the information of the spectral envelope of the signal. Method of Moments [53] is composed for the first five statistical moments (area, mean, power spectrum density, spectral skew and spectral kurtosis) describing the shape of the spectrograph of a given frame. Linear predictive coding (LPC) is used to estimate the basic parameters into a speech signal, such as the vocal tract transfer function and the formant frequencies. It has good intuitive interpretation both in time domain and in frequency domain. The cepstral representation (Linear Prediction Cepstral Coefficients (LPCC)) of its coefficients is more used because of its higher retrieval efficiency [54]. 2D Method of Moments (2DMM) [55] gives a spectrograph description and the variation of it during a short time frame. The feature is composed by spectral data in frames analyzed with two-dimensional method of moments. Strongest Frequency Via Zero Crossings [48] is an estimation of the highest frequency of the component of a signal, found through the number of zero-crossings. 2D Method of Moments of MFCCs [47] is the 2D statistical computation of the Mel Frequency Cepstral Coefficients (MFCCs), this feature com-

posed for a group of coefficients, allows recognizing the part of mid-frequencies from the signal. Fraction of Low Energy frame [56] is an indicator of the variability of the amplitude of frames; it is a fraction of previous frames, in which the Root Mean Square of each frame is less than the mean Root Mean Square. Strongest Frequency via FFT Maximum [48] is strongest frequency component in Hz of a signal. This is found by finding the highest bin (observations that fall into each of the disjoint categories) in the power spectrum. Strongest Frequency Via Spectral Centroid [48] is the strongest frequency in Hz in a signal related to the spectral centroid. The group of features conformed by Mel-Frequency Cepstral Coefficients, Linear Prediction Cepstral Coefficients, Method of Moments, 2D Method of Moments and 2D Method of Moments of MFCCs are matrices of, 4x13, 4x9, 4x5, 4x10, 4x10 respectively, that they will be transformed to vectors. Spectral Centroid, Spectral Roll off Point, Spectral Flux, Compactness, Spectral centroid Variability, Root Mean Square, Fraction of Low Energy frame, Zero Crossing rate, Strongest Frequency Via Zero Crossings, Strongest Frequency Via Spectral Centroid and Strongest Frequency Via FFT Maximum are conformed by 11 vectors of 8 features each one. Thus, the total feature vector contains 276 attributes that will be evaluated by the classifiers.

3) Machine learning techniques: The binary classification algorithm Support Vector Machine (SVM) which originated in statistical learning theory, offers robust classification to a very large number of variables and small samples [57]. SVM is capable of learning complex data from classification models applying mathematical principles to avoid overfitting. The more used kernels in SVM are polynomial and linear.

Another relatively fast classification model is the Decision tree, it works with a group of simple classification rules that are easy to understand. The rules represent the information in a tree based in a set of features. The classic decision tree is named ID3 based on growing and pruning [58], although C45 is other topdown decision trees inducers for continuous values [59], the last one is named as J48 in WEKA [66] and it uses the information gain as measure to select and split the nodes.

Within the connectionist techniques is also found the Artificial Neural Network (ANN). The ANN has a structure comparable to human neural networks where neurons located in layers process information. They have a graphical representation of an interconnected group of artificial neurons, in which the information resides in the weights from the arcs that connect the neurons. The ANN has two algorithms: feed-forward and recurrent neural network, in FF networks are supported over a directed acyclic graph, while RR networks have cycles. The most used feed-forward training algorithm is the Multilayer Perceptron named backpropagation [60]. The learning process covers two steps, the first step is a forward processing of input data by the neurons that produces a forecasted output, the second step is the adjustment of weights within the neuron layers, in order to minimize the errors of the forecasted solution compared with the correct output.

A graphical model (GMs) for probabilistic relationships

among a set of variables is bayesNet (Bayesian Network), it is used to represent knowledge the uncertainty [61]. The graph depicted in bayesNet is composed by nodes that represent random variables. In the graph, the edges between the nodes represent probabilistic dependencies among the corresponding random variables. Per each node there is a probability table specifying the conditional distribution of the variable given the values of its predecessors in the graph. These conditional dependencies in the graph are generally calculated by using known statistical and computational methods.

k Nearest Neighbors (kNN) is one of the simplest of classification algorithms available for supervised learning. The algorithm classifies unlabeled examples based on their similarity with examples in the training set. It is a lazy learning method that searches the closest match of the test data in feature space, based on distance function [62]. In this work is applied the Euclidean metric.

The supervised learning method naive Bayes [63] is a statistical method for classification, it is based on the well-known Bayes theorem with strong assumptions. The naive Bayes allows capturing the uncertainty about the model in a principled way by determining probabilities of the outputs. One of the advantages is the robustness to noise in input data. The classifier assumes that the presence (or absence) of a particular feature of a class is unrelated to the presence (or absence) of any other feature, given the class variable.

4) Validation techniques: Machine learning techniques have several measures in order to evaluate the performance of classifiers, which are principally focused on handling two-class problems. The performance of classifiers can be evaluated through several measures of machine learning techniques, which are principally focused in handling two-class problems. This research has faced a classification problem of six classes formed by six universal emotions. Most of the measures to evaluate binary problems could also apply to multi-class problem. In a problem with m classes, the performance of classifiers can be assessed based on an $m \times m$ confusion matrix, as shown in Table I. The groups of rows that describe the matrix represent the actual classes, while the columns are the predicted classes.

TABLE I: Confusion Matrix

	Predicted Class$_1$ CM_{11}	$\cdots$	Predicted Class$_m$ CM_{1m}
True Class$_1$			
$\vdots$	$\vdots$	$\ddots$	$\vdots$
True Class$_m$	CM_{m1}	$\cdots$	CM_{mm}

For example, the accuracy is the percentage of correctly classified cases of the dataset. Based on the confusion matrix, the accuracy can be computed as a sum of the main values from the diagonal of the matrix, which represents the correctly classified cases divided by the total number of instances in the dataset (Eq. 1).

$$Accuracy = \frac{\sum_{i=1}^{m} CM_{ii}}{\sum_{i=1}^{m} \sum_{j=1}^{m} CM_{ij}} \quad (1)$$

where CM_{ij} represents the elements in the row i and column j of the confusion matrix.

Some measures like accuracy do not represent the reality of the number of cases correctly classified per each class. In order to make a deeper analysis, the measure of recall has been calculated for each class. Recall provides the percentage of correctness of classification into each class. Eq. 2 represents the recall for class [64].

$$Recall_i = \frac{CM_{ii}}{\sum_{j=1}^{m} CM_{ij}} \quad (2)$$

A k-fold cross-validation with $k = 10$ was used to make validations over the classifiers. This technique allowed the evaluation of the model facing an unknown dataset. The group of data is randomly divided in k equal parts, one part of the group is used as a validation set and the rest $k - 1$ will be the training set. The process is repeated k times using a different group as a validation set, this process continues until each group can used once as validation test. Then, the k results obtained by groups can be averaged to a single result. The advantage of 10-fold cross-validation is that all examples of the database are used for both, training and testing stages [65].

III. RESULTS

The intent of this study was to provide the best classification of emotions contained in a speech signal, which might serve to feed the decision support system of a synthetic agent capable of supporting the societal participation of persons deprived of conventional modes of communication, in the context of socially intelligent systems. A 10-fold crossvalidation scheme was employed in the speech dataset for all the emotion classification experiments; this was done to validate the performances of the classifiers selected. Six classifiers were tested, the Support Vector Machine (SVM) has used three kernels, linear and polynomial (with degrees 2 and 3), k Nearest Neighbors (kNN) has used k from 1 to 15 (showed the best result with $k=5$), Multilayer Perceptron (MLP) with hidden neurons from 2 to 20 (showed the best result with 10 neurons), bayesNet (BN), NaiveBayes (NB) and decision tree (J48). All the several classification tests were conducted using the WEKA [66] toolbox. The best performance was achieved with the decision tree (J48) reaching a 96.21 % of accuracy facing the other classifiers, as shown in Fig. 2. The accuracy and the recall results were compared. As you can see in Fig. 3, the percentage of most relevant results per emotion positively classified (recall) was raised by the decision tree (J48).

The decision tree has reached the best result in accuracy and recall facing the remaining classifiers; therefore, this result is obtained with only a few features selected, taking into account their information gain. As can be seen the decision tree is composed of six nodes, which correspond to six features of the dataset, which means the tree only needs these six features to predict the emotions. The features selected of the tree are

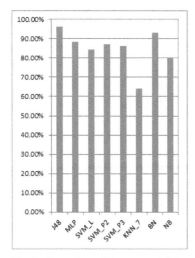

Fig. 2: Comparison of Accuracy of different classifiers

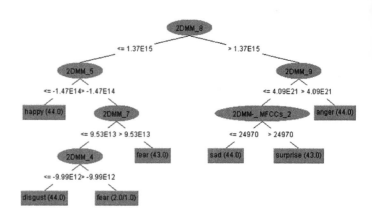

Fig. 4: Decision tree

to misclassification with poorer decoding accuracy [67].

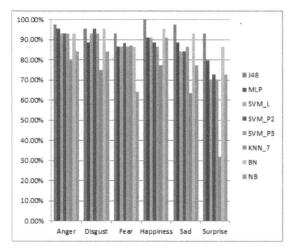

Fig. 3: Comparison of Recall of different classifiers

TABLE II: Confusion Matrix J48

	Anger	Disgust	Fear	Happiness	Sad	Surprise
Anger	43	0	0	0	0	1
Disgust	0	42	2	0	0	0
Fear	0	2	41	1	0	0
Happiness	0	0	0	44	0	0
Sad	0	0	0	0	43	1
Surprise	1	1	0	0	1	41

based on the 2D Method of Moments ($2DMM$) and the 2D Method of Moments of MFCCs ($2DMM - MFCCs$). Fig. 4 illustrates the graphical rendition obtained of the classification tree in which $2DMM_n$ is the n-th element within $n = 1, 2, 3...10$ from the feature vector of 2D Method of Moments. Similarly, $2DMM - MFCCs_m$ is the m-th element within $m = 1, 2, 3...10$ from the feature vector of 2D Method of Moments of MFCCs.

The confusion matrix of J48, as depicted in Table II, shows a balanced distribution of misclassifications rates in the group of emotions. For all six emotions, "happiness" is not confused at all with the rest of emotions and it is recognized with 100 %. Further analysis of the confusion matrix shows that the emotions "fear" and "surprise" attained a higher number of misclassifications and lower percentages of recall (both of them 93.20 %), as shown in Table II and Table III respectively. In case of emotions "disgust" and "fear", this speech signals could be interpreted from the psychophysiological framework, some acoustic cues in discrete emotions could lead listeners

Comparing the results of the decision tree with the remaining classifiers, it seems likely that the learning mechanism in the tree is essential in this problem. An important process in the algorithm, is how to determine which attribute to split on. The attributes are selected based on information gain, resulting in a set of selected relevant features. This can only lead to conclude that dataset probably has noisy and redundant features. Then, the information gain is visualized as a heuristic to select features as is done for the decision tree. Taking into account the features selected by the tree, the data set were reconstructed with the selection of the 2D Method of Moments and the 2D Method of Moments of MFCCs. The same classifiers with the same parameters were trained and compared. Also it is shown the best result for each classifier, where the best result for MLP was with 12 neurons and kNN for $k=6$. Comparisons between the accuracy previously obtained and the results are showed in Fig. 5. As can be seen, the features selected by the decision tree have highlighted improvements in performance of all classifiers in a range of 3.5 % to 31.8 %. The accuracies of MLP, BN and SVM (with polynomial kernel of degree 2) were superior to the decision tree. The MLP and BN have achieved 96.97 % and the SVM 96.59 %.

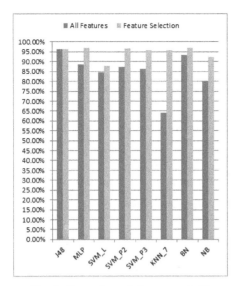

Fig. 5: Comparison of Accuracy of different classifiers with all features and with feature selection

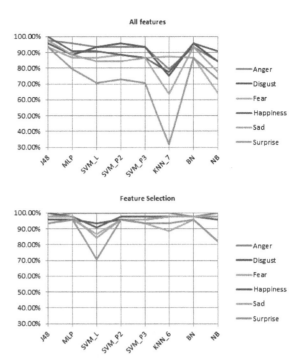

Fig. 6: Comparison of Recall of different classifiers with all features and with feature selection for each emotion

In order to analyze the results in each emotion, the recall can be analyzed in Fig. 6. It is clear that all emotions show improvements over the results obtained before. The best results also are obtained by MLP, BN and polynomial SVM with degree 2. In order to see more detail, Table III shows the recall of the four classifiers (J48, MLP, SVM with degree 2 (SVM-P2) and BN). The three last rows illustrate the average of recall for each classifier (Average) and the range of the recall (Min and Max). The differences between the classifiers are not significant in accuracy. However, the three classifiers are superior to J48 based on the range of the recall and the average. BN and MLP show similitudes in average, while MLP has a better range. The range of SVM is equal to MLP, while the average is lower than MLP. This comparison means that MLP has achieved the best results.

TABLE III: Comparison of recall for J48, MLP, SVM with degree 2 (SVM-P2) and BN

	J48	MLP	SVM-P2	BN
Anger	97.70 %	100.00 %	100.00 %	97.70 %
Disgust	95.50 %	95.50 %	95.50 %	97.70 %
Fear	93.20 %	95.50 %	95.50 %	95.50 %
Happiness	100.00 %	97.70 %	97.70 %	97.70 %
Sad	97.70 %	97.70 %	95.50 %	97.70 %
Surprise	93.20 %	95.50 %	95.50 %	95.50 %
Average	96.22 %	96.98 %	96.62 %	96.97 %
Min	93.20 %	95.50 %	95.50 %	95.50 %
Max	100.00 %	100.00 %	100.00 %	97.70 %

Keeping in mind that the building of a system for real time probably is applicable to a robot, the time consumption of the algorithm is relevant. Taking into account that the classifiers (SVM, BN and MLP) have a little difference in the results, it can be suggested the use of BN or SVM instead of MLP to consume less computational resources.

The emotions "anger" and "happiness" have achieved the best performance from the beginning, while surprise is the lowest results and shows the same behavior for all classifiers.

IV. CONCLUSIONS

The purpose of this research was to perform parameterization of audio data for the purpose of automatic recognition of emotions in speech. A collection of audio data from several videos related to human emotional expressions were gathered and turned into a data set. A group of six classifiers in order to identify the best of them to predict emotions in humans were selected. The outputs from a decision tree have been used as a feature selection technique to remove redundant and noisy features. The features provided by the decision tree were 2D Method of Moments and 2D Method of Moments of MFCCs. The feature selection increases the efficiency of the accuracy and the recall. The feature selection also allows reduction of the dimensionality of the data in turn leading to less computation processes in the robot memory.

After the selection of features, a group of experiments in order to select the best classifiers were conducted. Multilayer Perceptron, Support Vector Machine and bayesNet have achieved the best results. Support Vector Machine and bayesNet could be good candidates to build the emotional recognition system of a robot, because of their easily implementation and the less computational complexity.

This simple system with the classifiers is easy to understand and implement because of the utilization from a small group of features would work remarkably well on real-world data, making it possible to develop a real-time system in which the robot can make a fast decision in accordance with the emotional feedback provided from humans. As a real application, it could be considered a real-time system that can serve like a motor of emotional knowledge in order to understand the autistic children, to describe accurately their internal state and show the real content of their emotions. The system is not only applied to companion robots it could also be applicable to diverse smart sources (smart devices), this could be the case of healthcare, telemedicine or smart well-being systems that can be seen more often. This type of emotional devices working with emotional feedback will have the potential to reveal more about emotional state and the early detection of crisis, balanced lifestyle including and regulated stress level.

ACKNOWLEDGMENT

The authors greatly appreciate the financial support provided by the institution VINNOVA Swedish Governmental Agency for Innovation Systems through the ICT project The Next Generation (TNG). We also grateful to Antioquia School of Engineering "EIA" (Colombia) and Pontifical Catholic University (Perú) in a joint effort for collaborative research.

REFERENCES

[1]   R. Picard, Affective Computing. The MIT Press, United States, 1998.

[2]   T. Ziemke and R. Lowe, "On the role of emotion in embodied cognitive architectures: From organisms to robots," Cognitive computation, vol. 1, no. 1, pp. 104–117, 2009.

[3]   H.A. Samani and E. Saadatian, "A Multidisciplinary Artificial Intelligence Model of an Affective Robot," international Journal of Advanced Robotic Systems, vol. 9, pp. 1–11, 2012.

[4]   J.G. Rázuri, P.G. Esteban and D.R. Insua, "An adversarial risk analysis model for an autonomous imperfect decision agent," In T.V. Guy, M. Kárný and D.H. Wolpert, Eds. Decision Making and Imperfection. SCI, vol. 474, pp. 165–190. Springer, Heidelberg, 2013.

[5]   M. Pantic and L.J.M. Rothkrantz, "Automatic analysis of facial expressions: The state of the art," IEEE Trans. on Pattern Analysis and Machine Intelligence, vol. 22, no. 12, pp. 1424–1445, 2000.

[6]   D. Filko and G. Martinovic, "Emotion recognition system by a neural network based facial expression analysis, AutomatikaJournal for Control," Measurement, Electronics, Computing and Communications, vol. 54, no. 2, 2013.

[7]   R. Cowie and E. Douglas-Cowie, "Automatic statistical analysis of the signal and prosodic signs of emotion in speech," In Proc. International Conf. on Spoken Language Processing, pp. 1989–1992, 1996.

[8]   T. Sobol-Shikler, P. Robinson, "Classification of complex information: Inference of co-occurring affective states from their expressions in speech," IEEE Trans. Pattern Anal. Mach. Intell, vol. 32, no. 7, pp. 1284–1297, 2010.

[9]   K. Han, D. Yu and I. Tashev, "Speech Emotion Recognition Using Deep Neural Network and Extreme Learning Machine," Interspeech 2014, pp. 223–227, 2014.

[10] N. Bianchi-Berthouze and A. Kleinsmith, "A categorical approach to affective gesture recognition," Connection Science, vol. 15, no. 4, pp. 259–269, 2003.

[11] G. Castellano, S.D. Villalba and A. Camurri, "Recognising Human Emotions from Body Movement and Gesture Dynamics," In Proc. of 2nd International Conference on Affective Computing and Intelligent Interaction, Berlin, Heidelberg, 2007.

[12] K. Schindler, L. van Gool, and B. de Gelder, "Recognizing emotions expressed by body pose: a biologically inspired neural model," Neural Networks, vol. 21, no. 9, pp. 1238–1246, 2008.

[13] A. Kleinsmith and N. Bianchi-Berthouze, "Recognizing affective dimensions from body posture," In Affective Computing and Intelligent, Lecture Notes in Computer Science, pp. 48–58, Springer, Berlin, Germany, 2007.

[14] H. K. M. Meeren, C. van Heijnsbergen and B. de Gelder, "Rapid perceptual integration of facial expression and emotional body language," Proc. National Academy of Sciences of the USA, vol. 102, no. 45, pp. 16518–16523, 2005.

[15] A. Metallinou, A. Katsamanis and S. Narayanan, "Tracking changes in continuous emotion states using body language and prosodic cues," In IEEE International Conference on Acoustics, Speech and Signal Processing (ICASSP), pp. 2288–2291, 2011.

[16] C. Busso, Z. Deng, S. Yildirim, M. Bulut, C.M. Lee, A. Kazemzaeh. S. Lee, U. Neumann and S. Narayanan, "Analysis of Emotion Recognition using Facial Expressions, Speech and Multimodal information," In Proc. of ACM 6th int'l Conf. on Multimodal Interfaces (ICMI2004), State College, PA, pp. 205–211, 2004.

[17] J.Q. Wang, N. Trent, E. Skoe, M. Sams and N. Kraus, "Emotion and the auditory brainstem response to speech," Neuroscience Letters, vol. 469, no. 3, pp. 319–323, 2010.

[18] D.A. Abrams, N. Trent, S. Zecker and N. Kraus, "Rapid acoustic processing in the auditory brainstem is not related to cortical asymmetry for the syllable rate of speech," Clinical Neurophysiology, vol. 121, no. 8, pp. 1343–1350, 2010.

[19] M. Drolet, R.I Schubotz and J. Fischer, "Authenticity affects the recognition of emotions in speech: behavioral and fMRI evidence," Cognitive, Affective, and Behavioral Neuroscience, vol. 12, no. 1, pp. 140–150, 2012.

[20] J. Ang, R. Dhillon, A. Krupski, E. Shriberg and A. Stolcke, "Prosody-based automatic detection of annoyance and frustration in human-computer dialog," Proc. International Conference on Spoken Language Processing (ICSLP 2002), pp. 2037–2040, 2002.

[21] V. Hozjan and Z. Kacic, "Context-independent multilingual emotion recognition from speech signals," International journal of Speech Technology, vol. 6, pp. 311–320, 2003

[22] B. Schuller, D. Seppi, A. Batliner, A. Maier, and S. Steidl, "Towards more reality in the recognition of emotional speech," In International Conference on Acoustics, Speech, and Signal Processing, vol. 4, pp. 941–944, Honolulu, HI, USA, 2007.

[23] R. Cowie,E. Douglas-Cowie, N. Tsapatsoulis, S. Kollias, W. Fellenz and J. Taylor, "Emotion recognition in humancomputer interaction," IEEE Signal Process, vol. 18, pp.32–80, 2001.

[24] I. Murray, J. Arnott, "Toward a simulation of emotions in synthetic speech: A review of the literature on human vocal emotion," J. Acoust. Soc. Am, vol. 93, no. 2, pp. 1097–1108, 1993.

[25] D. Ververidis and C. Kotropoulos, "Emotional speech recognition: Resources, features, and methods," Speech Communication, pp. 1162–1181, 2006.

[26] D. Neiberg, K. Elenius, I. Karlsson and K. Laskowski, "Emotion Recognition in Spontaneous Speech," pp. 101–104, 2006.

[27] C. Busso, S. Lee, and S.S. Narayanan, "Using neutral speech models for emotional speech analysis," In Interspeech 2007-Eurospeech, pp. 2225–2228, 2007.

[28] K.P. Truong and D.A. van Leeuwen, "Automatic discrimination between laughter and speech," Speech Commun, vol. 49, no. 2. pp. 144–158, 2007.

[29] P. Alku, "Glottal inverse filtering analysis of human voice production A review of estimation and parameterization methods of the glottal excitation and their applications," Sadhana, vol. 36, no. 5, pp. 623–650, 2011.

[30] M. Lugger and B. Yang, "The relevance of voice quality features in speaker independent emotion recognition," IEEE International Conference on Acoustics, Speech and Signal Processing (ICASSP 2007), Honolulu, HI, USA, vol. 4, pp. 17–20, 2007.

[31] C.M. Lee and S.S. Narayanan, S. S, "Toward Detecting Emotions in Spoken Dialogs," IEEE Transactions on Speech and Audio Processing, vol. 13, no. 2, pp. 293–303, 2005.

[32] S. Steidl, Automatic Classification of Emotion-Related User States in Spontaneous Children's Speech, Logos-Verlag, 2009.

[33] B. Schuller, A. Batliner, S. Steidl and D. Seppi, "Emotion Recognition

from Speech: Putting ASR in the Loop," Proc. ICASSP 2009, IEEE, Taipei, Taiwan, pp. 4585–4588, 2009.

[34] D. Le and E. M. Provost, "Emotion recognition from spontaneous speech using Hidden Markov models with deep belief networks," in Automatic Speech Recognition and Understanding (ASRU), 2013 IEEE Workshop on, pp. 216–221, 2013.

[35] J. Wagner, T. Vogt, and E. André, "A systematic comparison of different HMM designs for emotion recognition from acted and spontaneous speech," in Proceedings of the 2nd International Conference on Affective Computing and Intelligent Interaction (ACII), Lisbon, Portugal, pp. 114–125, 2007.

[36] T. Hao, S.M. Chu, M. Hasegawa-Johnson and T.S. Huang, "Emotion recognition from speech VIA boosted Gaussian mixture models," in Multimedia and Expo, 2009. ICME 2009. IEEE International Conference, pp. 294–297, 2009.

[37] S.A. Rieger, R. Muraleedharan and R.P. Ramachandran, "Speech based emotion recognition using spectral feature extraction and an ensemble of kNN classifiers," in Chinese Spoken Language Processing (ISCSLP), pp. 589–593, 2014.

[38] S.A. Firoz, S.A. Raj and A.P. Babu, "Automatic Emotion Recognition from Speech Using Artificial Neural Networks with Gender-Dependent Databases," in Advances in Computing, Control and Telecommunication Technologies, ACT '09, pp. 162–164, 2009.

[39] C. Yu, Q. Tian, F. Cheng and S. Zhang, "Speech Emotion Recognition Using Support Vector Machines," in Advanced Research on Computer Science and Information Engineering. vol. 152, G. Shen and X. Huang, Eds., ed: Springer Berlin Heidelberg, pp. 215–220, 2011.

[40] M. Feraru and M. Zbancioc, "Speech emotion recognition for SROL database using weighted KNN algorithm," in Electronics, Computers and Artificial Intelligence (ECAI) , pp. 1–4, 2013.

[41] C.-C. Lee, E. Mower, C. Busso, S. Lee and S. Narayanan, "Emotion recognition using a hierarchical binary decision tree approach," Speech Commun, vol. 53, pp. 1162–1171, 2011.

[42] O. Martin, I. Kotsia, B. Macq and I. Pitas, "The eNTERFACE' 05 Audio-Visual Emotion Database," in Data Engineering Workshops, Proceedings. 22nd International Conference, pp. 8–8, 2006.

[43] P. Ekman and W.V. Friesen, "A new pan-cultural facial expression of emotion," Motivation and Emotion, vol. 10, no. 2, pp. 159–168, 1986.

[44] MathWorks, (2014). DSP System Toolbox: User's Guide (R2014b). http://fr.mathworks.com/help/pdf_doc/dsp/dsp_ug.pdf, 2014.

[45] T. Giannakopoulos and A. Pikrakis. Introduction to Audio Analysis. Elsevier Academic Press, 2014.

[46] P. Masri, "Computer modelling of sound for transformation and synthesis of musical signal," Ph.D. dissertation, University of Bristol, UK, 1996.

[47] G. Peeters, "Large Set of Audio Features for Sound Description," Technical report published by IRCAM, 2004.

[48] C. McKay and I. Fujinaga, "Automatic music classification and similarity analysis," International Conference on Music Information Retrieval, 2005.

[49] K.V. Cartwright, "Determining the Effective or RMS Voltage of Various Waveforms without Calculus," Technology Interface, vol. 8, no. 1, pps. 20, 2007.

[50] J. Kim, E. Andre, M. Rehm, T. Vogt and J. Wagner, "Integrating information from speech and physiological signals to achieve emotional sensitivity," In Proc. Interspeech, Lisbon, Portugal, pp. 809812, 2005.

[51] F. Pachetand P. Roy, P, "Analytical features: A knowledge-based approach to audio feature generation," EURASIP Journal on Audio, Speech, and Music Processing, 2009.

[52] B. Bogert, M. Healy, and J. Tukey, "The quefrency alanysis of time series for echoes: cepstrum, pseudo-autocovariance, cross- cepstrum, and saphe-cracking," Proceedings of the Symposium on Time Series Analysis, Wiley, 1963.

[53] I. Fujinaga, "Adaptive Optical Music Recognition," Ph.D. thesis, Department of Theory, Faculty of Music, McGill University, Montreal, Canada, 1997.

[54] X. Changsheng, M. C. Maddage and S. Xi, "Automatic music classification and summarization," Speech and Audio Processing, IEEE Transactions on, vol. 13, pp. 441-450, 2005.

[55] R. Mittra and V. Varadarajan, "A technique for solving 2D methodof-moments problems involving large scatterers," Microwave and Optical Technology Letters, vol. 8 no. 3, 2007.

[56] C. McKay and I. Fujinaga, jMIR: Tools for automatic music classification. Ann Arbor, MI: MPublishing, University of Michigan Library, 2009

[57] V. Vapnik, The Nature of Statistical Learning Theory ed.; Springer-Verlag, New York, 1995.

[58] J. R. Quinlan, "Induction of decision trees," Mach. Learn, vol. 1, no. 1, pp. 81-106, 1986.

[59] J. R. Quinlan, C4.5: Programs for Machine Learning, 1st ed.; Morgan Kaufmann Publishers: San Francisco, CA, USA, 1993.

[60] D. E. Rumelhart, G. E. Hinton and R. J. Williams, "Parallel distributed processing: explorations in the microstructure of cognition," D. E. Rumelhart and J.L. McClelland, eds, MIT Press: Cambridge, MA, USA, vol. 1, pp. 318-362, 1986

[61] J. Pearl, Probabilistic Reasoning in Intelligent Systems: Networks of Plausible Inference, Morgan Kaufmann, San Mateo, CA, 1988.

[62] T. M. Mitchell, Machine Learning; McGraw-Hill: New York, NY, p. 432, 1997.

[63] H. Zhang, "The Optimality of Naive Bayes," Proc. the 17th International FLAIRS conference, Florida, USA, pp. 17-19, 2004

[64] P. Flach, "Machine Learning: The Art and Science of Algorithms that Make Sense of Data," Cambridge University Press, 2012.

[65] B. Efron and R. J. Tibshirani, "An introduction to the Bootstrap," Chapman and Hall: New York, USA, 1993.

[66] M. Hall, E. Frank, G. Holmes, B. Pfahringer, P. Reutemann, and I. H. Witten, "The WEKA data mining software: an update," ACM SIGKDD Explorations Newsletter, vol. 11, no. 1, pp. 10-18, 2009.

[67] P. Laukkaand P. Juslin, "Similar patterns of age-related differences in emotion recognition from speech and music, " Motivation and Emotion, vol. 31, no. 3, pp. 182-191. 2007.

Application of Machine Learning Approaches in Intrusion Detection System

Nutan Farah Haq
Department of Computer Science and Engineering
Ahsanullah University of Science and Technology
Dhaka, Bangladesh

Abdur Rahman Onik
Department of Computer Science and Engineering
Ahsanullah University of Science and Technology
Dhaka, Bangladesh

Md. Avishek Khan Hridoy
Department of Computer Science and Engineering
Ahsanullah University of Science and Technology
Dhaka, Bangladesh

Musharrat Rafni
Department of Computer Science and Engineering
Ahsanullah University of Science and Technology
Dhaka, Bangladesh

Faisal Muhammad Shah
Department of Computer Science and Engineering
Ahsanullah University of Science and Technology
Dhaka, Bangladesh

Dewan Md. Farid
Department of Computer Science and Engineering
United International University
Dhaka, Bangladesh

Abstract—Network security is one of the major concerns of the modern era. With the rapid development and massive usage of internet over the past decade, the vulnerabilities of network security have become an important issue. Intrusion detection system is used to identify unauthorized access and unusual attacks over the secured networks. Over the past years, many studies have been conducted on the intrusion detection system. However, in order to understand the current status of implementation of machine learning techniques for solving the intrusion detection problems this survey paper enlisted the 49 related studies in the time frame between 2009 and 2014 focusing on the architecture of the single, hybrid and ensemble classifier design. This survey paper also includes a statistical comparison of classifier algorithms, datasets being used and some other experimental setups as well as consideration of feature selection step.

Keywords—Intrusion detection; Survey; Classifiers; Hybrid; Ensemble; Dataset; Feature Selection

I. INTRODUCTION

The Internet has become the most essential tool and one of the best sources of information about the current world. Internet can be considered as one of the major components of education and business purpose. Therefore, the data across the Internet must be secure. Internet security is one of the major concerns now-a-days. As Internet is threatened by various attacks it is very essential to design a system to protect those data, as well as the users using those data. Intrusion detection system (IDS) is therefore an invention to fulfill that requirement. Network administrators adapt intrusion detection system in order to prevent malicious attacks. Therefore, intrusion detection system became an essential part of the security management. Intrusion detection system detects and reports any intrusion attempts or misuse on the network. IDS can detect and block malicious attacks on the network, retain the performance normal during any malicious outbreak, perform an experienced security analysis.

Intrusion detection system approaches can be classified in 2 different categories. One of them is anomaly detection and the other one is signature based detection, also known as misuse detection based detection approach [4, 41]. The misuse detection is used to identify attacks in a form of signature or pattern. As misuse detection uses the known pattern to detect attacks the main disadvantage is that it will fail to identify any unknown attacks to the network or system. On the other hand, anomaly detection is used to detect unknown attacks. There are different ways to find out the anomalies. Different machine learning techniques are introduced in order to identify the anomalies.

Over the years, many researchers and scholars have done some significant work on the development of intrusion detection system. This paper reviewed the related studies in intrusion detection system over the past six years. This paper enlisted 49 papers in total from the year 2009 to 2014. This paper enlisted the proposed architecture of the classification techniques, algorithms being used. A Statistical comparison has been added to show classifier design, chosen algorithms, used datasets as well as the consideration of feature selection step.

This paper is organized as follows: Section 2 provides the research topic overview where a number of techniques for intrusion detection have been described. Section 3 represents a statistical overview of articles over the years on the algorithms that were frequently used, the datasets for each experiment and the consideration of feature selection step. Section 4 includes the discussion and conclusion as well as some issues which have been highlighted for future research in intrusion detection system using machine learning approaches.

II. RESEARCH PAPER OVERVIEW

A. Machine Learning Approach

Machine learning is a special branch of artificial intelligence that acquires knowledge from training data based on known facts. Machine learning is defined as a study that allows computers to learn knowledge without being programmed mentioned by Arthur Samuel in 1959.Machine learning mainly focuses on prediction. Machine learning techniques are classified into three broad categories such as – supervised learning, unsupervised learning, and reinforcement learning.

1) Supervised Learning

Supervised learning is also known as classification. In supervised learning data, instances are labeled in the training phase. There are several supervised learning algorithms. Artificial Neural Network, Bayesian Statistics, Gaussian Process Regression, Lazy learning, Nearest Neighbor algorithm, Support Vector Machine, Hidden Markov Model, Bayesian Networks, Decision Trees(C4.5,ID3, CART, Random Forrest), K-nearest neighbor, Boosting, Ensembles classifiers (Bagging, Boosting), Linear Classifiers (Logistic regression, Fisher Linear discriminant, Naive Bayes classifier, Perceptron, SVM), Quadratic classifiers are some of the most popular supervised learning algorithms.

2) Unsupervised Learning

In unsupervised learning data instances are unlabeled. A prominent way for this learning technique is clustering.

Some of the common unsupervised learners are Cluster analysis (K-means clustering, Fuzzy clustering), Hierarchical clustering, Self-organizing map, Apriori algorithm, Eclat algorithm and Outlier detection (Local outlier factor).

3) Reinforcement Learning

Reinforcement learning means computer interacting with an environment to achieve a certain goal. A reinforcement approach can ask a user (e.g., a domain expert) to label an instance, which may be from a set of unlabeled instances.

B. Single Classifiers

One machine learning algorithm or technique for developing an intrusion detection system can be used as a standalone classifier or single classifier. Some of the machine learning techniques have been discussed in this study which have been found as frequently used single classifiers in our studied 49 research papers.

1) Decision Tree

Creating a classifier for predicting the value of a target class for an unseen test instance, based on several already known instances is the task of Decision tree (DT). Through a sequence of decisions, an unseen test instance is being classified by a Decision tree [11]. Decision tree is very much popular as a single classifier because of its simplicity and easier implementation [14]. Decision tree can be expanded in 2 types: (i) Classification tree, with a range of symbolic class labels and (ii) Regression tree, with a range of numerically valued class labels [11].

2) Naive Bayes

On the basis of the class label given Naive Bayes assumes that the attributes are conditionally independent and thus tries to estimate the class-conditional probability[15]. Naive Bayes often produces good results in the classification where there exist simpler relations. Naive Bayes requires only one scan of the training data and thus it eases the task of classification a lot.

3) K-nearest neighbor

Various distance measure techniques are being used in K-nearest neighbor. K-nearest neighbor finds out k number of samples in training data that are nearest to the test sample and then it assigns the most frequent class label among the considered training samples to the test sample. For classifying samples, K-nearest neighbor is known as an approach which is the most simple and nonparametric[8]. K-nearest neighbor can be mentioned as an instance-based learner, not an inductive based [35].

4) Artificial Neural Network

Artificial Neural Network (ANN) is a processing unit for information which was inspired by the functionality of human brains [23]. Typically neural networks are organized in layers which are made up of a number of interconnected nodes which contain a function of activation. Patterns are presented to the network via the input layer, which communicates to one or more hidden layers where via a system of weighted connections the actual processing is done. The hidden layers then link to an output layer for producing the detection result as output.

5) Support Vector Machines

Support vector machine (SVM) was introduced in mid-1990's [5]. The concept behind SVM for intrusion detection basically is to use the training data as a description of only the normal class of objects or which is known as non-attack in intrusion detection system, and thus assuming the rest as anomalies [51]. The classifier constructed by support vector machines methodology discriminates the input space in a finite region where the normal objects are contained and all the rest of the space is assumed to contain the anomalies [9].

6) Fuzzy Logic

For reasoning purpose, dual logic's truth values can be either absolutely false (0) or absolutely true (1), but in Fuzzy logic these kinds of restrictions are being relaxed [60]. That means in Fuzzy logic the range of the degree of truth of a statement can hold the value between 0 and 1 along with '0' and '1'[11].

C. Hybrid Classifiers

A hybrid classifier offers combination of more than one machine learning algorithms or techniques for improving the intrusion detection system's performance vastly. Using some clustering-based techniques for preprocessing samples in training data for eliminating non-representative training samples and then, the results of the clustering are used as training samples for pattern recognition in order to design a classifier. Thus, either supervised or unsupervised learning approaches can be the first level of a hybrid classifier [11].

D. Ensemble Classifiers

The classifiers performing slightly better than a random classifier are known as weak learners. When multiple weak learners are combined for the greater purpose of improving the performance of a classifier significantly is known as Ensemble classifier [11].Majority vote, bagging and boosting are some common strategies for combining weak learners [15].Though it is known that the disadvantages of the component classifiers get accumulated in the ensemble classifier, but it has been producing a very efficient performance in some combination. So researchers are becoming more interested in ensemble classifiers day by day.

III.   STATISTICAL COMPARISONS OF RELATED WORK

A. Distribution of Papers by Year of Publication

The survey comprises 49 research papers in the time frame between 2009 and 2014. It discussed 8 papers from each of the year 2009, 2010 and 2012.The highest number of papers are studied from the year 2011.The number of papers from that year is 11. 10 papers are enlisted for the year 2013 and 4 papers from 2014.Fig.1 depicts the percentage of distribution of papers by year of publication.

B. Classifier design

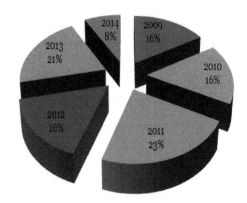

Fig. 1.   Year-wise distribution of papers

Intrusion detection method can be categorized in 3 categories namely single, hybrid and ensemble[11] .Fig.2 depicts the number of research papers in terms of single, hybrid and ensemble classifiers used in each year. According

TABLE I.   TOTAL NUMBERS OF RESEARCH PAPERS FOR THE Types Of CLASSIFIER DESIGN

Classifier design type	No. of research paper	References
Single	20	(D. Sa´nchez, 2009)[12], (Su-Yun Wua, 2009)[50], (Jun Ma, 2009)[27], (Mao Ye, 2009)[31], (Feng Jiang, 2009)[16], (Yung-Tsung Hou, 2010)[58], (Min Seok Mok, 2010)[34], (Han-Ching Wu, 2010)[22], (Chengpo Mua, 2010)[10], (Wang Dawei, 2011)[53], (G. Davanzo, 2011)[17], (Levent Koc, 2012)[29], (Carlos A. Catania, 2012)[9], (Inho Kang, 2012)[26], (Prabhjeet Kaur, 2012)[38], (Yusuf Sahin, 2013)[59], (S. Devaraju, 2013)[42], (Guillermo L. Grinblat, 2013)[21], (Mario Poggiolini, 2013)[32], (Adel Sabry Eesa, 2014)[2].
Hybrid	22	(Kamran Shafi, 2009)[28], (M. Bahrololum, 2009)[30], (Gang Wang, 2010)[18], (Woochul Shim, 2010)[55], (Muna Mhammad T. Jawhar, 2010)[37], (Ilhan Aydin, 2010)[25], (Seung Kim, 2011)[45], (I.T. Christou, 2011)[24], (Mohammad Saniee Abadeh, 2011)[36], (Shun-Sheng Wang, 2011)[47], (Su, 2011)[49], (Seungmin Lee, 2011)[46], (Yinhui Li, 2012)[57], (Bose, 2012)[6], (Prof. D.P. Gaikwad, 2012)[39], (A.M.Chandrashekhar, 2013)[1], (Mazyar Mohammadi Lisehroodi, 2013)[33], (Dahlia Asyiqin Ahmad Zainaddin, 2013)[13], (Seongjun Shin, 2013)[44], (Gisung Kim, A novel hybrid intrusion detection method integrating anomaly detection with misuse detection, 2013)[19], (Wenying Feng, 2014)[54], (Ravi Ranjan, 2014)[40].
Ensemble	7	(Tich Phuoc Tran, 2009)[52], (C.A. Laurentys, 2011)[7], (Dewan Md. Farid M. Z., 2011)[15], (Yang Yi, 2011)[56], (Siva S. Sivatha Sindhu, 2012)[48], (Dewan Md. Farid L. Z., 2013)[14], (Akhilesh Kumar Shrivas, 2014)[3]

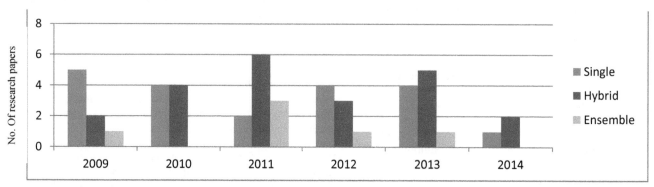

Fig. 2.   Year wise distribution of research papers for the types of classifier design

to the statistical comparison between the enlisted papers, hybrid classifiers have the highest number of literatures in the time frame mentioned earlier with a total number of 22. What comes later in terms of study is single classifiers which have been studied in 20 papers.

C. Single classifiers

Fig. 3 depicts the number of single learning algorithms used as classifiers. The number of research papers in the single classifier architecture using different classification techniques, e.g. Bayesian, SVM, DT, ANN, KNN, Fuzzy Logic enlisted in this survey paper is twenty. Table II enlists the proposed algorithms used in all the articles reviewed in this paper. Table IV shows Year wise distribution of single classifiers regarding results and citation of each article.

Support vector machine and Artificial neural network are the most popular approaches for single learning algorithm classifiers. Though we have taken 49 related papers and number of comparative samples is less but the comparison result implies that Support Vector machine is by far the most common and considered single classification technique. On the contrary, Fuzzy logic seems to be less considerable among the single classifiers over the enlisted literatures.

D. Ensemble classifiers

Multiple weak learners are combined in Ensemble classifiers. Table III depicts the articles using ensemble classifiers in intrusion detection system. Statistics shows AdaBoost is the most commonly used learning algorithm along with majority voting. Table III also enlists the detection rate of each of the classifier and the citation of each article throughout the time period.

E. Hybrid classifiers

Table V depicts Year wise distribution of Hybrid classifiers regarding results and citation of each article. Hybrid classifiers in intrusion detection have established in the mainstream study due to the performance accuracy in recent times Statistics shows hybrid classifiers have the highest number of articles in the Year of 2011. The table also shows the used algorithms in each article and their performance in intrusion detection system.

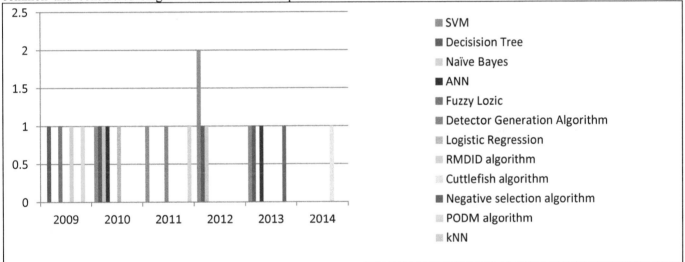

Fig. 3. Distribution of Single classifiers over the Years

TABLE II. ALGORITHMS USED IN SINGLE TYPE OF CLASSIFIER DESIGNED BASED RESEARCH PAPERS

Algorithm	Research paper Title	Reference
Naive Bayes	• A network intrusion detection system based on a hidden naive bayes multiclass classifier. • Malicious web content detection by machine learning.	(Levent Koc, 2012)[29] ; (Yung-Tsung Hou, 2010)[58]
Support Vector Machine	• An autonomous labeling approach to support vector machines algorithms for network traffic anomaly detection. • A differentiated one-class classification method with applications to intrusion detection. • Abrupt change detection with One-Class Time Adaptive Support Vector Machines. • Malicious web content detection by machine learning. • Anomaly detection techniques for a web defacement monitoring service.	(Carlos A. Catania, 2012)[9] ; (Inho Kang, 2012)[26] ; (Guillermo L. Grinblat, 2013)[21] ; (Yung-Tsung Hou, 2010)[58]; (G. Davanzo, 2011)[17].
Decision Tree	• Madam id for intrusion detection using data mining. • A cost-sensitive decision tree approach for fraud detection. • Data mining-based intrusion detectors. • Malicious web content detection by machine learning.	(Prabhjeet Kaur, 2012)[38]; (Yusuf Sahin, 2013)[59] ; (Su-Yun Wua, 2009)[50] ; (Yung-Tsung Hou, 2010)[58].
Artificial Neural Network	• Detection of accuracy for intrusion detection system using neural network classifier. • Neural networks-based detection of stepping-stone intrusion.	(S. Devaraju, 2013)[42] ; (Han-Ching Wu, 2010)[22].

Fuzzy Logic	• Data mining-based intrusion detectors.	(Su-Yun Wua, 2009)[50].
Detector Generation Algorithm	• Evolving boundary detector for anomaly detection	(Wang Dawei, 2011)[53].
Negative Selection algorithm	• Application of the feature-detection rule to the Negative Selection Algorithm	(Mario Poggiolini, 2013)[32].
Logistic regression	• Random effects logistic regression model for anomaly detection	(Min Seok Mok, 2010)[34].
RMDID	• Projected outlier detection in high-dimensional mixed-attributes data set.	(Mao Ye, 2009)[31].
PODM	• Information inconsistencies detection using a rule-map technique	(Jun Ma, 2009)[27]
Cuttlefish algorithm	• A novel feature-selection approach based on the cuttlefish optimization algorithm for intrusion detection systems.	(Adel Sabry Eesa, 2014)[2].
Sequence-based Outlier Detection algorithm	• Some issues about outlier detection in rough set theory.	(Feng Jiang, 2009)[16].
K-nearest neighbour (KNN)	• Anomaly detection techniques for a web defacement monitoring service.	(G. Davanzo, 2011)[17].

TABLE III.   YEAR WISE DISTRIBUTION OF ENSEMBLE CLASSIFIERS REGARDING RESULTS AND CITATION OF EACH ARTICLE

Year	Research Paper Title	Reference	Algorithm used	Result (%)	Citation
2009	Novel intrusion detection using probabilistic neural network & adaptive boosting	(Tich Phuoc Tran, 2009)[52]	• NN • AdaBoost • BSPNN	DR : 94.31	14
2011	A novel artificial immune system for fault behavior detection	(C.A. Laurentys, 2011)[7]	• GA • Majority Vote	DR : 97.85	17
	Adaptive intrusion detection based on boosting & naive Bayesian classifier	(Dewan Md. Farid M. Z., 2011)[15]	• NB • AdaBoost	DR : 99.75	14
	Incremental SVM based on reversed set for network intrusion detection	(Yang Yi, 2011)[56]	• SVM • ISVM	DR : 81.377	30
2012	Decision tree based light weight intrusion detection using a wrapper approach	(Siva S. Sivatha Sindhu, 2012)[48]	• Neural ensemble decision tree	DR : 98.38	44
2013	An adaptive ensemble classifier for mining concept drifting data streams	(Dewan Md. Farid L. Z., 2013)[14]	• NB • C4.5 • AdaBoost	DR : 92.65	13
2014	An ensemble model for classification of attacks with feature selection based on KDD-99 & NSL-KDD data set	(Akhilesh Kumar Shrivas, 2014)[3]	• ANN • Bayesian Network • Gain ratio FS	DR : 97.53 (using NSL-KDD) DR: 99.41 (using KDD-99)	[a]

[a]Not cited yet.

TABLE IV.   YEAR WISE DISTRIBUTION OF SINGLE CLASSIFIERS REGARDING RESULTS AND CITATION OF EACH ARTICLE

Year	Research Paper Title	Reference	Algorithm used	Result (%)	Citation
2009	Association rules applied to credit card fraud detection	(D. Sa´nchez, 2009)[12]	• Association rule methodology	Certainty factor : 80.08	64
	Data Mining based intrusion detectors	(Su-Yun Wua, 2009)[50]	• C4.5	DR : 70.62 FAR: 1.44	67
	Some issues about outlier detection in rough set theory	(Feng Jiang, 2009)[16]	• Outlier Detection algorithm	DR: SEQ based : 90 DIS based : 92	30
	Projected outlier detection in high dimensional mixed attributes data set	(Mao Ye, 2009)[31]	• PODM algorithm	DR: Credit approval data : 70 Breast Cancer Data : 80 Mushroom Data : 96 Synthetic Data : 97	24
	Information inconsistencies detection using a rule map technique	(Jun Ma, 2009)[27]	• RMDID algorithm	Error scales = 5.0% Inconsistent entries in Train Set = 5, Test Set = 4	1
2010	Malicious web content detection by machine learning	(Yung-Tsung Hou, 2010)[58]	• Naïve Bayes • DT • SVM • AdaBoost	Accuracy : NB : 58.28 DT : 94.74 SVM: 93.51 Boosted DT: 96.14	39
	Random effect logistic regression model for anomaly detection	(Min Seok Mok, 2010)[34]	• Logistic regression model.	Classification accuracy : Training dataset : 79.43 (Normal) 20.57(Attack) Validation dataset: 79.17 (Normal) 20.83 (Attack)	8
	An intrusion response decision making model based on hierarchical task network planning	(Chengpo Mua, 2010)[10]	• Hierarchical task network planning	Roc curve : excellent	20
	Neural Networks based detection of stepping stone intrusion	(Han-Ching Wu, 2010)[22]	• Neural Network	Accuracy : 99.0	13

2011	Evolving boundary detectors for anomaly detection	(Wang Dawei, 2011)[53]	• Detector Generation algorithm	DR : Iris Dataset : 99.28 considering Self radius = 0.08 Boundary threshold = 0.04 KDD dataset : DOS : 94.5 Probing : 93.64 U2R: 78.85 R2L: 50.69 considering Self radius = 0.05 Boundary threshold = 0.025	6
	Anomaly detection techniques for a web defacement monitoring service	(G. Davanzo, 2011)[17]	• K nearest neighbor • Support Vector machine	FPR: K nearest neighbor : 19.43 SVM :6.45	3
2012	A network intrusion detection system based on Hidden Naïve bayes multiclass classifier	(Levent Koc, 2012)[29]	• Hidden Naïve Bayes	Accuracy : 93.73 Error rate: 6.28	45
	An autonomous labeling approach to support vector machines algorithms for network traffic anomaly detection	(Carlos A. Catania, 2012)[9]	• Support Vector machine	DR : 88.64 (80% attack) 98.37 (1% attack)	11
	A differentiated one-class classification method with applications to intrusion detection	(Inho Kang, 2012)[26]	• Support Vector machine	DR : M=200* Targeted attack : 96.9 (4.7 % more than ordinary detection)	17
	Madam id for intrusion detection using Data mining	(Prabhjeet Kaur, 2012)[38]	• Decision Tree (J48)	FP rate :75.00 Precession : 1.7 Recall: 66.7	7
2013	A cost sensitive Decision tree approach for fraud detection	(Yusuf Sahin, 2013)[59]	• Decision Tree	TPR: Direct cost : 74.6 Class Probability : 92.1 CS-Gini : 92.8 Cs-IG: 92.6	9
	Detection of accuracy for intrusion detection system using neural network classifier	(S. Devaraju, 2013)[42]	• Neural Network	Accuracy : FFNN : 79.49 ENN : 78.1 GRNN: 58.74 PNN:85.50 RBNN: 83.51	4
	Abrupt change detection with one class time adaptive Support Vector Machine	(Guillermo L. Grinblat, 2013)[21]	• Support Vector Machine	495.9 sequences correctly classified within 500 sequences.	3
	Application of feature –detection rule to the negative selection algorithm	(Mario Poggiolini, 2013)[32]	• Negative Selection algorithm	Feature Detection rule : 0.9375 HD rule : 0.7686 RCHK(No MHC rule):0.8258 RCHK(Global MHC rule) : 0.5155 RCHK(MHC) rule : 0.9482	3
2014	A novel feature-selection approach based on the cuttlefish optimization algorithm for intrusion detection system	(Adel Sabry Eesa, 2014)[2]	• Cuttlefish algorithm	AR : 73.267 DR: 71.067 FPR: 17.685	b

[b]Not cited yet.

TABLE V.   A Detailed Information on Research Papers Designed with Hybrid Classifier

Year	Research Paper Title	Reference	Algorithm(s) used	Result (%)	Citation
2009	Anomaly intrusion detection design using hybrid of unsupervised and supervised neural network	(M. Bahrololum, 2009)[30]	• NN	TP rate : 97.00(Dos) 71.65(Probe) 26.69(R2L)	11
	An adaptive genetic-based signature learning system for intrusion detection	(Kamran Shafi, 2009)[28]	• GA	Accuracy : 92 FA rate : 0.84	31
2010	A new approach to intrusion detection using Artificial Neural Networks and fuzzy clustering	(Gang Wang, 2010)[18]	• ANN. • Fuzzy clustering.	Accuracy : 96.71 Precision : 99.91(Dos) 48.12(Probe) 93.18(R2L) 83.33(U2R)	114
	A distributed sinkhole detection method using cluster analysis	(Woochul Shim, 2010)[55]	• Hierarchical cluster analysis.	DR : 96.61	7
	Design Network Intrusion Detection System using hybrid Fuzzy-Neural Network	(Muna Mhammad T. Jawhar, 2010)[37]	• Fuzzy C-means clustering. • NN	Accuracy : 100(Dos) 100(U2R) 99.8(Probe) 40(R2L) 68.6(Unknown)	21
	Chaotic-based hybrid negative selection algorithm and its applications in fault and anomaly detection	(Ilhan Aydin, 2010)[25]	• Negative selection. • Clonal selection. • KNN.	Accuracy : 97.65	51
2011	Detecting fraud in online games of chance and lotteries	(I.T. Christou, 2011)[24]	• LOF. • K-means clustering. • EXAMCE.	DR : 98	3
	Fast outlier detection for very large log data	(Seung Kim, 2011)[45]	• Kd-tree indexing. • Approximated KNN. • LOF.	Gained time efficiency : 293-8727	11
	Design and analysis of genetic fuzzy systems for intrusion detection in computer networks	(Mohammad Saniee Abadeh, 2011)[36]	• Fuzzy genetic based machine learning methods: (i)Michigan,(ii)Pitsburg,(iii)IRL.	DR : 88.13 (Mitchigan) 99.53 (Pitsburg) 93.2 (IRL)	21

	An Integrated Intrusion Detection System for Cluster-based Wireless Sensor Networks	(Shun-Sheng Wang, 2011)[47]	• BPN. • ART. • Rule based method.	Accuracy: 95.13	24
	Real-time anomaly detection systems for Denial-of-Service attacks by weighted k-nearest-neighbor classifiers	(Su, 2011)[49]	• GA. • KNN.	Accuracy : 97.42 (with known attack) Accuracy : 78 (with unknown attack)	16
	Self-adaptive and dynamic clustering for online anomaly detection	(Seungmin Lee, 2011)[46]	• SOM. • K-means clustering	DR : 83.4 (offline) 86.4 (online)	14
2012	An efficient intrusion detection system based on support vector machines and gradually feature removal method	(Yinhui Li, 2012)[57]	• K-means clustering. • SVM. • Ant colony.	DR : 98.6249	40
	The combined approach for anomaly detection using neural networks & clustering techniques	(Bose, 2012)[6]	• SOM. • K-means clustering.	DR : 98.5 (Dos)	2
	Anomaly Based Intrusion Detection System Using Artificial Neural Network and Fuzzy Clustering	(Prof. D.P. Gaikwad, 2012)[39]	• ANN. • Fuzzy clustering.	*	6
2013	Fortification of hybrid intrusion detection system using variants of neural networks & support vector machines	(A.M.Chandrashekhar, 2013)[1]	• Fuzzy C-means clustering. • Fuzzy neural network. • SVM.	Accuracy : 98.94 (Dos) 97.11 (Probe) 97.80 (U2R) 97.78 (R2L)	2
	Hybrid of fuzzy clustering Neural network over NSL data set for intrusion detection system	(Dahlia Asyiqin Ahmad Zainaddin, 2013)[13]	• Fuzzy clustering.	Recall : 99.1 (Dos) 94.1 (Prob) 78 (U2R) 89 (R2L)	4
	A hybrid framework based on neural network MLP & K-means clustering for intrusion detection system	(Mazyar Mohammadi Lisehroodi, 2013)[33]	• K-means clustering. • MLP	DR : 99.99 (Dos) 99.97 (Probe) 99.99 (U2R) 99.98 (R2L)	c
	Advanced probabilistic approach for network intrusion forecasting and detection	(Seongjun Shin, 2013)[44]	• Markov chain. • K-means clustering. • APAN.	DR : 90	9
	A novel hybrid intrusion detection method integrating anomaly detection with misuse detection	(Gisung Kim, A novel hybrid intrusion detection method integrating anomaly detection with misuse detection, 2013)[19]	• C4.5. • 1-class SVM.	DR : 99.98 (with known attack) 97.4 (with unknown attack) Training time : 21.375 sec Testing time : 10.13 sec	9
2014	Mining network data for intrusion detection through combining SVMs with ant colony networks	(Wenying Feng, 2014)[54]	• CSOACN (self organized ant colony network) • SVM • CSVAC(combining support vectors with ant colony)	DR : 94.86 FP : 6.01 FN : 1.00	10
	A new clustering approach for anomaly intrusion detection	(Ravi Ranjan, 2014)[40]	• C4.5. • SVM. • K-means clustering.	DR : 96.12 (Dos) 90.10 (R2L) 70.51 (U2R) 70.13 (R2L). Accuracy : 96.38 False alarm rate : 3.2	4

cNot cited yet

F. Used Dataset in Researches

Datasets are assigned for default tasks e.g., Classification, Regression, Function learning, Clustering. Datasets reviewed by this paper is for classification purpose. As Fig.4 depicts, by far the most common dataset being used is KDD cup 1999 dataset. This dataset contains 4,000,000 instances and 42 attributes. The number of papers using KDD cup 1999 data set yields a peak in 2011 and in total 20 research papers has mentioned KDD Cup 1999 as their dataset.

Car evolution dataset [32] contains 1,728 instances with 6 attributes, attribute types are categorical. Wisconsin Breast cancer [16] has multivariate data types, all 10 attributes are integer types and it has 699 instances. Glass [32] dataset with multivariate data types and 214 instances It has 10 real attributes. Mushroom dataset [32] contains 22 categorical attributes and 8,124 instances. Lympography dataset [16] contains 18 categorical attributes and 148 instances. Yeast dataset [24] have 8 real attributes with 1,484 instances. Fisher-Iris dataset [25] contains 4 real attributes with 150 instances.Bicup2006 dataset and CO2 dataset [27] have 1,323 and 296 instances respectively. Public datasets like DARPA 1998, DARPA 2000, Fisher-Iris dataset, NSL KDD datasets are used in many related studies. Study also shows that few private or non-public datasets used over the time frame. Although the study briefly highlights public datasets like KDD

cup 99, DARPA 1998, DARPA 2000 being considered as standard datasets for intrusion detection system. DARPA dataset contains around 1.5 million traffic instances [36]. NSL-KDD dataset was proposed by removing all redundant instances from KDD'99. Thus, NSL-KDD dataset is more efficient than KDD'99 in getting more accurate evaluation of different learning techniques [19]. Some of the datasets were randomly used by the researchers. Table VI shows the year-wise distribution of randomly used dataset.

TABLE VI.   YEAR-WISE DISTRIBUTION OF RANDOMLY USED DATASET

Data Set	2009	2010	2011	2012	2013	2014	Total
Car Evaluation					1		1
Glass					1		1
DAMADICS			1				1
Yeast			1				1
Ionosphere			1				1
Musk			1				1
Malicious Web pages		1	1				2
Bicup2006	1						1
CO2	1						1
Lympography	1						1

G. Feature Selection

Feature Selection is an important step for the improvement of the system performance. Feature selection is considered before the training phase. Feature selection points out the best features and eliminates the redundant and irrelevant features. Table VII shows the year-wise distribution of feature selection step consideration. Table VII implies that out of 49 studies, 21 used feature selection step for their proposed architecture. It also shows that the number of papers using feature selection

yields a peak in the year 2012, where out of 8 papers in that year 7 used feature selection step. On the contrary, in 2009 the scenario was completely opposite. Though we have taken 49 related papers and number of differences in those papers are trivial but the comparison result implies that 21 experiments used feature selection where 28 experiments did not. It implies that feature selection is not a popular procedure in intrusion detection. Table VII and VIII overview the year-wise distribution of feature selection considered in related studies and the count of paper.

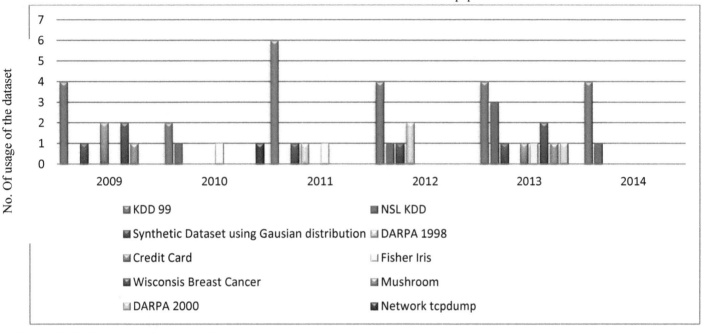

Fig. 4.   Distribution of popular datasets over the years

TABLE VII.   YEAR-WISE DISTRIBUTION OF FEATURE SELECTION CONSIDERED

Feature Selection Considered	2009	2010	2011	2012	2013	2014	Total
YES	1	3	4	7	4	2	21
NO	7	5	7	1	6	2	28

TABLE VIII.   DISTRIBUTION OF RESEARCH PAPERS CONSIDERING THE FEATURE SELECTION STEP

Feature Selection	No. of research papers	Research papers
YES	21	A.m.chandrashekhar, k. (2013)[1]. adel sabryeesa, z. o. (2014)[2]. Akhilesh Kumar Shrivas, A. K. (2014)[3]. Bose, A. A. (2012)[6] Carlos A. Catania, F. B. (2012)[9]. Inho Kang, M. K. (2012)[26]. Levent Koc, T. A. (2012)[29].M. Bahrololum, E. S. (2009)[30]. Mario Poggiolini, A. E. (2013)[32]. Min Seok Mok, S. Y. (2010)[34]. Prabhjeet Kaur, A. K. (2012)[38]. S. Devaraju, S. R. (2013)[42]. Seongjun Shin, S. L. (2013)[44]. Shun-Sheng Wang, K.-Q. Y.-C.-W. (2011)[47]. Siva S. Sivatha Sindhu, S. G. (2012)[48]. Su, M.-Y. (2011)[49]. Woochul Shim, G. K. (2010)[55]. Yang Yi, J. W. (2011)[56]. Yinhui Li, J. X. (2012)[57]. Yung-Tsung Hou, Y. C.-S.-M. (2010)[58]. Yusuf Sahin, S. B. (2013)[59].
NO	28	C.A. Laurentys, R. P. (2011)[7] Chengpo Mua, Y. L. (2010)[10] D. Sa´nchez, M. V. (2009)[12] Dahlia Asyiqin Ahmad Zainaddin, Z. M. (2013)[13]. Dewan Md. Farid, L. Z. (2013)[14] Dewan Md. Farid, M. Z. (2011)[15] Feng Jiang, Y. S. (2009)[16] G. Davanzo, E. M. (2011)[17] Gang Wang, J. H. (2010)[18] Gisung Kim, S. L. (2013)[19](Ravi Ranjan, 2014)[40] Guillermo L. Grinblat, L. C. (2013)[21] Han-Ching Wu, S.-H. S. (2010)[22] I.T. Christou, M. B. (2011)[24] Ilhan Aydin, M. K. (2010)[25]. Jun Ma, J. L. (2009)[27] Kamran Shafi, H. A. (2009)[28]Mao Ye, X. L. (2009)[31]. Mazyar Mohammadi Lisehroodi, Z. M. (2013)[33]. Mohammad Saniee Abadeh, H. M. (2011)[36]. Muna Mhammad T. Jawhar, M. M. (2010)[37]. Prof. D.P. Gaikwad, S. J. (2012)[39] Seung Kim, N. W.-H. (2011)[45]. Seungmin Lee, G. K. (2011)[46]. Su-Yun Wua, E. Y. (2009)[50]. Tich Phuoc Tran, L. C. (2009)[52]. Wang Dawei, Z. F. (2011)[53]. Wenying Feng, Q. Z. (2014)[54].

IV.   DISCUSSION, FUTURE WORK AND CONCLUSION

Uses of different classifier techniques in intrusion detection system is an emerging study in machine learning and artificial intelligence. It has been the attention of researchers for a long period of time. This paper has identified 49 research papers related to application of using different classifiers for intrusion detection published between 2009 and 2014. Though this survey paper cannot claim to be an in-depth study of those studies, but it presents a reasonable perspective and shows a valid comparison of works in this field over those years. The following issues could be useful for future research:

- Removal of redundant and irrelevant features for the training phase is a key factor for system performance. Consideration of feature selection will play a vital role in the classification techniques in future work.

- Feature selection has many algorithms to work with. Using different feature selection algorithms and working with the best possible one will be helpful for the classification techniques and also increase the consideration of feature selection step in intrusion detection.

- Uses of single classifiers or baseline classifiers in performance measurement can be replaced by hybrid or ensemble classifiers.

REFERENCES

[1]   A.M.Chandrashekhar, K. (2013). Fortification of hybrid intrusion detection system using variants of neural networks & support vector machines. International Journal of Network Security & Its Applications (IJNSA) .

[2]   Adel Sabry Eesa, Z. O. (2014). A novel feature-selection approach based on the cuttlefish optimization algorithm for intrusion detection systems. Expert Systems with Applications,ELSEVIER .

[3]   Akhilesh Kumar Shrivas, A. K. (2014). An Ensemble Model for Classification of Attacks with Feature Selection based on KDD99 and NSL-KDD Data Set. International Journal of Computer Applications .

[4]   Anderson, J. (1995). An introduction to neural networks. Cambridge: MIT Press.

[5]   Bernhard E Boser, I. M. (1992). A Training Algorithm for Optimal Margin Classiers. Proceedings of the 5th Annual ACM Workshop on Computational , 144-152.

[6]   Bose, A. A. (2012). THE COMBINED APPROACH FOR ANOMALY detection using neural networks & clustering techniques. Computer Science & Engineering: An International Journal (CSEIJ) .

[7]   C.A. Laurentys, R. P. (2011). A novel Artificial Immune System for fault behavior detection. Expert Systems with Applications,ELSEVIER .

[8]   C.M.Bishop. (1995). Neural networks for pattern recognition. England: Oxford University.

[9]   Carlos A. Catania, F. B. (2012). An autonomous labeling approach to support vector machines algorithms for network traffic anomaly detection. Expert Systems with Applications,ELSEVIER .

[10] Chengpo Mua, Y. L. (2010). An intrusion response decision-making model based on hierarchical. Expert Systems with Applications,ELSEVIER .

[11] Chih-Fong Tsai, Y.-F. H.-Y.-Y. (2009). Intrusion detection by machine learning: A review. expert systems with applications,ELSEVIER .

[12] D. Sa´nchez, M. V. (2009). Association rules applied to credit card fraud detection. Expert Systems with Applications,ELSEVIER .

[13] Dahlia Asyiqin Ahmad Zainaddin, Z. M. (2013). HYBRID OF FUZZY CLUSTERING NEURAL NETWORK OVER NSL DATASET FOR INTRUSION DETECTION SYSTEM. Journal of Computer Science.

[14] Dewan Md. Farid, L. Z. (2013). An Adaptive Ensemble Classifier for Mining Concept-Drifting Data Streams. Expert systems with Applications,ELSEVIER .

[15] Dewan Md. Farid, M. Z. (2011). Adaptive Intrusion Detection based on Boosting and. International Journal of Computer Applications .

[16] Feng Jiang, Y. S. (2009). Some issues about outlier detection in rough set theory. expert systems with application,ELSEVIER .

[17] G. Davanzo, E. M. (2011). Anomaly detection techniques for a web defacement monitoring service. Expert Systems with Applications,ELSEVIER .

[18] Gang Wang, J. H. (2010). A new approach to intrusion detection using Artificial Neural Networks and. Expert Systems with Applications,ELSEVIER .

[19] Gisung Kim, S. L. (2013). A novel hybrid intrusion detection method integrating anomaly detection with misuse detection. Expert Systems with Applications,ELSEVIER .

[20] Gisung Kim,J.C,S.K. (2012). A congestion-aware IDS node selection method for wireless sensor networks. IJDSN.

[21] Guillermo L. Grinblat, L. C. (2013). Abrupt change detection with One-Class Time-Adaptive Support Vector Machines. Expert Systems with Applications,ELSEVIER .

[22] Han-Ching Wu, S.-H. S. (2010). Neural networks-based detection of stepping-stone intrusion. Expert Systems with Applications,ELSEVIER .

[23] Haykin, S. (1999). Neural networks: A comprehensive foundation (2nd Edition). New Jersey: Prentice Hall.

[24] I.T. Christou, M. B. (2011). Detecting fraud in online games of chance and lotteries. Expert Systems with Applications,ELSEVIER .

[25] Ilhan Aydin, M. K. (2010). Chaotic-based hybrid negative selection algorithm and its applications in fault. expert systems with applications,ELSEVIER .

[26] Inho Kang, M. K. (2012). A differentiated one-class classification method with applications to intrusion detection. Expert Systems with Applications,ELSEVIER .

[27] Jun Ma, J. L. (2009). Information inconsistencies detection using a rule-map technique. expert systems with applications,ELSEVIER .

[28] Kamran Shafi, H. A. (2009). An adaptive genetic-based signature learning system for intrusion detection. Expert Systems with Applications, ELSEVIER .

[29] Levent Koc, T. A. (2012). A network intrusion detection system based on a Hidden Naïve Bayes multiclass classifier. Expert Systems with Applications,ELSEVIER .

[30] M. Bahrololum, E. S. (2009). ANOMALY INTRUSION DETECTION DESIGN USING. International Journal of Computer Networks & Communications (IJCNC) .

[31] Mao Ye, X. L. (2009). Projected outlier detection in high-dimensional mixed-attributes data set. Expert systems with applications,ELSEVIER .

[32] Mario Poggiolini, A. E. (2013). Application of the feature-detection rule to the Negative Selection Algorithm. Expert Systems with Applications,ELSEVIER .

[33] Mazyar Mohammadi Lisehroodi, Z. M. (2013). A HYBRID FRAMEWORK BASED ON NEURAL NETWORK MLP AND K-MEANS CLUSTERING FOR INTRUSION DETECTION SYSTEM. Proceedings of the 4th International Conference on Computing and Informatics, ICOCI 2013 (p. Paper No. 020). Sarawak, Malaysia: Universiti Utara Malaysia.

[34] Min Seok Mok, S. Y. (2010). Random effects logistic regression model for anomaly detection. Expert Systems with Applications,ELSEVIER .

[35] Mitchell, T. (1997). Machine learning. New york: MacHraw Hill.

[36] Mohammad Saniee Abadeh, H. M. (2011). Design and analysis of genetic fuzzy systems for intrusion detection in. Expert Systems with Applications,ELSEVIER .

[37] Muna Mhammad T. Jawhar, M. M. (2010). Design Network Intrusion Detection System using hybrid. International Journal of Computer Science and Security.

[38] Prabhjeet Kaur, A. K. (2012). MADAM ID FOR INTRUSION DETECTION USING DATA MINING. International Journal of Research in IT & Management,IJRIM .

[39] Prof. D.P. Gaikwad, S. J. (2012). Anomaly Based Intrusion Detection System Using Artificial Neural Network & Fuzzy clustering. International Journal of Engineering Research & Technology (IJERT) .

[40] Ravi Ranjan, G. S. (2014). A NEW CLUSTERING APPROACH FOR ANOMALY INTRUSION DETECTION. International Journal of Data Mining & Knowledge Management Process (IJDKP) .

[41] Rhodes, B. M. (2000). Multiple self-organizing maps for intrusion detection. Baltimore, MD.

[42] S. Devaraju, S. R. (2013). DETECTION OF ACCURACY FOR INTRUSION DETECTION SYSTEM USING NEURAL NETWORK CLASSIFIER. International Journal of Emerging Technology and Advanced Engineering .

[43] Sahoo, R. R. (2014). A NEW CLUSTERING APPROACH FOR ANOMALY INTRUSION DETECTION. International Journal of Data Mining & Knowledge Management Process (IJDKP) .

[44] Seongjun Shin, S. L. (2013). Advanced probabilistic approach for network intrusion forecasting and detection. Expert Systems with Applications,ELSEVIER .

[45] Seung Kim, N. W.-H. (2011). Fast outlier detection for very large log data. Expert Systems with Applications,ELSEVIER .

[46] Seungmin Lee, G. K. (2011). Self-adaptive and dynamic clustering for online anomaly detection. Expert Systems with Applications,ELSEVIER .

[47] Shun-Sheng Wang, K.-Q. Y.-C.-W. (2011). An Integrated Intrusion Detection System for Cluster-based Wireless. Expert Systems with Applications.

[48] Siva S. Sivatha Sindhu, S. G. (2012). Decision tree based light weight intrusion detection using a wrapper approach. Expert Systems with Applications,ELSEVIER .

[49] Su, M.-Y. (2011). Real-time anomaly detection systems for Denial-of-Service attacks by weighted. Expert Systems with Applications,ELSEVIER .

[50] Su-Yun Wua, E. Y. (2009). Data mining-based intrusion detectors. Expert Systems with Applications,ELSEVIER .

[51] Tax, D. &. (1999). Data domain description using support vectors. Proceedings of the european symposium on artificial neural networks, 251-256.

[52] Tich Phuoc Tran, L. C. (2009). Novel Intrusion Detection using Probabilistic Neural. (IJCSIS) International Journal of Computer Science and Information Security.

[53] Wang Dawei, Z. F. (2011). Evolving boundary detector for anomaly detection. Expert Systems with Applications.

[54] Wenying Feng, Q. Z. (2014). Mining network data for intrusion detection through combining SVMs with ant colony networks. Future Generation Computer Systems,ELSEVIER .

[55] Woochul Shim, G. K. (2010). A distributed sinkhole detection method using cluster analysis. Expert Systems with Applications,ELSEVIER .

[56] Yang Yi, J. W. (2011). Incremental SVM based on reserved set for network intrusion detection. Expert Systems with Applications .

[57] Yinhui Li, J. X. (2012). An efficient intrusion detection system based on support vector machines and gradually feature removal method. Expert Systems with Applications,ELSEVIER .

[58] Yung-Tsung Hou, Y. C.-S.-M. (2010). Malicious web content detection by machine learning. expert systems with applications,ELSEVIER .

[59] Yusuf Sahin, S. B. (2013). A cost-sensitive decision tree approach for fraud detection. Expert Systems with Applications,ELSEVIER.

[60] Zimmermann, H.-J. (2010). Fuzzy set theory. Advanced Review John Wiley & Sons, Inc

Appropriate Tealeaf Harvest Timing Determination Referring Fiber Content in Tealeaf Derived from Ground based Nir Camera Images

Kohei Arai 1
Graduate School of Science and Engineering
Saga University
Saga City, Japan

Shihomi Kasuya [2],
2 Sasaki Green Tea Company,
Kakegawa city – Japan

Yoshihiko Sasaki [2],
2 Sasaki Green Tea Company,
Kakegawa city – Japan

Hideto Matusura [2]
2 Sasaki Green Tea Company,
Kakegawa city – Japan

Abstract—Method for most appropriate tealeaves harvest timing with the reference to the fiber content in tealeaves which can be estimated with ground based Near Infrared (NIR) camera images is proposed. In the proposed method, NIR camera images of tealeaves are used for estimation of nitrogen content and fiber content in tealeaves. The nitrogen content is highly correlated to Theanine (amid acid) content in tealeaves. Theanine rich tealeaves taste good. Meanwhile, the age of tealeaves depend on fiber content. When tealeaves are getting old, then fiber content is increased. Tealeaf shape volume also is increased with increasing of fiber content. Fiber rich tealeaves taste not so good, in general. There is negative correlation between fiber content and NIR reflectance of tealeaves. Therefore, tealeaves quality of nitrogen and fiber contents can be estimated with NIR camera images. Also, the shape volume of tealeaves is highly correlated to NIR reflectance of tealeaf surface. Therefore, not only tealeaf quality but also harvest amount can be estimated with NIR camera images. Experimental results show the proposed method works well for estimation of appropriate tealeaves harvest timing with fiber content in the tealeaves in concern estimated with NIR camera images.

Keywords—Tealeaves; Nitrigen content; Amino accid; Leaf volume; NIR images; Fiber content; Theanine; Amid acid; Regressive analysis

I. INTRODUCTION

There is a strong demand for monitoring of the vitality of crops in agricultural areas automatically with appropriate measuring instruments in order to manage agricultural area in an efficient manner. It is also required to monitor not only quality but also quantity of vegetation in the farmlands. Vegetation monitoring is attempted with red and photographic cameras [1]. Grow rate monitoring is also attempted with spectral observation [2].

Total nitrogen content corresponds to amid acid which is highly correlated to Theanine: 2-Amino-4-(ethylcarbamoyl) butyric acid for tealeaves so that total nitrogen is highly correlated to tea taste. Meanwhile, fiber content in tealeaves has a negative correlation to tea taste. Near Infrared: NIR camera data shows a good correlation to total nitrogen and fiber contents in tealeaves so that tealeaves quality can be monitored with network NIR cameras. It is also possible to estimate total nitrogen and fiber contents in tealeaves with remote sensing satellite data, in particular, Visible and near infrared: VNIR radiometer data. Moreover, VC, NDVI, BRDF of tealeaves have a good correlation to grow index of tealeaves so that it is possible to monitor expected harvest amount and quality of tealeaves with network cameras together with remote sensing satellite data. BRDF monitoring is well known as a method for vegetation growth [3],[4]. On the other hand, degree of polarization of vegetation is attempted to use for vegetation monitoring [5], in particular, Leaf Area Index: LAI together with new tealeaves growth monitoring with BRDF measurements [6].

Theanine in new tealeaves are changing to Catechin [7],[8],[9] with sunlight. In accordance with increasing of sunlight, also, new tealeaves grow up so that there is a most appropriate time for harvest of tealeaves in order to maximize the amount and the taste of new tealeaves simultaneously.

Optical properties of tealeaves and methods for estimation of tealeaves quality and harvest amount estimation accuracy are well reported [10]-[17].

The method proposed here is to determine tealeaves harvest timing by using NIR camera images through the estimation of fiber content in tealeaves together with meteorological data. This paper, also, deals with automatic monitoring of a quality of tealeaves with earth observation satellite, network cameras together with a method that allows estimation of total nitrogen and fiber contents in tealeaves as an example.

The following section describes the proposed method together with some research background. Then experimental results are described followed by some tealeaves harvest timing related discussions. Finally, conclusions are described together with some discussions.

II.   PROPOSED METHOD

The proposed method for most appropriate tealeaves harvest timing is based on Near Infrared (NIR) camera images of tealeaves which are acquired in tea farm areas. The most appropriate tealeaves harvest timing is difficult to define. Tea farmers determine the timing empirically. It may be defined as maximizing Theanine in tealeaves and harvest amount as well as softness of tealeaves (less fiber content).

Theanine is highly correlated to the nitrogen content in tealeaves. Nitrogen content in tealeaves is proportional to NIR reflectance of tealeaf surfaces. Meanwhile, harvest amount is proportional to leaf area of tealeaves which depends on NIR reflectance of tealeaf surfaces. Also, tealeaf thickness, length and width can be measured results in shape volume of tealeaves measurements. It is also crossly related to the harvest amount. For time being, tealeaves are getting old and fiber content is also increased accordingly. Such old tealeaves taste not so good. New fresh tealeaves which have less fiber and have much Theanine taste good. Due to the fact that there is relations between NIR reflectance of tealeaves and Theanine (Nitrogen), fiber (age), Theanine and fiber content in tealeaves and harvest amount can be estimated with NIR camera imagery data.

NIR reflectance can be obtained from the acquired NIR camera images if a standard reflectance panel or plaque is acquired simultaneously with tealeaves in concern. Standard panel, for instance Spectralon, is not so cheap. In order to minimize a required cost for acquisition of NIR camera images, typical print sheets are used as standard panel. Therefore, cross comparison of reflectance between Spectralon and the print sheet is needed. Fig.1 shows acquired NIR image of Spectralon and the print sheet. Correction curve for conversion of NIR reflectance with the print sheet to Spectralon based NIR reflectance can be obtained from the acquired reflectance measured for Spectralon and the print sheet.

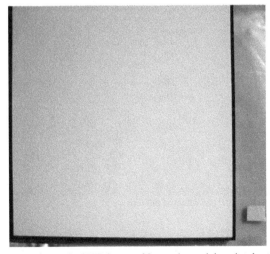

Fig. 1.   Acquired NIR image of Spectralon and the print sheet

Fig.2 shows the ratio of the Spectralon based NIR reflectance and the print sheet based NIR reflectance.

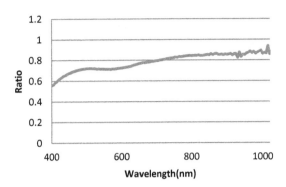

Fig. 2.   Ratio of the Spectralon based NIR reflectance and the print sheet based NIR reflectance

III.   EXPERIMENTS

A.   Intensive Study Test Sites

Experiments are conducted at the tea farm areas of Kakegawa in Shizuoka prefecture, Japan in April and May time frame (first harvesting time period a year). Outlook of one of the typical tea farm areas which is called "Pilot farm area" is shown in Fig.3. Longitude and latitude of the Pilot tea farm area is as follows,

34°44'30.7"N 138°01'27.6"E

Fig. 3.   Outlook of one of the typical tea farm areas which is called "Pilot farm area"

Other than Pilot tea farm area, there are other test sites, "Front" and "Back" which are situated the following longitude and latitude, respectively.

34°45'44.3"N 138°03'11.3"E(Front)

34°44'40.0"N 138°03'11.9"E(Back)

Usually, tealeaves are harvested three times, (1) begging in May, (2) middle in July, (3) middle in October. Therefore, tealeaves are growing in April and then tealeaves are harvested in the begging of May in general for the first harvesting time period. Fig.4 shows examples of NIR camera image of photos which are acquired before and after the harvest. Fig.3 (a) and (b) shows an example of NIR camera image of tealeaves which is acquired just before harvest together with the print sheet and of tealeaves which is acquired just after harvest together with the same print sheet, respectively. In the experiments, the center wavelength of NIR camera is 800nm with band width of 100nm.

(a)Before harvest

(b)After harvest

Fig. 4.   Examples of NIR camera images which are acquired just before (a) and after (b) harvest of tealeaves together with the print sheet as secondly standard panel

Fig.5 shows examples of tealeaves and nadir view of tea farm areas of the intensive test sites, "Pilot", "Front" and "Back".

(a)Tealeaves in Pilot

(b)Nadir view in Pilot

(c)Tealeaves in Front

(d)Nadir view in Front

(e)Tealeaves in Back

(f)Nadir view in Back

Fig. 5. Example of the acquired NIR images for the three intensive test sites, Pilot, Front and Back of tea farm areas which are taken just before harvest, April 23, Mar 2 and May 3 in 2015, respectively

B. Relation between NIR Reflectance and TN and Fiber

Through the comparison of NIR reflectance measured in both time frames, it is found that Theanine which is proportional to the measured NIR reflectance is increased for the time being. Also, it is found that the Theanine is decreased for the short term periods (around one month). During the short term periods, tealeaf thickness (T), length (L), and width (W) are increased as shown in Fig.6 (a) and (b) shows the relation between NIR reflectance which is calculated with nadir view camera data and nitrogen and fiber contents in tealeaves. Although R square values are not so large, there are not so bad correlations between NIR reflection and nitrogen and fiber contents. There are positive and negative correlations between NIR reflectance and nitrogen as well as fiber contents, respectively. Trends of nitrogen and fiber contents in April 2015 are shown in Fig.7 (a) and (b), respectively. Fiber content is getting large while nitrogen content is getting small for the time being, obviously. These trends are different from each other tea farm areas.

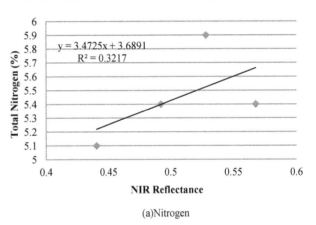

(a)Nitrogen

(b)Fiber

Fig. 6.   Relation   between   NIR   reflect

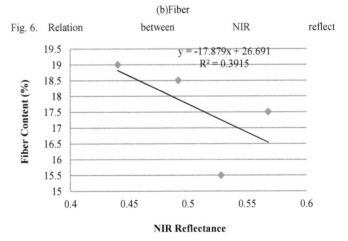

ance which is calculated with nadir view camera data and nitrogen and fiber contents in tealeaves

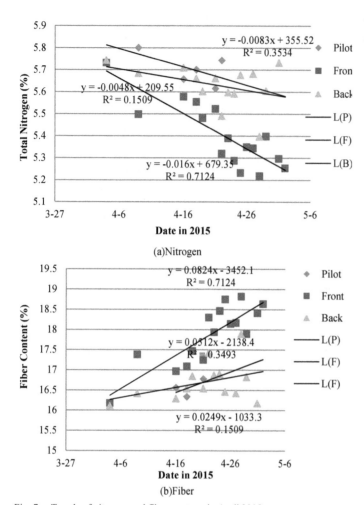

(a)Nitrogen

(b)Fiber

Fig. 7.   Trends of nitrogen and fiber contents in April 2015

C. Relation between TN and Fiber as well as Meteorological Data

If the appropriate harvest timing is relating to the meteorological data, it is very convenient to determine the timing with the meteorological data. The harvest dates for the test sites of Pilot, Front and Back are April 23, May 2 and May3, respectively. These are determined by tea farmers in subjective manner. They used to determine the best harvest days through touching the typical top of tealeaves (softness of the tealeaves), and looking the shape of tealeaves, length, width, and thickness.

Fig.8 (a) shows the trends of the maximum air temperature on the ground a day (triangle), the sunshine duration time a day (cross) while Fig.8 (b) shows cumulative maximum air temperature (blue square) and cumulative sunshine duration time a day (brown square) since March 23 2015, respectively. Also, Fig.8 (c) shows the measured irradiance at the three test sites. Due to the fact that the harvest dates for the test sites of Pilot, Front and Back are April 23, May 2 and May3, respectively, these dates are coincident to the dates when the irradiance reaches to 10000 for each test site. These measured irradiances vary by day by day. 10000 arbitrary unit of measured irradiance is the maximum of the measurable irradiance range. In this case, the irradiance is measured when

the tealeaf shape is measured together with the NIR camera data is taken.

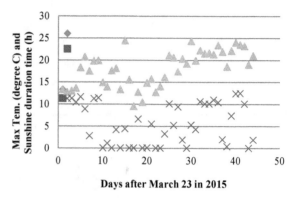

(a)Maximum air temperature a day and sunshine duration time a day

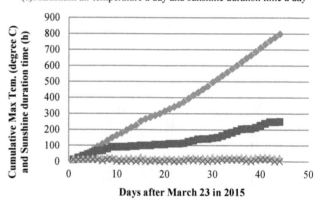

(b)Cumulative

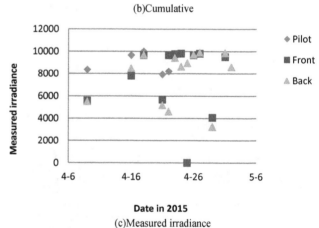

(c)Measured irradiance

Fig. 8.   Maximum air temperature a day and sunshine duration time a day and cumulative those since March 23 2015

Fig.9 (a) and (b), meanwhile, shows nitrogen and fiber contents in tealeaves at the three test sites, Pilot, Front and Back as a function of cumulative maximum air temperature a day since March 23 2015, respectively. Also, Fig.9 (c) shows fiber content in tealeaves and the shape volume (width by length by thickness) at the test sites, Pilot, Front and Back. Due to the fact that the harvest day at the test site, Pilot is April 23 2015, there is only one data point of the measured shape volume while there are two data points for the test sites, Front and Back.

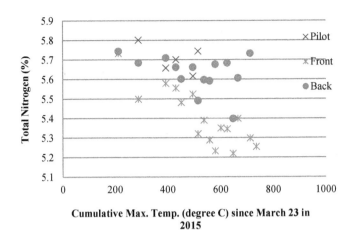

(a)Nitrogen

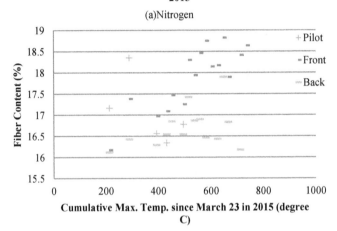

(b)Fiber

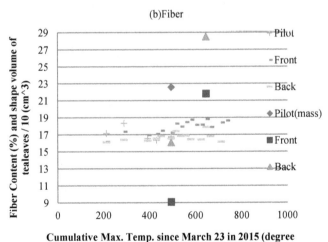

(c)Shape Volume

Fig. 9. Nitrogen and fiber contents in tealeaves at the test sites, Pilot, Front, Back tea farm areas

Although the harvest days for the different test sites, Pilot, Front, Back are different, the shape volumes at the harvest days are almost same (greater than 230 cm³). It is not easy to measure the shape volume. Tea farmers used to measure the shape of tealeaves not so frequently because it takes time. On the other hand, it is not so difficult to take a NIR camera image for tea farmers. Also, as is aforementioned, the shape

volume is proportional to fiber content and is negatively proportional to nitrogen content. Therefore, it seems reasonable to determine harvest day with nitrogen and fiber contents in tealeaves. The relations between leaf shape volume and nitrogen and fiber contents in tealeaves are shown in Fig.10. It is quite obvious that sensitivity of fiber content is greater than that of nitrogen content.

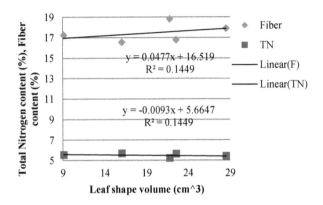

Fig. 10. relations between leaf shape volume and nitrogen and fiber contents in tealeaves

A relation between fiber content and cumulative maximum air temperature a day since March 23 2015 is estimated by using this relation of regressive equation. Fig.11 shows the relation between both.

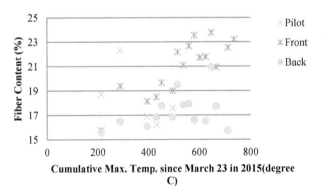

Fig. 11. Relation between fiber content and cumulative maximum air temperature a day since March 23 2015

From this figure, it is found that the harvest day can be determined by using fiber content in tealeaves with the threshold at around 23 (%). If the fiber content is greater than 23 (%), such tealeaves are no longer new fresh tealeaves and are getting down in terms of tealeaf quality (it tastes getting bad).

D. Estimation of Harvest Amount

Harvest amounts (Kg/a) of the test sites, Pilot, Front, and Back are 60.79, 93.4, 85.0, respectively. Harvest amount may be estimated with shape volume. Namely, it seems reasonable that large size of tealeaf shape means a great harvest amount. The other factor would be fiber content in tealeaves. Namely, in accordance with tealeaf age, fiber content is increased. This implies that harvest amount is increased with increasing of fiber content. Fig.12 shows the relations between harvest amount and shape volume as well as fiber content in tealeaves.

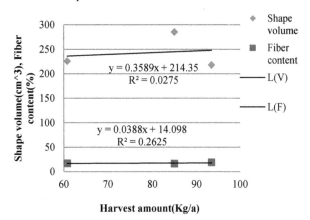

Fig. 12. Relations between harvest amount and shape volume as well as fiber content in tealeaves

It is found that the relation between harvest amount and fiber content is greater than that shape volume.

IV. CONCLUSION

Method for most appropriate tealeaves harvest timing with the reference to the fiber content in tealeaves which can be estimated with ground based Near Infrared (NIR) camera images is proposed. In the proposed method, NIR camera images of tealeaves are used for estimation of nitrogen content and fiber content in tealeaves. The nitrogen content is highly correlated to Theanine (amid acid) content in tealeaves. Theanine rich tealeaves taste good. Meanwhile, the age of tealeaves depend on fiber content. When tealeaves are getting old, then fiber content is increased. Tealeaf shape volume also is increased with increasing of fiber content. Fiber rich tealeaves taste not so good, in general. There is negative correlation between fiber content and NIR reflectance of tealeaves. Therefore, tealeaves quality of nitrogen and fiber contents can be estimated with NIR camera images. Also, the shape volume of tealeaves is highly correlated to NIR reflectance of tealeaf surface. Therefore, not only tealeaf quality but also harvest amount can be estimated with NIR camera images.

Experimental results show the proposed method works well for estimation of appropriate tealeaves harvest timing with fiber content in the tealeaves in concern estimated with NIR camera images. It is found that the appropriate tealeaf harvest day can be determined by using fiber content in tealeaves which is estimated with NIR camera data. Also, it is found that harvest amount can be estimated with fiber content in tealeaves which is derived from NIR camera data.

Further study is required for a relation between micro meteorological condition and the best harvest timing. Pilot tea farm area is the best solar illumination condition among three test sites followed by Front and Back. The other conditions, fertilizer amount, water supply, insect damage management, etc. are almost same for each test sites. Therefore, weather condition; in particular, solar illumination condition must be key issues for determination of the best harvest timing.

REFERENCES

[1] J.T.Compton, Red and photographic infrared linear combinations for monitoring vegetation, Journal of Remote Sensing of Environment, 8, 127-150, 1979.

[2] C.Wiegand, M.Shibayama, and Y.Yamagata, Spectral observation for estimating the growth and yield of rice, Journal of Crop Science, 58, 4, 673-683, 1989.

[3] S.Tsuchida, I.Sato, and S.Okada, BRDF measurement system for spatially unstable land surface-The measurement using spectro-radiometer and digital camera- Journal of Remote Sensing, 19, 4, 49-59, 1999.

[4] K.Arai, Lecture Note on Remote Sensing, Morikita-shuppan Co., Ltd., 2000.

[5] K.Arai and Y.Nishimura, Degree of polarization model for leaves and discrimination between pea and rice types of leaves for estimation of leaf area index, Abstract, COSPAR 2008, A3.10010-08#991, 2008.

[6] K.Arai and Long Lili, BRDF model for new tealeaves and new tealeaves monitoring through BRDF monitoring with web cameras, Abstract, COSPAR 2008, A3.10008-08#992, 2008.

[7] Greivenkamp, John E., *Field Guide to Geometrical Optics*. SPIE Field Guides vol. FG01. SPIE. ISBN 0-8194-5294-7, 2004.

[8] Seto R H. Nakamura, F. Nanjo, Y. Hara, *Bioscience, Biotechnology, and Biochemistry,* Vol.61 issue9 1434-1439 1997.

[9] Sano M, Suzuki M ,Miyase T, Yoshino K, Maeda-Yamamoto, M.,J.Agric.Food Chem., 47 (5), 1906-1910 1999.

[10] Kohei Arai, Method for estimation of grow index of tealeaves based on Bi-Directional reflectance function: BRDF measurements with ground based network cameras, International Journal of Applied Science, 2, 2, 52-62, 2011.

[11])Kohei Arai, Wireless sensor network for tea estate monitoring in complementarily usage with Earth observation satellite imagery data based on Geographic Information System(GIS), International Journal of Ubiquitous Computing, 1, 2, 12-21, 2011.

[12] Kohei Arai, Method for estimation of total nitrogen and fiber contents in tealeaves with ground based network cameras, International Journal of Applied Science, 2, 2, 21-30, 2011.

[13] Kohei Arai, Monte Carlo ray tracing simulation for bi-directional reflectance distribution function and grow index of tealeaves estimation, International Journal of Research and Reviews on Computer Science, 2, 6, 1313-1318, 2011.

[14] K.Arai, Monte Carlo ray tracing simulation for bi-directional reflectance distribution function and grow index of tealeaves estimations, International Journal of Research and Review on Computer Science, 2, 6, 1313-1318, 2012.

[15] K.Arai, Fractal model based tea tree and tealeaves model for estimation of well opened tealeaf ratio which is useful to determine tealeaf harvesting timing, International Journal of Research and Review on Computer Science, 3, 3, 1628-1632, 2012.

[16] Kohei Arai, Method for tealeaves quality estimation through measurements of degree of polarization, leaf area index, photosynthesis available radiance and normalized difference vegetation index for characterization of tealeaves, International Journal of Advanced Research in Artificial Intelligence, 2, 11, 17-24, 2013.

[17] K.Arai, Optimum band and band combination for retrieving total nitrogen, water, and fiber in tealeaves through remote sensing based on regressive analysis, International Journal of Advanced Research in Artificial Intelligence, 3, 3, 20-24, 2014.

22

Human Lips-Contour Recognition and Tracing

Md. Hasan Tareque[1]
Department of Computer Science and Engineering
IBAIS University
Dhaka, Bangladesh

Ahmed Shoeb Al Hasan[2]
Department of Computer Science and Engineering
Bangladesh University of Business & Technology
Dhaka, Bangladesh

Abstract—**Human-lip detection is an important criterion for many automated modern system in present day. Like computerized speech reading, face recognition etc. system can work more precisely if human-lip can detect accurately. There are many processes for detecting human-lip. In this paper an approach is developed so that the region of a human-lip can be detected, we called it lip contour. For this a region-based Active Contour Model (ACM) is introduced with watershed segmentation. In this model we used global energy terms instead of local energy terms because, global energy gives better convergence rate for malicious environment. At the time of ACM initialization by using H∞ based on Lyapunov stability theory, the system gives more accurate and stable result.**

Keywords—Watershed Model; Active contour models (ACM); H ∞ filter Contour model; Lypunov stability theory

I. INTRODUCTION

Aut;omatic lips contour detection and tracking is always a crucial prerequisite process for various kinds of applications. It has been extensively utilized in the state-of-the art of audio-visual speech recognition. [1]

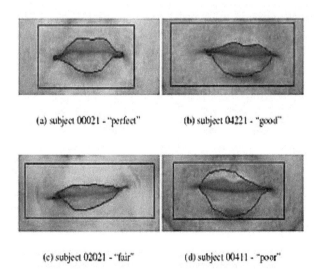

(a) subject 00021 - "perfect"   (b) subject 04221 - "good"

(c) subject 02021 - "fair"   (d) subject 00411 - "poor"

Fig. 1.   Different shape of lips contour

There are several uses of lips contour detection like Audio-visual speech authentication [3], intelligent human–computer interaction Human expression recognition [2] etc. Automatic Speech Recognition (ASR) [4] systems use only Acoustic Information, for this system show poor performance in noisy environment. Bimodal Audio-Visual Systems signal often contains information that is complementary to audio

information. Again visual information is not affected by acoustic noise. For this overall performance of the combined system is better [5]-[7].

Moreover, when working with the video sequence, an automatic and precise initial contour placement for the ACM is a crucial issue. Therefore, lips tracking algorithm is adopted into the system to keep track on the lips feature points at the subsequent incoming video frame as the initialization of the ACM lips contour process. In this paper, the conventional H∞ filtering is modified according to the LST as to assure that the system is always at the stable condition. Additionally, by properly selecting the Lyapunov function V (k), during the system design, as the time approaches to infinity, the tracking error of the proposed H ∞ filtering would asymptotically converge to zero. This is because the Lyapunov function [11] selected for the proposed system,

$$V (k) = e^2 (k)$$

This consists of a unique global minimum point at the system origin. From the simulation results demonstrated, the modified H∞ shows an appreciable improvement in terms of reducing the lips feature points tracking error compared to the conventional H∞ approach

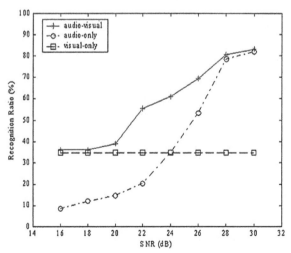

Fig. 2.   SNR ratio of different system

In this paper some of the terminology will be discussed like

A. *Contour:*

A contour line of a function of two variables is a curve along which the function has a constant value. Minimal Cost

Path: The optimal path from every pixel in the image to the seed point is determined by using Dijkstra's algorithm.[9]

B. Active contour model:

Active contour model, also called snakes, is a framework for delineating an object outline from a possibly noisy 2D image. This framework attempts to minimize an energy associated to the current contour as a sum of an internal and external energy.[12]

Fig. 3.   Active contour model.

C. H∞ filter:

After a decade or so of reappraising the nature and role of Kalman filters, engineers realized they needed a new filter that could handle system modelling errors and noise uncertainty. State estimators that can tolerate such uncertainty are called robust. The H∞ filter [8] was designed for robustness.

D. Lyapunov stability theory:

The most important type is that concerning the stability of solutions near to a point of equilibrium. If all solutions of the dynamical system that start out near an equilibrium point xe stay near xe forever, then xe is Lyapunov stable. [12]

II.   LIPS CONTOUR DETECTION & TRACKING

In this paper, instead of utilizing the global region model, the active contour energy with the local information interpretation corresponds to the watershed segmentation is proposed. This gives better performance on the contour detection compared to the global energies calculation, particularly when the inner and outer region shares the similar image statistic. Performs is better compared to the localized ACM algorithm.

Compared to the aforementioned approach, the integration of ACM and watershed segmentation presented in this paper not only provides high deformability but is also better in handling the unclear boundary between lips and facial skin.

For video sequence, an automatic and precise initial contour placement for the ACM is a crucial issue. Keep track on the lips feature points at the subsequent incoming video frame as the initialization of the ACM [6] lips contour process.

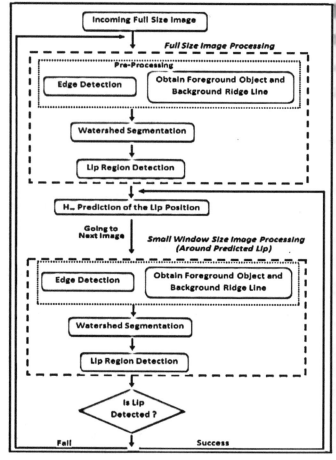

Fig. 4.   Follow chart of the system.

H∞ filtering is modified according to the LST as to assure that the system is always at the stable condition. By properly selecting the Lyapunov function, V (k), as the time approaches to infinity, the tracking error of the proposed H∞ filtering would asymptotically converge to zer This is because the Lyapunov function selected for the proposed system,

$$V(k) = e^2(k).$$

The overall process of the proposed lips contour detection and tracking is demonstrated in Figure 5. The input image is first sent to the lips localization to roughly obtain the lips location. Subsequently, the watershed region segmentation and at the same time, also the ACM initialization are working on the lips image. The output from the watershed and the ACM initial contour position are then passed to the localized region-based ACM for further lips contour detection. Once the lips contour is successfully detected, the feature points from the detected contour are extracted and used by the modified. H∞ lips tracking process to keep track on the initial contour position of the succeeding image.   The lips localization is triggered only once per 20 rounds of the overall lips detection process.

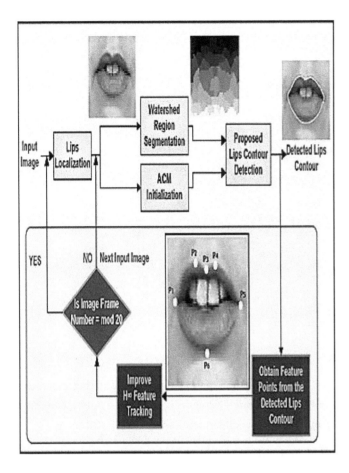

Fig. 5.   Overview of the overall system.

III.   LIPS CONTOUR DETECTION WITH LOCALIZED WATERSHED-BASED ACTIVE CONTOUR MODEL

Each pixel is situated at a certain altitude levels. Where the white pixel (intensity value = 255) is referred to as the maximum altitude while black pixel (intensity value = 0) is known as the minimum altitude.

The watershed algorithm which applied in this paper is based on the rain-flow simulation. By employing the falling rain concept, the raindrop falls from the highest altitude to the lowest (known as catchment basin) according to the steepest descent order.

The coloured watershed regions are known as the local mask region M, for a particular point within the narrow band. The local exterior region is illustrated in orange colour while the red region is known as the local interior region.

The local mask region can be mathematically construed:

$$M(x, y) = \begin{cases} 1, & \text{if} \quad \|x - y\| < w \\ 0, & \text{if} \quad \text{otherwise} \end{cases}$$

Where w is known as the segmented watershed region situated at the local point to be analyzed, and including the segmented watershed regions located at the points of north, east, south and west from the local point. The energy function,

F of an image, i(x, y) could be mathematically written as follows:

$$F(C, u, v) = \alpha \cdot Length(C) + \lambda \int_{in} |i(x, y) - u|^2 dx dy \\ + \lambda \int_{out} |i(x, y) - v|^2 dx dy$$

Where $\alpha \geq 0$, $\lambda > 0$, u and v are the average intensities levels inside and outside of the curve C, respectively.

By applying the level set approach onto the variation of the ACM, the unknown C is substituted with the unknown value φ, $\varphi(x, y)$ is taken as the signed distance function, where $\varphi(x, y) > 0$ inside the curve, $\varphi(x, y) < 0$ outside the curve, and $\varphi(x, y) = 0$ on the curve.

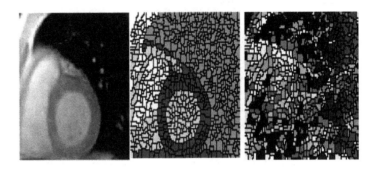

Fig. 6.   a) Original image (b) expected watershed segmentation (c) over-segmentation.

So by applying Sobel filtering & watershed segmentation the system can gain expected segmentation.

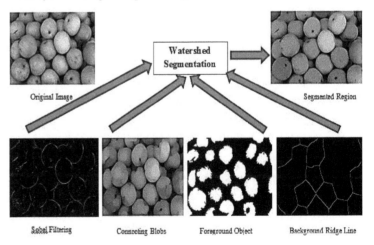

Fig. 7.   The outcomes of the pre-processing procedures.

IV.   MODIFIED H∞ BASED ON LYAPUNOV STABILITY THEORY FOR LIPS TRACKING SYSTEM

An active shape model sample grey level perpendicular to the lip contour and centered at the model points.

After obtaining the lips contour from the proposed localized ACM system as discussed in the previous section, six feature points from the detected lips region is tracked using the modified H∞ filtering. [9]

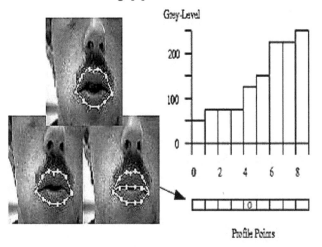

Fig. 8. Profile points & grey-level.

After obtaining the lips contour from the proposed localized ACM system as discussed in the previous section, six feature points from the detected lips region are tracked using the modified H∞ filtering which is elaborated in this section. The feature points include the right and left lip corners, the lower central point, and three points of the Cupidon's bow as depicted in Fig. 1. The six feature points (P_1 –P_6) have to be first extracted from the detected contour. P_1 and P_5 are, respectively, obtained from the right and left most of the contour. Whereas P_6 situated at the bottom of the lips contour while P_3 at the upper contour that has the similar vertical coordination as P_6. P_2, which is the left Cupidon's bow is obtained by calculating the maximum altitude between P_1 and P_3, while P_4 (the right Cupidon's bow) is the maximum altitude between P_5 and P_3.

H∞ filtering is mathematically represented according to the state-space concept. State equation:

$$X_k = AX_{k-1} + w_k$$

Where $X_k = \left[x_{k-p+1}^T\, x_{k-p+2}^T \cdots x_k^T \right]^T$, A is known as the P ×P dimensional state transition matrix, which bring forward the state value, Xk from time k to k + 1, wk is the P-dimensional model error that analogous covariance matrix is known as $Q_k = E\left[w_k w_k^T \right]$.

Observation equation: $y_k = CX_{k-1} + v_k$

Where y_k is known as the N-dimensional sequential observation vector, C is the N ×P dimensional observation function, v_k is the observation error that analogous covariance matrix is known as $R_k = E\left[v_k v_k^T \right]$.

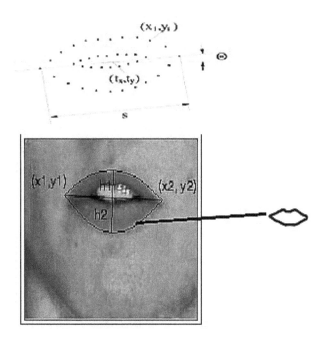

Fig. 9. Lip feature extraction and tracking.

In this paper a new gain for the H∞ filtering parameter updating rules is proposed.

The modified H∞ is interpreted as follows.

Theorem: With the given linear parameter vector, H_k and desired output, d_k, the state vector, x_k is updated as follows:

$$x_k = x_{k-1} + g_k \alpha_k$$

The modified H∞ adaptation gain that fulfills the condition of $\Delta v < 0$ is proposed as follows:

$$g_k = \left[I - \frac{F_k P_{k-1} L_k F^T e_{k-1} - P_k}{\alpha_k} e_{k-1} \right] \frac{H_k}{\|H_k\|^2}$$

where

$$L_k = I - \gamma Q P_k + H_k^T V^{-1} H_k P_k$$

$$P_k = F P_k L_k F^T + W.$$

Where γ is known as the user-defined performance bound, Q, W and V are the respective weighting matrices for the estimation error, process noise, and measurement noise. The priori estimation error, α_k, is denoted as follows:

$$\alpha_k = d_k - x_{k-1}^T H_k.$$

When the time, k, approaches infinity, the tracking error, e_k, would be asymptotically converged to the value of zero.

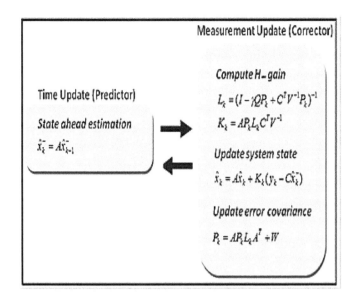

Fig. 10. The process flow of the H ∞ filter algorithm

V. FUTURE WORK

This method can be applied in 3-D system by some modification. Face recognition, emotion detection, speech recognition from a video etc. area can achieve a high percentage of accuracy by applying lips detection method.

VI. CONCLUSION

It is well known that visual information from lip shapes and movements helps improve the accuracy and robustness of a speech recognition system. Hence, the development of an accurate and robust algorithm for extracting lip features relevant to speech information becomes vital. Watershed-based active contour model and modified H∞ filtering is illustrated in this paper. Compared to the global region-based ACM, the performance of the proposed localized ACM is successfully improved. The modified H ∞ tracking approach validates improved lips feature points tracking compared to the conventional H∞ algorithm [11].

By applying Shape Constraint factor this method could be more robust. Given an image, the region of the lip may be located by using Shape Models. The overall system would be implemented into the audio-visual speech recognition for further research in the future. This system can be used to identify face or face recognition system.

But the system has High time complexity. Tracking through large number of frames point P1-P6 sometimes difficult to identify in robust situation. Sometime the skin & lip color are too similar that hard to distinguish. As a result contour detection might be tuff. For different expression like smiley, sad etc. finding accurate contour of the lip is a challenge.

VII. ACKNOWLEDGMENT

A special thanks to Prof. Dr. Mohammad Mahfuzul Islam for giving us a chance to work with this topic. Also a superior thanks goes to Siew Wen Chin, Kah Phooi Seng, Li-Minn Ang, King Hann Lim, there works really was very helpful.

REFERENCES

[1] B. J. Borgstrom and A. Alwan, "A low-complexity parabolic lip contour model with speaker normalization for high-level feature extraction in noise-robust audiovisual speech recognition," IEEE Trans. Syst. Man Cybern. Part A: Syst. Hum., vol. 38, no. 6, pp. 1273–1280, Nov. 2008.

[2] Y. Zhang, Q.-J. Liu, Y.-H. Li, and Z. Li, "Intelligent wheelchair multi-modal human-machine interfaces in lip contour extraction based on PMM," in Proc. IEEE Int. Conf. ROBIO, Dec. 2009, pp. 2108–2113.

[3] M.-I. Faraj and J. Bigun, "Audio-visual person authentication using lip-motion from orientation maps," Pattern Recognit. Lett., vol. 28, no. 11, pp. 1368–1382, Aug. 2007.

[4] J. A. Dargham, A. Chekima, and S. Omatu, "Lip detection by the use of neural networks," Artif. Life Robot., vol. 12, nos. 1–2, pp. 301–306, 2008.

[5] K. S. Jang, "Lip contour extraction based on active shape model and snakes," Int. J. Comp. Sci. Network Security, vol. 7, no. 10, pp. 148–153, Oct. 2007.

[6] T. F. Chan and L. A. Vese, "Active contours without edges," IEEE Trans. Image Process., vol. 10, no. 2, pp. 266 277, Feb. 2001.

[7] S. Lankton and A. Tannenbaum, "Localizing region-based active contours," IEEE Trans. Image Process., vol. 17, no. 11, pp. 2029–2039, Nov. 2008.

[8] N. Eveno, A. Caplier, and P.-Y. Coulon, "Accurate and quasi-automatic lip tracking," IEEE Trans. Circuits Syst. Video Technol., vol. 14, no. 5, pp. 706–715, May 2004.

[9] C. S. Wen et al., "Lips detection for audio-visual speech recognition system," in Proc. Int. Symp. ISPACS, 2009, pp. 1–4.

[10] W.-Y. Chang, C.-S. Chen, and Y.-D. Jian, "Visual tracking in highdimensional state space by appearance-guided particle filtering," IEEE Trans. Image Process., vol. 17, no. 7, pp. 1154–1167, Jul. 2008.

[11] Siew Wen Chin, Kah Phooi Seng, and Li-Minn Ang, "Lips Contour Detection and Tracking Using Watershed Region-Based Active Contour Model and Modified H∞," IEEE Transactions On Circuits And Systems For Video Technology, Vol. 22, No. 6, June 2012.

[12] N. Widynski, S. Dubuisson, and I. Bloch, "Integration of fuzzy spatial information in tracking based on particle filtering," IEEE Trans. Syst. Man Cybern. Part B: Cybern., vol. 41, no. 3, pp. 635–649, Jun. 2011.

23

Some more results on fuzzy k-competition graphs

Sovan Samanta

Department of Applied
Mathematics with Oceanology
and Computer Programming,
Vidyasagar University,
Midnapore - 721 102, India.
email: ssamantavu@gmail.com

Madhumangal Pal

Department of Applied
Mathematics with Oceanology
and Computer Programming,
Vidyasagar University,
Midnapore - 721 102, India.
email: mmpalvu@gmail.com

Anita Pal

Department of Mathematics,
National Institute of
Technology Durgapur,
Durgapur-713209, India.
e-mail: anita.buie@gmail.com

Abstract—**Fuzzy competition graph as the generalization of competition graph is introduced here. Fuzzy k-competition graph as a special type of fuzzy competition graph is defined here along with fuzzy isolated vertices. The fuzzy competition number is also introduced and investigated several properties. Isomorphism properties on fuzzy competition graphs are discussed.**

Keywords: Fuzzy graphs, fuzzy competition graphs, fuzzy k-competition graphs, fuzzy competition number.

I. INTRODUCTION

The concept of fuzzy graph was introduced by Rosenfeld [36] in 1975. Fuzzy graph theory has a vast area of applications. It is used in evaluation of human cardiac function, fuzzy neural networks, etc. Fuzzy graphs can be used to solve traffic light problem, time table scheduling, etc. In fuzzy set theory, there are different types of fuzzy graphs which may be a graph with crisp vertex set and fuzzy edge set or fuzzy vertex set and crisp edge set or fuzzy vertex set and fuzzy edge set or crisp vertices and edges with fuzzy connectivity, etc. A lot of works have been done on fuzzy graphs [5], [6], [14], [20], [22], [26].

In 1968, Cohen [12] introduced the notion of competition graphs in connection with a problem in ecology. Let $\overrightarrow{D} = (V, \overrightarrow{E})$ be a digraph, which corresponds to a food web. A vertex $x \in V(\overrightarrow{D})$ represents a species in the food web and an arc $\overrightarrow{(x, s)} \in \overrightarrow{E}(\overrightarrow{D})$ means that x preys on the species s. If two species x and y have a common prey s, they will compete for the prey s. Based on this analogy, Cohen defined a graph which represents the relations of competition among the species in the food web. The competition graph $C(\overrightarrow{D})$ of a digraph $\overrightarrow{D} = (V, \overrightarrow{E})$ is an undirected graph $G = (V, E)$ which has the same vertex set V and has an edge between two distinct vertices $x, y \in V$ if there exists a vertex $s \in V$ and arcs $\overrightarrow{(x, s)}, \overrightarrow{(y, s)} \in \overrightarrow{E}(\overrightarrow{D})$. The competition graph is also applicable in channel assignment, coding, modelling of complex economic and energy systems, etc. [35].

A lot of works have been done on competition graphs and its variations. In all these works, it is assumed that the vertices and edges of the graphs are precisely defined. But, in reality we observe that sometimes the vertices and edges of a graph can not be defined precisely. For example, in ecology, species

may be of different types like vegetarian, non-vegetarian, strong, weak, etc. Similarly in ecology, preys may be tasty, digestive, harmful, etc. The terms tasty, digestive, harmful, etc. have no precise meanings. They are fuzzy in nature and hence the species and preys may be assumed as fuzzy sets and inter-relationship between the species and preys can be designed by a fuzzy graph. This motivates the necessity of fuzzy competition graphs.

The competition graphs and fuzzy graphs are well known topics. In this article, fuzzy competition graphs are defined as motivated from fuzzy food web. Also, the generalization of it, the fuzzy k-competition graphs is introduced. Fuzzy neighbourhood graphs and their properties are investigated. Fuzzy competition number and fuzzy isolated vertex are exemplified. Isomorphism of fuzzy competition graphs is discussed.

A. Review of literature

In 1968, Cohen [12] presented a nice technique for food webs to find minimum number of dimensions of niche spaces and defined competition graphs. After Cohen's introduction of competition graph, several variations of it are found in literature. These are p-competition graphs of a digraph [16], [19], tolerance competition graphs [9], m-step competition graphs of a digraph [11], competition hypergraphs [46], common enemy graph of a digraph [21], competition-common enemy graph of a digraph [45], etc. Surveys of the large literature related to competition graphs can be found in [10], [15], [43]. All these representations are crisp in sense and do not include all real field competitions. The competition graphs describe about the common prey and related species, but they do not measure the strength of competitions. These graphs do not show that how much the species depend on a common prey compared to other species. On the other hand fuzzy graph theory, after Rosenfeld [36], is increased with a large number of branches [2], [3], [4], [27], [28], [29], [30], [32], [34], [37], [38].

To include the representations of all real world competitions, fuzzy competition graphs are introduced. The relations among neighbourhoods of any species are described in fuzzy neighbourhood graphs. Sometimes the preys are so valuable that the species (competitors) compete strongly. Different level of competitions can shown by "fuzzy k-competition graphs", a generalization of fuzzy competition graph which is introduced here.

II. Preliminaries

A. Competition graphs

A *directed graph (digraph)* $\overrightarrow{G}$ is a graph which consists of non-empty finite set $V(\overrightarrow{G})$ of elements called vertices and a finite set $\overrightarrow{E}(\overrightarrow{G})$ of ordered pairs of distinct vertices called arcs. We will often write $\overrightarrow{G} = (V, \overrightarrow{E})$. For an arc $\overrightarrow{(u, v)}$, u is the tail and v is the head. The *order (size)* of $\overrightarrow{G}$ is the number of vertices (arcs) in $\overrightarrow{G}$. The *out-neighbourhood* [17] of a vertex v is the set $N^+(v) = \{u \in V - v : \overrightarrow{(v, u)} \in \overrightarrow{E}\}$. Similarly, the *in-neighbourhood* [17] $N^-(v)$ of a vertex v is the set $\{w \in V - v : \overrightarrow{(w, v)} \in \overrightarrow{E}\}$. The *open neighbourhood* of a vertex is the union of out-neighbourhood and in-neighbourhood of the vertex. A *walk* in $\overrightarrow{G}$ is an alternating sequence $W = x_1 \overrightarrow{e_1} x_2 \overrightarrow{e_2} \ldots x_{k-1} \overrightarrow{e_k} x_k$ of vertices x_i and arcs $\overrightarrow{e_i}$ of $\overrightarrow{G}$ such that tail of $\overrightarrow{e_i}$ is x_i and head is x_{i+1} for every $i = 1, 2, \ldots, k-1$. A walk is closed if $x_1 = x_k$. A *trail* is a walk in which all arcs are distinct. A *path* is a walk in which all vertices are distinct. A path $x_1, x_2, \ldots, x_k$ with $k \geq 3$ is a cycle if $x_1 = x_k$.

For an undirected graph, *open-neighbourhood* [1] $N(x)$ of the vertex x is the set of all vertices adjacent to x in the graph. *Open neighbourhood graph* [1] $N(G)$ of G is a graph whose vertex set is same as G and has an edge between two vertices x and y in $N(G)$ if and only if $N(x) \cap N(y) \neq \phi$ in G. *Closed neighbourhood* $N[x]$ of x is the set $N(x) \cup \{x\}$. *Closed neighbourhood graph* $N[G]$ of a graph G is similarly defined, except has an edge in $N[G]$ if and only if $N[x] \cap N[y] \neq \phi$ in G. *(p)-neighbourhood graph* (read as open p-neighbourhood graph) [8], $N_p(G)$ of a graph G is a graph whose vertex set is same as G and has an edge between two vertices x and y if and only if $|N(x) \cap N(y)| \geq p$ (note that $|X|$ is the number of elements in the crisp set X) in G. Similarly *[p]-neighbourhood graph* (closed p-neighbourhood graph) $N_p[G]$ [8] is defined similar in $N_p(G)$ except there is an edge between x and y if and only if $|N[x] \cap N[y]| \geq p$ in G.

Definition 1: [12] The competition graph $C(\overrightarrow{G})$ of a digraph $\overrightarrow{G} = (V, \overrightarrow{E})$ is an undirected graph $G = (V, E)$ which has the same vertex set V and has an edge between distinct two vertices $x, y \in V$ if there exist a vertex $a \in V$ and arcs $\overrightarrow{(x, a)}, \overrightarrow{(y, a)} \in \overrightarrow{E}$ in $\overrightarrow{G}$. We say that a graph G is a competition graph if there exists a digraph $\overrightarrow{G}$ such that $C(\overrightarrow{G}) = G$.

Many variations of competition graph have been available in literature [9], [11], [16], [19], [46]. One of the important graphs, known as p-competition graphs, is defined below .

Definition 2: [19] If p is a positive integer, the p-competition graph $C_p(\overrightarrow{G})$ corresponding to the digraph $\overrightarrow{G}$ is defined to have a vertex set V with an edge between x and y in V if and only if, for some distinct vertices $a_1, a_2, \ldots, a_p$ in

V, $\overrightarrow{(x, a_1)}, \overrightarrow{(y, a_1)}, \overrightarrow{(x, a_2)}, \overrightarrow{(y, a_2)}, \ldots, \overrightarrow{(x, a_p)}, \overrightarrow{(y, a_p)}$ are arcs in $\overrightarrow{G}$.

If $\overrightarrow{G}$ is thought of as a food web whose vertices are the species in some ecosystem, (x, y) is an edge of $C_p(\overrightarrow{G})$ if and only if x and y have at least p common preys. So $C_1(\overrightarrow{G})$ is the competition graph.

B. Fuzzy graphs

A fuzzy set A on a set X is characterized by a mapping $m : X \to [0, 1]$, which is called the *membership function*. A fuzzy set is denoted by $A = (X, m)$. The *support* of A is supp $A = \{x \in X \mid m(x) \neq 0\}$. The *core* of A is the crisp set of all members whose membership values are 1. A is non trivial if supp A is non-empty. The *height* of A is $h(A) = max\{m(x) \mid x \in X\}$. A is *normal* if $h(A) = 1$. The membership function of the intersection of two fuzzy sets A and B with membership functions m_A and m_B respectively is defined as the minimum of the two individual membership functions. $m_{A \cap B} = min\{m_A, m_B\}$. We write $A = (X, m) \leq B = (X, m')$ (fuzzy subset) if $m(x) \leq m'(x)$ for all $x \in X$. The family of all fuzzy subsets is denoted by $\mathcal{F}(x)$. *The cardinality of a fuzzy set $A = (X, m)$ is a positive real number $c(A)$ or $|A|$ is the sum of membership values of the elements of X* [?]. Now the fuzzy graph is defined below.

Definition 3: [36] A fuzzy graph $\xi = (V, \sigma, \mu)$ is a non-empty set V together with a pair of functions $\sigma : V \to [0, 1]$ and $\mu : V \times V \to [0, 1]$ such that for all $x, y \in V$, $\mu(x, y) \leq \sigma(x) \wedge \sigma(y)$ and μ is a symmetric fuzzy relation on σ. Here $\sigma(x)$ and $\mu(x, y)$ represent the membership values of the vertex x and of the edge (x, y) in ξ.

Since μ is well defined, a fuzzy graph has no multiple edges. A loop at a vertex x in a fuzzy graph is represented by $\mu(x, x) \neq 0$. The fuzzy set (V, σ) is called fuzzy vertex set of ξ and the elements of the fuzzy set are called fuzzy vertices. $(V \times V, \mu)$ is called the fuzzy edge set of ξ and the elements of the fuzzy set are called fuzzy edge. An edge is non-trivial if $\mu(x, y) \neq 0$. The fuzzy graph $\xi' = (V', \tau, \nu)$ is called a *fuzzy subgraph* [24] of ξ if $\tau(x) \leq \sigma(x)$ for all $x \in V'$ and $\nu(x, y) \leq \mu(x, y)$ for all $x, y \in V'$ where $V' \subset V$.

For the fuzzy graph $\xi = (V, \sigma, \mu)$, an edge $(x, y), x, y \in V$ is called strong [13] if $\frac{1}{2} min\{\sigma(x), \sigma(y)\} \leq \mu(x, y)$ and it is called weak otherwise.

Like crisp digraphs there is fuzzy digraphs which are defined below.

Definition 4: [25] Directed fuzzy graph (fuzzy digraph) $\overrightarrow{\xi} = (V, \sigma, \overrightarrow{\mu})$ is a non-empty set V together with a pair of functions $\sigma : V \to [0, 1]$ and $\overrightarrow{\mu} : V \times V \to [0, 1]$ such that for all $x, y \in V$, $\overrightarrow{\mu}(x, y) \leq \sigma(x) \wedge \sigma(y)$.

Since $\overrightarrow{\mu}$ is well defined, a fuzzy digraph has at most two directed edges (which must have opposite directions) between any two vertices. Here $\overrightarrow{\mu}(u, v)$ is denoted by the membership value of the edge $\overrightarrow{(u, v)}$. The loop at a vertex x is represented by $\overrightarrow{\mu}(x, x) \neq 0$. Here $\overrightarrow{\mu}$ need not be symmetric as $\overrightarrow{\mu}(x, y)$ and $\overrightarrow{\mu}(y, x)$ may have different values. The *underlying crisp graph of directed fuzzy graph* is the graph similarly obtained except the directed arcs are replaced by undirected edges.

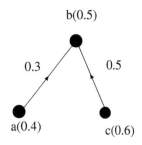

Fig. 1. Example of fuzzy out-neighbourhood and in-neighbourhood of a vertex

A *homomorphism* [27] between fuzzy graphs ξ and ξ' is a map $h : S \to S'$ which satisfies $\sigma(x) \le \sigma'(h(x))$ for all $x \in S$ and $\mu(x, y) \le \mu'(h(x), h(y))$ for all $x, y \in S$ where S is set of vertices of ξ and S' is that of ξ'.

A *weak isomorphism* [27] between fuzzy graphs is a bijective homomorphism $h : S \to S'$ which satisfies $\sigma(x) = \sigma'(h(x))$ for all $x \in S$.

A *co-weak isomorphism* [27] between fuzzy graphs is a bijective homomorphism $h : S \to S'$ which satisfies $\mu(x, y) = \mu'(h(x), h(y))$ for all $x, y \in S$.

An *isomorphism* [27] between fuzzy graphs is a bijective homomorphism $h : S \to S'$ which satisfies $\sigma(x) = \sigma'(h(x))$ for all $x \in S$ and $\mu(x, y) = \mu'(h(x), h(y))$ for all $x, y \in S$.

III. FUZZY COMPETITION GRAPHS

Now, we come to our main objective of the paper, the fuzzy competition graph. Like crisp graph, fuzzy out-neighbourhood and fuzzy in-neighbourhood of a vertex in directed fuzzy graph are defined below.

Definition 5: Fuzzy out-neighbourhood of a vertex v of a directed fuzzy graph $\overrightarrow{\xi} = (V, \sigma, \overrightarrow{\mu})$ is the fuzzy set $\mathcal{N}^+(v) = (X_v^+, m_v^+)$ where $X_v^+ = \{u | \overrightarrow{\mu}(v, u) > 0\}$ and $m_v^+ : X_v^+ \to [0, 1]$ defined by $m_v^+(u) = \overrightarrow{\mu}(v, u)$. Similarly, fuzzy in-neighbourhood of a vertex v of a directed fuzzy graph $\overrightarrow{\xi} = (V, \sigma, \overrightarrow{\mu})$ is the fuzzy set $\mathcal{N}^-(v) = (X_v^-, m_v^-)$ where $X_v^- = \{u | \overrightarrow{\mu}(u, v) > 0\}$ and $m_v^- : X_v^- \to [0, 1]$ defined by $m_v^-(u) = \overrightarrow{\mu}(u, v)$.

Example 1: Let $\overrightarrow{\xi}$ be a directed fuzzy graph. Let the vertex set be $\{a, b, c\}$ with membership values $\sigma(a) = 0.4$, $\sigma(b) = 0.5$, $\sigma(c) = 0.6$. The membership values of arcs are $\overrightarrow{\mu}(a, b) = 0.3$, $\overrightarrow{\mu}(c, b) = 0.5$. So $\mathcal{N}^+(a) = \{(b, 0.3)\}$. $\mathcal{N}^-(b) = \{(a, 0.3), (c, 0.5)\}$. (Note that $(a, \sigma(a))$ represents the vertex a with membership value $\sigma(a)$). It is shown in Figure 1.

Now, we define fuzzy competition graph.

Definition 6: The fuzzy competition graph $\mathcal{C}(\overrightarrow{\xi})$ of a fuzzy digraph $\overrightarrow{\xi} = (V, \sigma, \overrightarrow{\mu})$ is an undirected fuzzy graph $\xi = (V, \sigma, \mu)$ which has the same fuzzy vertex set as in $\overrightarrow{\xi}$ and has a fuzzy edge between two vertices $x, y \in V$ in $\mathcal{C}(\overrightarrow{\xi})$ if and only if $\mathcal{N}^+(x) \cap \mathcal{N}^+(y)$ is non-empty fuzzy set in $\overrightarrow{\xi}$. The edge membership value between x and y in $\mathcal{C}(\overrightarrow{\xi})$ is $\mu(x, y) = (\sigma(x) \wedge \sigma(y))h(\mathcal{N}^+(x) \cap \mathcal{N}^+(y))$.

Example 2: Let $\overrightarrow{\xi}$ be a directed fuzzy graph. Let the vertices with membership values of $\overrightarrow{\xi}$ be $(a, 0.3)$, $(b, 0.6)$,

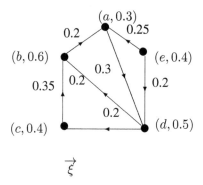

$(a) :$ Directed fuzzy graph

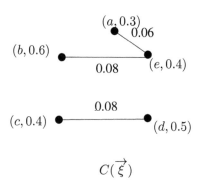

$(b) :$ Fuzzy competition graph

Fig. 2. Example of fuzzy competition graph

$(c, 0.4)$, $(d, 0.5)$, $(e, 0.4)$ with membership values of arcs be $\overrightarrow{\mu}(b, a) = 0.2$, $\overrightarrow{\mu}(c, b) = 0.35$, $\overrightarrow{\mu}(d, c) = 0.2$, $\overrightarrow{\mu}(e, d) = 0.2$, $\overrightarrow{\mu}(e, a) = 0.25$, $\overrightarrow{\mu}(a, d) = 0.3$ (shown in Figure 2(a)). The corresponding fuzzy competition graph is shown in Figure 2(b).

Edge in fuzzy competition graphs indicates that there is a competition between two vertices (species) for at least one prey. So the strengths of the edges are important to characterize the competitions.

Now an extension of fuzzy competition graph, called fuzzy k-competition graph is defined in the following.

Definition 7: Let k be a non-negative number. The fuzzy k-competition graph $\mathcal{C}_k(\overrightarrow{\xi})$ of a fuzzy digraph $\overrightarrow{\xi} = (V, \sigma, \overrightarrow{\mu})$ is an undirected fuzzy graph $\xi = (V, \sigma, \mu)$ which has the same fuzzy vertex set as $\overrightarrow{\xi}$ and has a fuzzy edge between two vertices $x, y \in V$ in $\mathcal{C}_k(\overrightarrow{\xi})$ if and only if $|\mathcal{N}^+(x) \cap \mathcal{N}^+(y)| > k$. The edge membership value between x and y in $\mathcal{C}_k(\overrightarrow{\xi})$ is $\mu(x, y) = \frac{(k'-k)}{k'}[\sigma(x) \wedge \sigma(y)]h(\mathcal{N}^+(x) \cap \mathcal{N}^+(y))$ where $k' = |\mathcal{N}^+(x) \cap \mathcal{N}^+(y)|$.

So fuzzy k-competition graph is simply fuzzy competition graph when $k = 0$. An example of fuzzy 0.15-competition graph is given below.

Example 3: Let $\overrightarrow{\xi}$ be a directed fuzzy graph (Figure 3(a)). Let vertices with membership values of $\overrightarrow{\xi}$ be $(x, 0.4)$, $(y, 0.6)$, $(a, 0.6)$, $(b, 0.7)$, $(c, 0.8)$, and the membership values of arcs be $\overrightarrow{\mu}(x, a) = 0.3$, $\overrightarrow{\mu}(x, b) = 0.35$, $\overrightarrow{\mu}(x, c) = 0.36$,

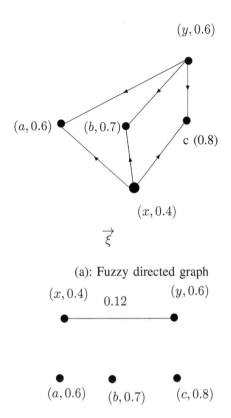

(y, 0.6)

(a, 0.6) (b, 0.7)

c (0.8)

(x, 0.4)

$\overrightarrow{\xi}$

(a): Fuzzy directed graph

(x, 0.4)   0.12   (y, 0.6)

(a, 0.6)   (b, 0.7)   (c, 0.8)

$\mathcal{C}_{0.2}(\overrightarrow{\xi})$

(b): Fuzzy 0.15-competition graph

Fig. 3. Example of fuzzy 0.15-competition graph

$\overrightarrow{\mu}(y,a) = 0.6$, $\overrightarrow{\mu}(y,b) = 0.5$, $\overrightarrow{\mu}(y,c) = 0.45$. The corresponding fuzzy 0.15-competition graph is shown in Figure 3(b).

Theorem 1: Let $\overrightarrow{\xi} = (V, \sigma, \overrightarrow{\mu})$ be a fuzzy digraph. If $\mathcal{N}^+(x) \cap \mathcal{N}^+(y)$ is singleton set, then the edge (x, y) of $C(\overrightarrow{\xi})$ is strong if and only if $|\mathcal{N}^+(x) \cap \mathcal{N}^+(y)| > 0.5$.

Proof. Here $\overrightarrow{\xi} = (V, \sigma, \overrightarrow{\mu})$ is a fuzzy digraph. Let $\mathcal{N}^+(x) \cap \mathcal{N}^+(y) = \{(a, m)\}$, where m is the membership value of the element a. Here, $|\mathcal{N}^+(x) \cap \mathcal{N}^+(y)| = m = h(\mathcal{N}^+(x) \cap \mathcal{N}^+(y))$. So, $\mu(x, y) = m \times \sigma(x) \wedge \sigma(y)$. Hence the edge (x, y) in $C(\overrightarrow{\xi})$ is strong if and only if $m > 0.5$. □

If all the edges of a fuzzy digraph are strong, then all the edges of the corresponding fuzzy competition graph may not be strong. This result is illustrated below. Let us consider two vertices x, y with $\sigma(x) = 0.3, \sigma(y) = 0.4$ in a fuzzy digraph such that the vertices have a common prey z with $\sigma(z) = 0.2$. Let $\overrightarrow{\mu}(x, z) = 0.2$, $\overrightarrow{\mu}(y, z) = 0.15$. Clearly, the edges $\overrightarrow{(x, z)}$ and $\overrightarrow{(y, z)}$ are strong. But membership value of the edge (x, y) in corresponding competition graph is $0.3 \times 0.15 = 0.045$. Hence the edge is not strong as $\frac{0.045}{0.3} = 0.15 < 0.5$. But if all the edges are strong of a fuzzy digraph, then a result can be found from the following theorem.

Theorem 2: If all the edges of a fuzzy digraph $\overrightarrow{\xi} = (V, \sigma, \overrightarrow{\mu})$ be strong, then $\frac{\mu(x,y)}{(\sigma(x) \wedge \sigma(y))^2} > 0.5$ for all edge (x, y)

in $C(\overrightarrow{\xi})$.

Proof. Let $\overrightarrow{\xi} = (V, \sigma, \overrightarrow{\mu})$ be a fuzzy digraph and every edge of $\overrightarrow{\xi}$ be strong i.e., $\frac{\overrightarrow{\mu}(x,y)}{\sigma(x) \wedge \sigma(y)} > 0.5$ for all edge (x, y) in $\overrightarrow{\xi}$. Let the corresponding fuzzy competition graph be $C(\overrightarrow{\xi}) = (V, \sigma, \mu)$.

Case 1: Let $\mathcal{N}^+(x) \cap \mathcal{N}^+(y)$ be a null set for all $x, y \in V$. Then there exist no edge in $C(\overrightarrow{\xi})$ between x and y.

Case 2: $\mathcal{N}^+(x) \cap \mathcal{N}^+(y)$ is not a null set. Let $\mathcal{N}^+(x) \cap \mathcal{N}^+(y) = \{(a_1, m_1), (a_2, m_2), \ldots, (a_z, m_z)\}$, where $m_i, i = 1, 2, \ldots, z$ are the membership values of $a_i, i = 1, 2, \ldots, z$, respectively. So $m_i = \min\{\overrightarrow{\mu}(x, a_i), \overrightarrow{\mu}(y, a_i)\}, i = 1, 2, \ldots, z$. Let $h(\mathcal{N}^+(x) \cap \mathcal{N}^+(y)) = \max\{m_i, i = 1, 2, \ldots, z\} = m_{max}$.

$\mu(x, y) = (\sigma(x) \wedge \sigma(y)) h(\mathcal{N}^+(x) \cap \mathcal{N}^+(y)) = m_{max} \times \sigma(x) \wedge \sigma(y)$. Hence $\frac{\mu(x,y)}{(\sigma(x) \wedge \sigma(y))^2} = \frac{m_{max}}{\sigma(x) \wedge \sigma(y)} > 0.5$. □

We have seen that if height of intersection between two out neighbourhoods of two vertices of a fuzzy digraph is greater than 0.5, the edge between the two vertices in corresponding fuzzy competition graph is strong. This result is not true in corresponding fuzzy k-competition graph. A related result is presented below.

Theorem 3: Let $\overrightarrow{\xi} = (V, \sigma, \overrightarrow{\mu})$ be a fuzzy digraph. If $h(\mathcal{N}^+(x) \cap \mathcal{N}^+(y)) = 1$ and $|\mathcal{N}^+(x) \cap \mathcal{N}^+(y)| > 2k$, then the edge (x, y) is strong in $C_k(\overrightarrow{\xi})$.

Proof. Let $\overrightarrow{\xi} = (V, \sigma, \overrightarrow{\mu})$ be a fuzzy digraph and $C_k(\overrightarrow{\xi}) = (V, \sigma, \mu)$ be the corresponding fuzzy k-competition graph. Also let, $h(\mathcal{N}^+(x) \cap \mathcal{N}^+(y)) = 1$ and $|\mathcal{N}^+(x) \cap \mathcal{N}^+(y)| > 2k$.

Now, $\mu(x, y) = \frac{k'-k}{k'} \sigma(x) \wedge \sigma(y) h(\mathcal{N}^+(x) \cap \mathcal{N}^+(y))$, where $k' = |\mathcal{N}^+(x) \cap \mathcal{N}^+(y)|$. So, $\mu(x, y) = \frac{k'-k}{k'} \sigma(x) \wedge \sigma(y)$. Hence $\frac{\mu(x,y)}{\sigma(x) \wedge \sigma(y)} = \frac{k'-k}{k'} > 0.5$ as $k' > 2k$. Hence the edge (x, y) is strong. □

A. *Fuzzy isolated vertex*

Isolated vertex in crisp graph is a vertex which has no incident edge i.e. a vertex of degree 0. In fuzzy graph, degree of a vertex is the sum of the membership values of incident edges. If the degree of a vertex in fuzzy graph is zero approximately, then the vertex can be assumed as isolated vertex. The formal definition of fuzzy isolated vertex is given below.

Definition 8: Let $\xi = (V, \sigma, \mu)$ be a fuzzy graph and ϵ be a pre-assigned positive real number. A vertex x is said to be fuzzy isolated vertex if the degree of x is less than ϵ i.e. $d(x) < \epsilon$.

An example of isolated vertex is given as follows.

Example 4: Let $\epsilon = 0.05$. In Fig. 4, the degree of the vertex b is $0.01 + 0.03 = 0.04$. As $d(b) < \epsilon$, b is said to be fuzzy isolated vertex. By similar reason, a and c are fuzzy isolated vertices.

The fuzzy isolated vertices are not same like crisp isolated vertices. Fuzzy isolated vertices may have an edge of small strength. The definition of strong isolated vertices are given below.

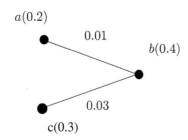

Fig. 4. An example of fuzzy isolated vertex.

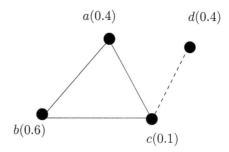

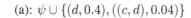

(a): $\psi \cup \{(d, 0.4), ((c, d), 0.04)\}$

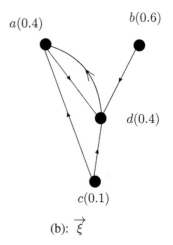

(b): $\overrightarrow{\xi}$

Fig. 5. Example of fuzzy competition number.

Definition 9: Let $\xi = (V, \sigma, \mu)$ be a fuzzy graph. A vertex $x \in V$ is said to be strong isolated vertex in ξ if $d(x) = 0$.

In fuzzy graph, vertices which are not connected to any other vertices are called strong isolated vertices. So, strong isolated vertices are comparable with crisp isolated vertices.

B. Fuzzy competition number

Fuzzy competition number is associated with a fuzzy competition graph which is defined below.

Definition 10: Let $\overrightarrow{\xi}$ be an arbitrary fuzzy digraph and $C(\overrightarrow{\xi})$ be the corresponding fuzzy competition graph. The fuzzy competition number of a fuzzy graph ψ is the minimum number of isolated fuzzy vertices such that ψ along with these isolated fuzzy vertices (say ξ) is the fuzzy competition graph of $\overrightarrow{\xi}$ i.e. $C(\overrightarrow{\xi}) = \xi$.

This concept is illustrated in the following example.

Example 5: Let $\psi = (V, \sigma, \mu)$ be a fuzzy graph (see Fig. 5(a)) with $V = \{(a, 0.4), (b, 0.6), (c, 0.1)\}$ and $E = \{((a, b), 0.2), ((b, c), 0.1), ((c, a), 0.1)\}$. Now this fuzzy graph along with an isolated fuzzy vertex $d(0.4)$ and an edge $((c, d), 0.04)$ is drawn. Let $\xi = \psi \cup \{(d, 0.4), ((c, d), 0.04)\}$. Now in Fig 5(b), a fuzzy multi-digraph $\overrightarrow{\xi}$ is shown. Clearly $C(\overrightarrow{\xi}) = \xi$. Also, d is said to be fuzzy isolated vertex if $\epsilon = 0.05$. Thus the fuzzy competition number of ψ is 1.

An important theorem related to strong isolated vertices is given below.

Theorem 4: A path of length m along with a strong isolated vertex may be a fuzzy competition graph of a fuzzy directed tree of $m + 1$ vertices.

Proof. Let $P_m = (V, \sigma, \mu)$ be a fuzzy path where $V = \{v_1, v_2, \ldots, v_m\}$. We take another vertex u which is distinct from the vertices of P_m. So, the fuzzy path together with the strong isolated vertex has $m + 1$ vertices. Now, we construct the directed fuzzy graph whose fuzzy competition graph is $P_m \cup \{u\}$. Keeping the concept of fuzzy competition graph of a fuzzy digraph, we construct the directed fuzzy edges $\overrightarrow{(v_i, u)}, i = 1, 2, \ldots, m$. Now this directed fuzzy graph is a directed fuzzy tree T_{m+1} of $m+1$ vertices. Also, $C(T_{m+1}) = P_m \cup \{u\}$. Hence the result. □

The above theorem can be extended as follows. Some disjoint paths along with a strong isolated vertex may be a fuzzy competition graph of a fuzzy directed tree. Let the disjoint paths be $P_{m_1}, P_{m_2}, \ldots, P_{m_n}$, then the number of vertices of the corresponding directed fuzzy graph is

$m_1 + m_2 + \ldots + m_n + 1$.

From Theorem 4, it is easy to observe that the fuzzy competition number of a path is 1. Like crisp graph the fuzzy competition numbers can be found as follows.

The competition number of a fuzzy chordal graph which has no strong isolated vertex is 1.

If a fuzzy graph is triangle free, then the fuzzy competition number is equal to two more than the difference between the number of edges and number of vertices.

The competition number of fuzzy complete graph is one.

C. Fuzzy neighbourhood graphs

The fuzzy open neighbourhood and fuzzy closed neighbourhood of a vertex in fuzzy graph are defined below.

Definition 11: Fuzzy open neighbourhood of a vertex v of a fuzzy graph $\xi = (V, \sigma, \mu)$ is the fuzzy set $\mathcal{N}(v) = (X_v, m_v)$ where $X_v = \{u | \mu(v, u) > 0\}$ and $m_v : X_v \to [0, 1]$ defined by $m_v(u) = \mu(v, u)$. For each vertex $v \in V$, we define fuzzy singleton set, $A_v = (\{v\}, \sigma')$ such that $\sigma' : \{v\} \to [0, 1]$ defined by $\sigma'(v) = \sigma(v)$. Fuzzy closed neighbourhood of a vertex v is $\mathcal{N}[v] = \mathcal{N}(v) \cup A_v$.

In this section, fuzzy open neighbourhood graphs are defined and then fuzzy closed neighbourhood graphs. Based on these fuzzy graphs fuzzy k-neighbourhood graphs of open and closed types are defined.

Definition 12: Let $\xi = (V, \sigma, \mu)$ be a fuzzy graph. Fuzzy open neighbourhood graph of ξ is a fuzzy graph $\mathcal{N}(\xi) = (V, \sigma, \mu')$ whose fuzzy vertex set is same as ξ and has a fuzzy edge between two vertices x and $y \in V$ in $\mathcal{N}(\xi)$ if and only if $\mathcal{N}(x) \cap \mathcal{N}(y)$ is non-empty fuzzy set in ξ and $\mu' : V \times V \to [0, 1]$ such that $\mu'(x, y) = [\sigma(x) \wedge \sigma(y)] h(\mathcal{N}(x) \cap \mathcal{N}(y))$.

Definition 13: Let $\xi = (V, \sigma, \mu)$ be a fuzzy graph. Fuzzy closed neighbourhood graph of ξ is a fuzzy graph $\mathcal{N}[\xi] = (V, \sigma, \mu')$ whose fuzzy vertex set is same as ξ and has a fuzzy edge between two vertices x and $y \in V$ in $\mathcal{N}[\xi]$ if and only if $\mathcal{N}[x] \cap \mathcal{N}[y]$ is non-empty fuzzy set in ξ and $\mu' : V \times V \to [0, 1]$ such that $\mu'(x, y) = [\sigma(x) \wedge \sigma(y)] h(\mathcal{N}[x] \cap \mathcal{N}[y])$.

Definition 14: Let $\xi = (V, \sigma, \mu)$ be a fuzzy graph. Fuzzy (k)-neighbourhood graph (read as open fuzzy k-neighbourhood graph) of ξ is a fuzzy graph $\mathcal{N}_k(\xi) = (V, \sigma, \mu')$ whose vertex set is same as ξ and has an edge between two vertices x and $y \in V$ in $\mathcal{N}_k(\xi)$ if and only if $|\mathcal{N}(x) \cap \mathcal{N}(y)| > k$ in ξ and $\mu' : V \times V \to [0, 1]$ such that $\mu'(x, y) = \frac{(k'-k)}{k'}[\sigma(x) \wedge \sigma(y)] h(\mathcal{N}(x) \cap \mathcal{N}(y))$ where $k' = |\mathcal{N}(x) \cap \mathcal{N}(y)|$.

Definition 15: Let $\xi = (V, \sigma, \mu)$ be a fuzzy graph. Fuzzy $[k]$-neighbourhood graph (read as fuzzy closed k-neighbourhood graph) of ξ is a fuzzy graph $\mathcal{N}_k[\xi] = (V, \sigma, \mu')$ whose fuzzy vertex set is same as ξ and has a fuzzy edge between two vertices x and $y \in V$ in $\mathcal{N}_k[\xi]$ if and only if $|\mathcal{N}[x] \cap \mathcal{N}[y]| > k$ in ξ and $\mu' : V \times V \to [0, 1]$ such that $\mu'(x, y) = \frac{(k'-k)}{k'}[\sigma(x) \wedge \sigma(y)] h(\mathcal{N}[x] \cap \mathcal{N}[y])$ where $k' = |\mathcal{N}[x] \cap \mathcal{N}[y]|$.

Theorem 5: For every edge of a fuzzy graph ξ, there exists one edge in $\mathcal{N}[\xi]$.

Proof. Let (x, y) be an edge of a fuzzy graph $\xi = (V, \sigma, \mu)$. Let the corresponding closed neighbourhood graph be $\mathcal{N}[\xi] = (V, \sigma, \nu)$. Then $x, y \in \mathcal{N}[x]$ and $x, y \in \mathcal{N}[y]$. So $x, y \in \mathcal{N}[x] \cap \mathcal{N}[y]$. Hence $h(\mathcal{N}[x] \cap \mathcal{N}[y]) \neq 0$. Now, $\nu(x, y) = \sigma(x) \wedge \sigma(y) h(\mathcal{N}[x] \cap \mathcal{N}[y]) \neq 0$. So for every edge (x, y) in ξ, there exists an edge (x, y) in $\mathcal{N}[\xi]$. $\square$

Definition 16: Let $\overrightarrow{\xi} = (V, \sigma, \overrightarrow{\mu})$ be a fuzzy digraph. The underlying fuzzy graph of $\overrightarrow{\xi}$ is denoted by $\mathcal{U}(\xi)$ and is defined as $\mathcal{U}(\xi) = (V, \sigma, \mu)$ where $\mu(u, v) = \min\{\overrightarrow{\mu}(u, v), \overrightarrow{\mu}(v, u)\}$ for all $u, v \in V$.

A relation between fuzzy (k)-neighbourhood graph and fuzzy k-competition graph is established below.

Theorem 6: If the symmetric fuzzy digraph $\overrightarrow{\xi}$ is loop less, $\mathcal{C}_k(\overrightarrow{\xi}) = \mathcal{N}_k(\mathcal{U}(\xi))$ where $\mathcal{U}(\xi)$ is the fuzzy graph underlying $\overrightarrow{\xi}$.

Proof. Let a directed fuzzy graph be $\overrightarrow{\xi} = (V, \sigma, \overrightarrow{\mu})$ and the corresponding underlying fuzzy graph be $\mathcal{U}(\xi) = (V, \sigma, \mu)$. Also let $\mathcal{C}_k(\overrightarrow{\xi}) = (V, \sigma, \nu)$ and $\mathcal{N}_k(\mathcal{U}(\xi)) = (V, \sigma, \nu')$. The fuzzy vertex set of $\overrightarrow{\xi}$ is equal to $\mathcal{C}_k(\overrightarrow{\xi})$. Also an underlying fuzzy graph has the same fuzzy vertex set as the directed fuzzy graph. Hence $\mathcal{N}_k(\mathcal{U}(\xi))$ has the same fuzzy vertex set as $\overrightarrow{\xi}$. Now we need to show that $\nu(x, y) = \nu'(x, y)$ for all $x, y \in V$. If $\nu(x, y) = 0$ in $\mathcal{C}_k(\overrightarrow{\xi})$ then $|\mathcal{N}^+(x) \cap \mathcal{N}^+(y)| \leq k$. As $\overrightarrow{\xi}$ is symmetric fuzzy set, $|\mathcal{N}[x] \cap \mathcal{N}[y]| \leq k$ in $\mathcal{U}(\xi)$. So $\nu'(x, y) = 0$.

If $|\mathcal{N}^+(x) \cap \mathcal{N}^+(y)| > k$ then $\nu(x, y) > 0$ in $\mathcal{C}_k(\overrightarrow{\xi})$. So $\nu(x, y) = \frac{(k'-k)}{k'}[\sigma(x) \wedge \sigma(y)] h(\mathcal{N}^+(x) \cap \mathcal{N}^+(y))$ where $k' = |\mathcal{N}^+(x) \cap \mathcal{N}^+(y)|$. As $\overrightarrow{\xi}$ is symmetric fuzzy digraph, $|\mathcal{N}[x] \cap \mathcal{N}[y]| > k$ in $\mathcal{U}(\xi)$. So $\nu' = \frac{(k''-k)}{k''}[\sigma(x) \wedge \sigma(y)] h(\mathcal{N}[x] \cap \mathcal{N}[y])$ where $k'' = |\mathcal{N}[x] \cap \mathcal{N}[y]|$. It is clear that $h(\mathcal{N}^+(x) \cap \mathcal{N}^+(y))$ in $\overrightarrow{\xi}$ is equal to $h(\mathcal{N}[x] \cap \mathcal{N}[y])$ in $\mathcal{U}(\xi)$ as $\overrightarrow{\xi}$ is symmetric. $k' = k''$ for similar reason. Hence $\nu(x, y) = \nu'(x, y)$ for all $x, y \in V$. $\square$

Similarly, a relation between fuzzy $[k]$-neighbourhood graph and fuzzy k-competition graph is established below.

Theorem 7: If the symmetric fuzzy digraph $\overrightarrow{\xi}$ has loop at every vertex, then $\mathcal{C}_k(\overrightarrow{\xi}) = \mathcal{N}_k[\mathcal{U}(\xi)]$ where $\mathcal{U}(\xi)$ is the loop less fuzzy graph underlying $\overrightarrow{\xi}$.

Proof. Let a directed fuzzy graph be $\overrightarrow{\xi} = (V, \sigma, \overrightarrow{\mu})$ and the corresponding underlying loop less graph be $\mathcal{U}(\xi) = (V, \sigma, \mu)$. Also let $\mathcal{C}_k(\overrightarrow{\xi}) = (V, \sigma, \nu)$ and $\mathcal{N}_k[\mathcal{U}(\xi)] = (V, \sigma, \nu')$. The fuzzy vertex set of $\overrightarrow{\xi}$ is equal to $\mathcal{C}_k(\overrightarrow{\xi})$. Also an underlying fuzzy graph has the same fuzzy vertex set as the directed fuzzy graph. Hence $\mathcal{N}_k[\mathcal{U}(\xi)]$ has the same fuzzy vertex set as $\overrightarrow{\xi}$. Now we need to show that $\nu(x, y) = \nu'(x, y)$ for all $x, y \in V$. As the directed fuzzy graph $\overrightarrow{\xi}$ has a loop at every vertex, out-neighbourhood of each vertex contains the vertex itself. Hence if $\nu(x, y) = 0$ in $\mathcal{C}_k(\overrightarrow{\xi})$ then $|\mathcal{N}^+(x) \cap \mathcal{N}^+(y)| \leq k$. As $\overrightarrow{\xi}$ is symmetric fuzzy set, $|\mathcal{N}(x) \cap \mathcal{N}(y)| \leq k$ in $\mathcal{U}(\xi)$. So $\nu'(x, y) = 0$.

If $|\mathcal{N}^+(x) \cap \mathcal{N}^+(y)| > k$ then $\nu(x, y) > 0$ in $\mathcal{C}_k(\overrightarrow{\xi})$. So $\nu(x, y) = \frac{(k'-k)}{k'}[\sigma(x) \wedge \sigma(y)] h(\mathcal{N}^+(x) \cap \mathcal{N}^+(y))$ where $k' = |\mathcal{N}^+(x) \cap \mathcal{N}^+(y)|$. As $\overrightarrow{\xi}$ is symmetric fuzzy digraph, $|\mathcal{N}(x) \cap \mathcal{N}(y)| > k$ in $\mathcal{U}(\xi)$. So $\nu' = \frac{(k''-k)}{k''}[\sigma(x) \wedge \sigma(y)] h(\mathcal{N}(x) \sqcap \mathcal{N}(v))$ where $k'' = |\mathcal{N}(x) \cap \mathcal{N}(y)|$. It is clear that $h(\mathcal{N}^+(x) \cap \mathcal{N}^+(y))$ in $\overrightarrow{\xi}$ is equal to $h(\mathcal{N}(x) \cap \mathcal{N}(v))$ in $\mathcal{U}(\xi)$ and $k' = k''$ as $\overrightarrow{\xi}$ is symmetric. Hence $\nu(x, y) = \nu'(x, y)$ for all $x, y \in V$. $\square$

IV. ISOMORPHISM IN FUZZY COMPETITION GRAPH

Isomorphism on fuzzy graphs are well known in literature. Here the isomorphism in fuzzy digraphs are introduced. A homomorphism of fuzzy digraphs $\overrightarrow{\xi}$ and $\overrightarrow{\xi'}$ is a map $h : S \to S'$ which satisfies $\sigma(x) \leq \sigma'(h(x))$ for all $x \in S$ and $\overrightarrow{\mu(x, y)} \leq \mu'\overrightarrow{(h(x), h(y))}$ for all $x, y \in S$ where S is the set of vertices of $\overrightarrow{\xi}$ and S' is that of $\overrightarrow{\xi'}$.

A weak isomorphism between fuzzy digraphs is a bijective homomorphism $h : S \to S'$ which satisfies $\sigma(x) = \sigma'(h(x))$ for all $x \in S$.

A co-weak isomorphism between fuzzy digraphs is a bijective homomorphism $h : S \to S'$ which satisfies $\overrightarrow{\mu(x, y)} = \mu'\overrightarrow{(h(x), h(y))}$ for all $x, y \in S$.

An isomorphism between fuzzy graphs of fuzzy digraphs is a bijective homomorphism $h : S \to S'$ which satisfies $\sigma(x) = \sigma'(h(x))$ for all $x \in S$ and $\overrightarrow{\mu(x, y)} = \mu'\overrightarrow{(h(x), h(y))}$ for all $x, y \in S$.

Isomorphism between fuzzy graphs is an equivalence relation. But, if there is an isomorphism between two fuzzy

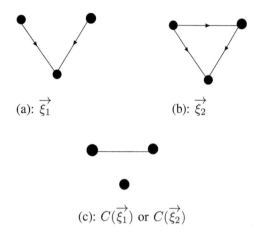

(a): $\overrightarrow{\xi_1}$ (b): $\overrightarrow{\xi_2}$

(c): $C(\overrightarrow{\xi_1})$ or $C(\overrightarrow{\xi_2})$

Fig. 6. Fuzzy competition graphs are isomorphic but corresponding digraphs are not isomorphic.

graph and one is fuzzy competition graph, then the other will be fuzzy competition graph. But, the corresponding digraphs may not be isomorphic. This result can be shown as follows.

Remark 1: Let $\overrightarrow{\xi_1}$ and $\overrightarrow{\xi_2}$ be two digraphs such that $C(\overrightarrow{\xi_1})$ and $C(\overrightarrow{\xi_2})$ are isomorphic. But $\overrightarrow{\xi_1}$ and $\overrightarrow{\xi_2}$ may not be isomorphic.

This remark can be illustrated from the following example. In Fig. 6(a) and Fig. 6(b), two fuzzy digraphs are shown. These fuzzy digraphs are not isomorphic as degree of vertices of these fuzzy digraphs are not same. But, the corresponding fuzzy competition graphs are isomorphic.

If two fuzzy digraphs are isomorphic, their corresponding fuzzy competition graphs must be isomorphic. This can established in the following theorem.

Theorem 8: If two fuzzy digraphs $\overrightarrow{\xi_1}$ and $\overrightarrow{\xi_2}$ are isomorphic, then $C(\overrightarrow{\xi_1})$ and $C(\overrightarrow{\xi_2})$ are isomorphic.

Proof. Here two fuzzy digraphs $\overrightarrow{\xi_1}$ and $\overrightarrow{\xi_2}$ are isomorphic. Then there exists one to one correspondence between vertices and edges. Also the membership values of the vertices and edges are preserved. So the adjacency of edges are preserved. Then it is easy to observe that $C(\overrightarrow{\xi_1})$ and $C(\overrightarrow{\xi_2})$ are isomorphic. □

Similarly, we can prove that if two fuzzy digraphs $\overrightarrow{\xi_1}$ and $\overrightarrow{\xi_2}$ are isomorphic, then $C_k(\overrightarrow{\xi_1})$ and $C_k(\overrightarrow{\xi_2})$ are isomorphic.

V. CONCLUSIONS

This study introduces fuzzy competition graphs and fuzzy k-competition graph. In these fuzzy graphs, if there is a common out neighbourhood of two vertices, then there will be an edge between the vertices. Thus the fuzzy competition graph can be stated as 1-step fuzzy competition graph. In future, m-step fuzzy competition graphs can be investigated as an extension of this study. Besides, fuzzy isolated vertices and fuzzy competition numbers are introduced. The competition numbers of paths and several graphs are investigated. But, fuzzy competition number of any fuzzy graphs is not calculated here. This can be investigated in future. Isomorphism in

fuzzy graphs is new in the research field. Isomorphism relation between two fuzzy competition graphs are established. Also, some fundamental results related to fuzzy competition graphs are presented. These results will be helpful in ecosystem, competitive markets, etc.

REFERENCES

[1] B. D. Acharya and M. N. Vartak, Open neighbourhood graphs, Research Report 07, IIT Bombay, 1973.

[2] M. Akram and W. A. Dudek, Interval-valued fuzzy graphs, *Computers and Mathematics with Applications*, 61(2), 289-299, 2011.

[3] M. Akram, W. A. Dudek, Intuitionistic fuzzy hypergraphs with applications, *Information Sciences*, 218, 182-193, 2013.

[4] M. Akram, Bipolar fuzzy graphs with applications, *Knowledge Based System*, 39, 1-8, 2013.

[5] K. R. Bhutani and A. Battou, On M-strong fuzzy graphs, *Information Sciences*, 155(12), 103-109, 2003.

[6] K. R. Bhutani and A. Rosenfeld, Strong arcs in fuzzy graphs, *Information Sciences*, 152, 319-322, 2003.

[7] K. R. Bhutani, J. Moderson and A. Rosenfeld, On degrees of end nodes and cut nodes in fuzzy graphs, *Iranian Journal of Fuzzy Systems*, 1(1), 57-64, 2004.

[8] R. C. Brigham and R. D. Dutton, On neighbourhood graphs, *Journal of Combinatories, Information, and System Sciences*, 12, 75-85, 1987.

[9] R.C. Brigham, F. R. McMorris and R. P. Vitray, Tolerance competition graphs, *Linear Algebra and its Application*, 217, 41-52, 1995.

[10] C. Cable, K. F. Jones, J. R. Lundgren and S. Seager, Niche graphs, *Discrete Apllied Mathematics*, 23(3), 231-241, 1989.

[11] H. H. Cho, S. R. Kim and Y. Nam, The m-step competition graph of a diagraph, *Discrete Applied Mathematics*, 105(1-3), 115-127, 2000.

[12] J. E. Cohen, Interval graphs and food webs: a finding and a problem, Document 17696-PR, RAND Corporation, Santa Monica, CA (1968).

[13] C. Eslahchi and B. N. Onaghe, Vertex Strength of Fuzzy Graphs, *International Journal of Mathematics and Mathematical Sciences*, Volume 2006, Article ID 43614, Pages 1-9, DOI 10.1155/IJMMS/2006/43614.

[14] P. Ghosh, K. Kundu and D. Sarkar, Fuzzy graph representation of a fuzzy concept lattice, *Fuzzy Sets and Systems*, 161(12), 1669-1675, 2010.

[15] M. C. Gulumbic and A. Trenk, *Tolerance Graphs*, Cambridge University Press, (2004).

[16] G. Isaak, S. R. Kim, T. A. McKee, F.R. McMorris and F. S. Roberts, 2-competition graphs, *SIAM J. Disc. Math.*, 5(4), 524-538, 1992.

[17] J. B. Jenson and G. Z. Gutin, *Digraphs: Theory, Algorithms and Applications*, Springer-verlag, 2009.

[18] S. R. Kim, Graphs with one hole and competition number one, *J. Korean Math. Soc*, 42(6), 1251-1264, 2005.

[19] S. R. Kim, T. A. McKee, F.R. McMorris and F. S. Roberts, p-competition graphs, *Linear Algebra and its Application*, 217, 167-178, 1995.

[20] L. T. Koczy, Fuzzy graphs in the evaluation and optimization of networks, *Fuzzy Sets and Systems*, 46, 307-319, 1992.

[21] J. R. Lundgren and J. S. Maybee, Food webs with interval competition graph, In Graphs and Applcations: Proceedings of the first colorado symposium on graph theory, Wiley, Newyork, 1984.

[22] S. Mathew and M.S. Sunitha, Types of arcs in a fuzzy graph, *Information Sciences*, 179, 1760-1768, 2009.

[23] S. Mathew and M.S. Sunitha, Node connectivity and arc connectivity of a fuzzy graph, *Information Sciences*, 180(4), 519-531, 2010.

[24] J. N. Mordeson and P. S. Nair, *Fuzzy graphs and hypergraphs*, Physica Verlag, 2000.

[25] J. N. Mordeson and P. S. Nair, Successor and source of (fuzzy) finite state machines and (fuzzy) directed graphs, *Information Sciences*, 95(1-2), 113-124, 1996.

[26] S. Muoz, M. T. Ortuo, J. Ramirez and J. Yez, Coloring fuzzy graphs, *Omega*, 33(3), 211-221, 2005.

[27] A. Nagoorgani and K. Radha, On regular fuzzy graphs, *Journal of Physical Sciences*, 12, 33-40, 2008.

[28] A. Nagoorgani and J. Malarvizhi, Isomorphism properties of strong fuzzy graphs, *International Journal of Algorithms, Computing and Mathematics*, 2(1), 39-47, 2009.

[29] A. Nagoorgani and P. Vadivel, Relations between the parameters of independent domination and irredundance in fuzzy graph, *International Journal of Algorithms, Computing and Mathematics*, 2(1), 15-19, 2009.

[30] A. Nagoorgani and P. Vijayalaakshmi, Insentive arc in domination of fuzzy graph, *Int. J. Contemp. Math. Sciences*, 6(26), 1303-1309, 2011.

[31] A. Nagoorgani and R. J. Hussain, Fuzzy effective distance k-dominating sets and their applications, *International Journal of Algorithms, Computing and Mathematics*, 2(3), 25-36, 2009.

[32] P. S. Nair and S. C. Cheng, Cliques and fuzzy cliques in fuzzy graphs, IFSA World Congress and 20th NAFIPS International Conference, 4, 2277 - 2280, 2001.

[33] P. S. Nair, Perfect and precisely perfect fuzzy graphs, *Fuzzy Information Processing Society*, 19-22, 2008.

[34] C. Natarajan and S. K. Ayyasawamy, On strong (weak) domination in fuzzy graphs, *World Academy of Science, Engineering and Technology*, 67, 247-249, 2010.

[35] A. Raychaudhuri and F. S. Roberts, Generalized competition graphs and their applications, *IX symposium on operations research*, Part I, Sections 14 (Osnabruck, 1984) Athenaum/Hain/Hanstein, Konigstein, Methods Oper. Res. 49,295-311, 1985.

[36] A. Rosenfeld, Fuzzy graphs, in: L.A. Zadeh, K.S. Fu, M. Shimura (Eds.), *Fuzzy Sets and Their Applications*, Academic Press, New York, 77-95, 1975.

[37] S. Samanta and M. Pal, Fuzzy tolerance graphs, *International Journal of Latest Trends in Mathematics*, 1(2), 57-67, 2011.

[38] S. Samanta and M. Pal, Fuzzy threshold graphs, *CIIT International Journal of Fuzzy Systems*, 3(12), 360-364, 2011.

[39] S. Samanta and M. Pal, Irregular bipolar fuzzy graphs, *Inernational Journal of Applications of Fuzzy Sets*, 2, 91-102, 2012.

[40] S. samanta and M. Pal, Bipolar fuzzy hypergraphs, *International Journal of Fuzzy Logic Systems*, 2(1), 17 − 28, 2012.

[41] S. Samanta, M. Pal and A. Pal, Some more results on bipolar fuzzy sets and bipolar fuzzy intersection graphs, To appear in *The Journal of Fuzzy Mathematics*.

[42] S. Samanta and M. Pal, A new approach to social networks based on fuzzy graphs, To appear in *Journal of Mass Communication and Journalism*.

[43] Y. Sano, The competition-common enemy graphs of digraphs satisfying Conditions $C(p)$ and $C'(p)$, arXiv:1006.2631v2 [math.CO], 2010.

[44] Y. Sano, Characterizations of competition multigraphs, *Discrete Applied Mathematics*, 157(13), 2978-2982, 2009.

[45] D. D. Scott, The competition-common enemy graph of a digraph, *Discrete Appl. Math.*, 17, 269-280, 1987.

[46] M. Sonnatag and H. M. Teichert, Competition hypergraphs, *Discrete Appl. Math.*, 143, 324-329, 2004.

[47] A. Somasundaram and S. Somasundaram, Domination in fuzzy graphs-1, *Pattern Recognition Letters*, 19(9), 787-791, 1998.

Locality of Chlorophyll-A Distribution in the Intensive Study Area of the Ariake Sea, Japan in Winter Seasons based on Remote Sensing Satellite Data

Kohei Arai 1

1Graduate School of Science and Engineering
Saga University
Saga City, Japan

Abstract—**Mechanism of chlorophyll-a appearance and its locality in the intensive study area of the Ariake Sea, Japan in winter seasons is clarified by using remote sensing satellite data. Through experiments with Terra and AQUA MODIS data derived chlorophyll-a concentration and truth data of chlorophyll-a concentration together with meteorological data and tidal data which are acquired for 6 years (winter 2010 to winter 2015), it is found that strong correlation between the chlorophyll-a concentration and tidal height changes. Also it is found that the relations between ocean wind speed and chlorophyll-a concentration. Meanwhile, there is a relatively high correlation between sunshine duration a day and chlorophyll-a concentration. Furthermore, it is found that there are different sources of chlorophyll-a in the three different sea areas of Ariake Sea area in the back, Isahaya bay area, and Kumamoto offshore area.**

Keywords—chlorophyl-a concentration; red tide; diatom; solar irradiance; ocean winds; tidal effect

I. Introduction

The Ariake Sea is the largest productive area of Nori (Porphyra yezoensis1) in Japan. In winters of 2012 and 2013, a massive diatom bloom occurred in the Ariake Sea, Japan [1]. In case of above red tides, bloom causative was Eucampia zodiacus2. This bloom has being occurred several coastal areas in Japan and is well reported by Nishikawa et al. for Harimanada sea areas [2]-[10]. Diatom blooms have recurrently occurred from late autumn to early spring in the coastal waters of western Japan, such as the Ariake Sea [11] and the Seto Inland Sea [12], where large scale "Nori" aquaculture occurs. Diatom blooms have caused the exhaustion of nutrients in the water column during the "Nori" harvest season. The resultant lack of nutrients has suppressed the growth of "Nori" and lowered the quality of "Nori" products due to bleaching with the damage of the order of billions of yen [3].

This bloom had been firstly developed at the eastern part of the Ariake Sea. However, as the field observation is time-consuming, information on the developing process of the red tide, and horizontal distribution of the red tide has not yet been clarified in detail. To clarify the horizontal distribution of red tide, and its temporal change, remote sensing satellite data is quite useful.

In particular in winter, almost every year, relatively large size of diatoms of *Eucampia zodiacus* appears in Ariake Sea areas. That is one of the causes for damage of *Porphyra yezoensis*. There is, therefore, a strong demand to prevent the damage from Nori farmers. Since 2007, *Asteroplanus karianus* appears in the Ariake Sea almost every year. In addition, *Eucampia zodiacus* appears in Ariake Sea since 2012. There is a strong demand on estimation of relatively large size of diatoms appearance, size and appearance mechanism).

The chlorophyll-a concentration algorithm developed for MODIS[3] has been validated [13]. The algorithm is applied to MODIS data for a trend analysis of chlorophyll-a distribution in the Ariake Sea area in winter during from 2010 to 2015 is made. Then chlorophyll-a distributions of three specific areas, Ariake Bay, Isahaya Bay and Kumamoto Offshore are compared. It is intended to confirm that the sources of the chlorophyll-a concentration are different each other of sea areas.

The major influencing factors of chlorophyll-a concentration are species, sea water temperature (sunshine duration a day), northern winds for convection of sea water, and tidal effect have to be considered. Therefore, the relations between chlorophyll-a concentration and tidal effects, ocean wind speed as well as sunshine duration a day are, then, clarified.

In the next section, the method and procedure of the experimental study is described followed by experimental data and estimated results. Then conclusion is described with some discussions.

---

[1] http://en.wikipedia.org/wiki/Porphyra

[2] http://www.eos.ubc.ca/research/phytoplankton/diatoms/centric/eucampia/e_zodiacus.html

[3] http://modis.gsfc.nasa.gov/

II. METHOD AND PROCEDURE

A. The Procedure

The procedure of the experimental study is as follows,

1) Gather MODIS data of the Ariake Sea areas together with the chlorophyll-a concentration estimation with the MODIS data,

2) Compare chlorophyll-a distribution of three different sea areas, Ariake Bay, Isahaya Bay and Kumamoto Offshore,

3) Gather the meteorological data which includes sunshine duration a day, ocean wind speed and direction, tidal heights,

4) Correlation analysis between MODIS derived chlorophyll-a concentration and geophysical parameters, ocean wind speed, sunshine duration a day, tidal heights is made.

B. The Intensive Study Areas

Fig.1 shows the intensive study areas in the Ariake Sea area, Kyushu, Japan.

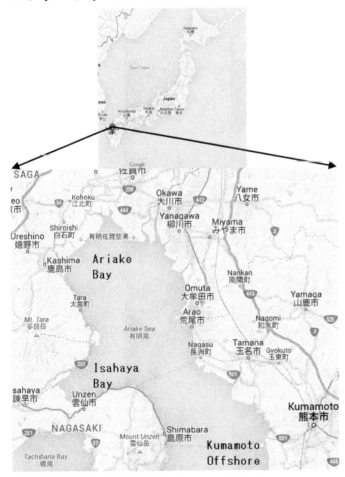

Fig. 1. Intensive study areas

III. EXPERIMENTS

A. The Data Used

MODIS derived chlorophyll-a concentrations which area acquired for the observation period of one month (January) in

2010 to 2015 are used for the experiments. Also, the meteorological data which includes sunshine duration a day, ocean wind speed and direction, tidal heights which are acquired for the same time periods as MODIS acquisitions mentioned above. In particular for 2015, two months January and February) data are used for trend analysis. Fig.2 shows the data used for two month period of time series MODIS derived chlorophyll-a concentrations in January and February, 2015. These data are acquired on January 4, 6, 7, 8, 9, 9[4], 10, 12, 17, 18, 20, 23, February 1, 3, 6, 9, 13, 14, 20, 27, and March 2 in 2015, respectively (from top left to bottom right in Fig.2). MODIS data are acquired on these days. MODIS data cannot be acquired on the rest of days due to cloudy condition. White portions in the chlorophyll-a concentration images are cloud covered areas.

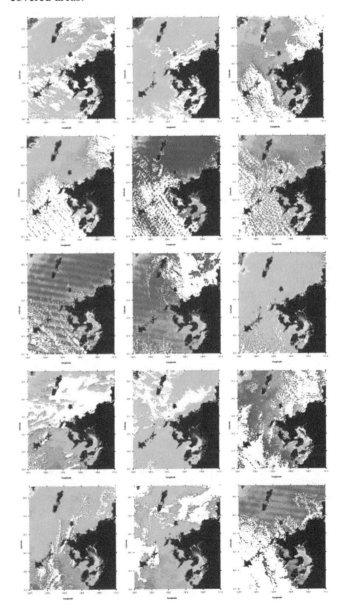

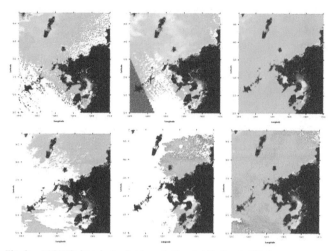

Fig. 2.   MODIS data derived chlorophyll-a distribution in 2015

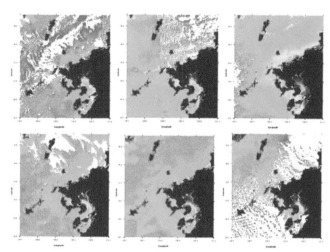

Fig. 3.   MODIS data derived chlorophyll-a concentrations in 2014

This time frame is red tide (Phytoplankton) blooming period. Such this MODIS derived chlorophyll-a concentration data are available almost every day except cloudy and rainy conditions.

Blooming is used to be occurred when the seawater becomes nutrient rich water, calm ocean winds, long sunshine duration after convection of seawater (vertical seawater current from the bottom to sea surface). Therefore, there must are relations between the geophysical parameters, ocean wind speed, sunshine duration, tidal heights and chlorophyll-a concentration.

As shown in Fig.2, it is clear that the diatom appeared at the back in the Ariake Sea, Ariake Bay and is not flown from somewhere else. Also, there is relatively low chlorophyll-a concentration sea areas between Isahaya Bay and Ariake Bay. Therefore, chlorophyll-a concentration variations are isolated each other (Isahaya Bay and Ariake Bay).

Fig.3 to Fig.7 also shows MODIS data derived chlorophyll-a concentrations in January 2014, 2013, 2012, 2011 and 2010, respectively. MODIS data are acquired on January 10, 13, 15, 16, 19, 23, 24, 26, 27, 29, 30 and February 4, respectively (from top left to bottom right in Fig.3).

Fig.4 shows the time series of MODIS data derived chlorophyll-a concentrations in 2013. MODIS data are acquired on January 4, 6, 10, 12, 15, 18, 19, 25, 28, 30, and 31, respectively (from top left to bottom right in Fig.4).

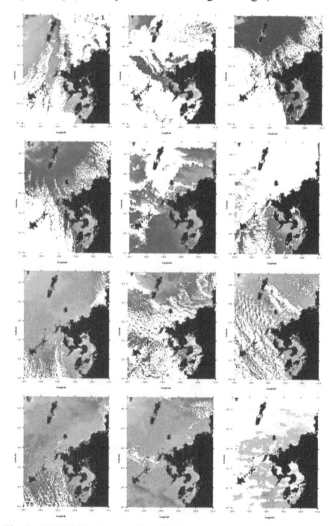

Fig. 4.   MODIS data derived chlorophyll-a concentrations in 2013

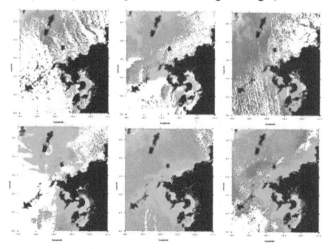

Fig.5 also shows MODIS data derived chlorophyll-a concentrations acquired in 2012. MODIS data are acquired on January 2, 6, 7, 12, 17, 20, 21, 23, 26, 29, 30, and 31, respectively (from top left to bottom right in Fig.5).

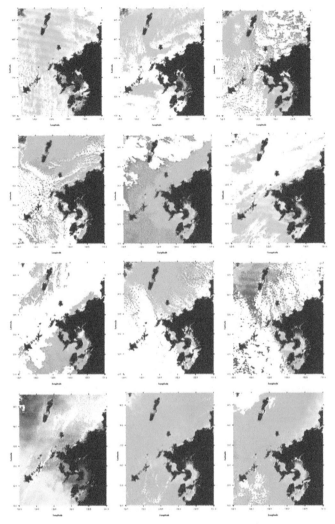

Fig. 5.   MODIS data derived chlorophyll-a concentrations in 2012

Fig.6 shows the time series of MODIS data derived chlorophyll-a concentrations in 2011. MODIS data are acquired on January 1, 2, 7, 8, 14, 17, 22, 26, and 27, respectively (from top left to bottom right in Fig.6).

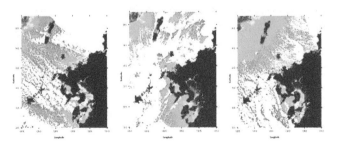

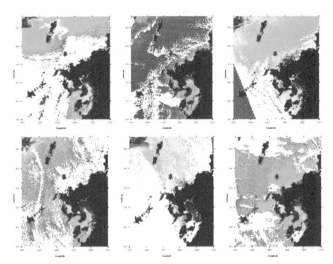

Fig. 6.   MODIS data derived chlorophyll-a concentrations in 2011

Fig.7 shows the time series of MODIS data derived chlorophyll-a concentrations in 2011. MODIS data are acquired on January 1, 2, 9, 14, 16, 17, 18, 22, 24, 26, 27, 29, respectively (from top left to bottom right in Fig.7).

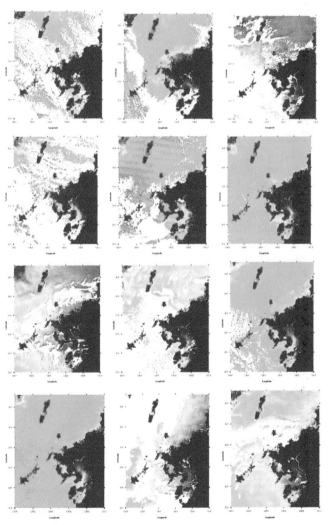

Fig. 7.   MODIS data derived chlorophyll-a concentrations in 2010

B. Trends of Chlorophyll-a Concentration in Ariake Bay Area as well as Tidal Hieght, Solar Direct Irradiation, and Wind Speed from North

Fig.8 (a) to (f) shows trends of chlorophyll-a concentration in Ariake Bay area as well as tidal height difference a day, solar direct irradiance and wind speed from the North in 2015, 2014, 2013, 2012, 2011 and 2010, respectively. Typical trend is that chlorophyll-a concentration is increased in accordance with the tidal height difference a day it is not always true though. The reason for this is the following, namely, (1) chlorophyll appears in around sea bottom because nutrition rich water is situated in around sea bottom, (2) chlorophyll moves up to sea surface due to tidal effect (from the neap to the spring tide).

This fact is not always true. For instance, chlorophyll is not increased at the spring tide (35 days in the begging of February in 2015 and 12 days in January in 2012). In such cases, wind speed from the north is relatively strong and solar irradiance is not so high. This implies that sea water is mixed up between sea surface and sea bottom due to convection caused by relatively strong wind. Also it is implied that sea surface temperature is not getting warm because solar irradiance is weak results in decreasing of chlorophyll-a concentration.

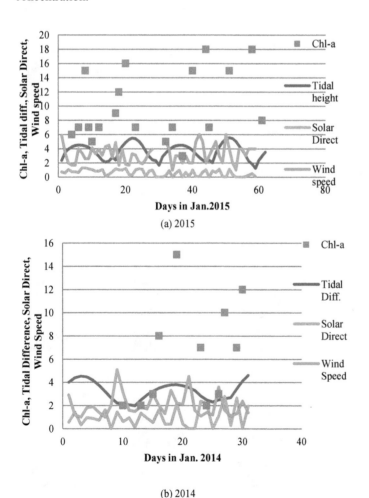

(a) 2015

(b) 2014

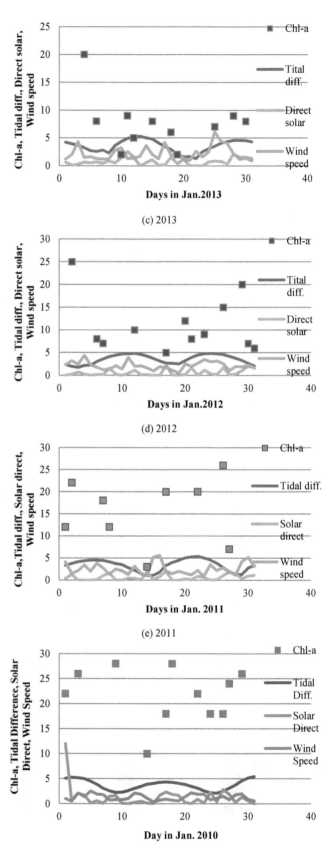

(c) 2013

(d) 2012

(e) 2011

(f) 2010

Fig. 8. Trends of chlorophyll-a concentration in Ariake Bay area as well as tidal height, solar direct irradiance and wind speed from the North

C. Locality of Chlorophyll-a Concentration in Ariake Bay, Isahaya Bay and Kumamoto Offshore Areas

In order to make sure that the sources of the chlorophyll-concentrations of the three different sea areas, Ariake Bay, Isahaya Bay and Kumamoto Offshore are different each other, trends of chlorophyll-a concentration of three sea areas are compared. Fig.9 (a) to (f) shows the calculated trends for the year of 2010 to 2015. Although these trends are very similar due to the fact that nutrition condition and weather condition are almost same in the Ariake Sea area, the chlorophyll-a concentrations of these sea areas shows somewhat different trends each other in detail. Therefore, it may say that the sources of the chlorophyll-a concentration are different.

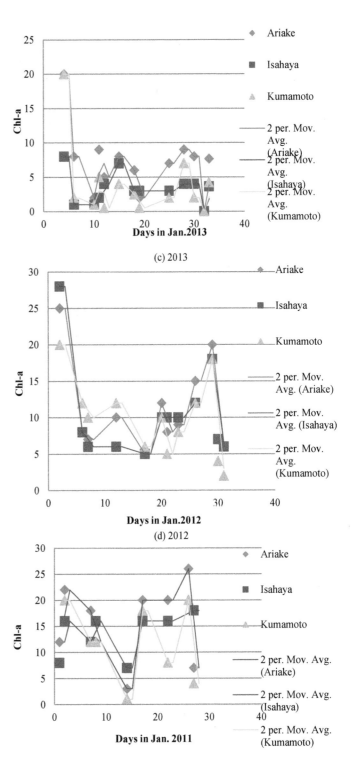

(c) 2013

(d) 2012

(e) 2011

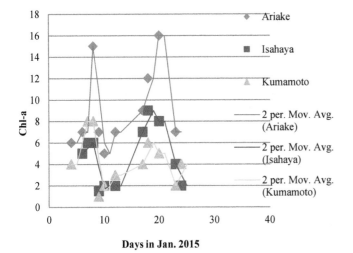

(a) 2015

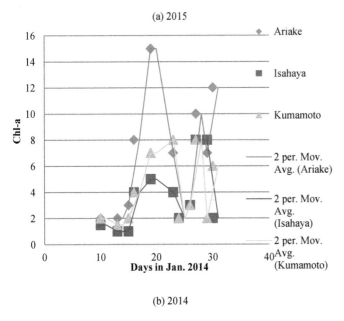

(b) 2014

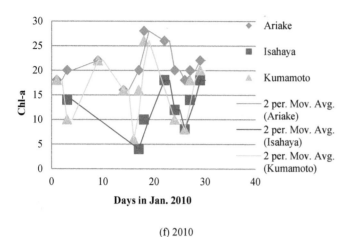

(f) 2010

Fig. 9.   Trends of chlorophyll-a concentrations at three intensive test sites

As is mentioned before, these three trends are very similar due to the fact that nutrition condition and weather condition are almost same in the Ariake Sea area. Therefore, monthly mean of chlorophyll-a concentrations of three sea areas show almost same trends as shown in Fig.10.

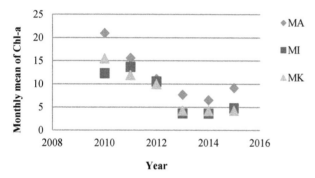

Fig. 10.   Monthly mean of chlorophyll-a concentrations of three sea areas

Also, correlations of chlorophyll-a concentration of three different sea areas are very similar as shown in Fig.11.

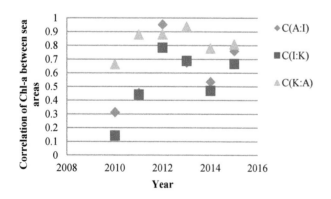

Fig. 11.   Correlations of chlorophyll-a concentrations of three different sea areas

D. Correlations Between Chlorophyll-a Concentration and Meteorological Conditions for Three Different Sea Areas

Fig.12 shows the correlations between chlorophyll-a concentration and the meteorological conditions, tidal difference a day, solar irradiance, wind speed from the north.

Correlation coefficients are calculated between chlorophyll-a concentration and the other data of tidal difference a day, sun shine time duration a day and wind speed from the north. The result shows that there is a strong relation between chlorophyll-a concentration and tidal difference a day, obviously followed by wind speed from the north.

It is not always true. The situation may change by year by year. In particular, there is clear difference between year of 2011 and the other years, 2012 to 2015. One of the specific reasons for this is due to the fact that chlorophyll-a concentration in 2011 is clearly greater than those of the other years. Therefore, clear relation between chlorophyll-a concentration and the other data of tidal difference a day, sun shine time duration a day and wind speed from the north cannot be seen. That is because of the fact that there is time delay of chlorophyll-a increasing after the nutrient rich bottom seawater is flown to the sea surface.

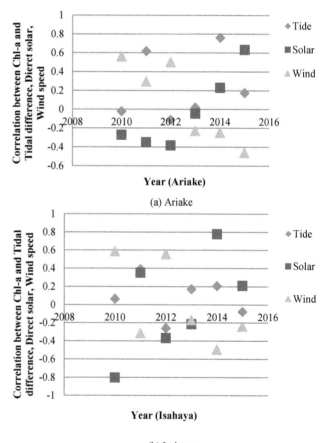

(a) Ariake

(b) Isahaya

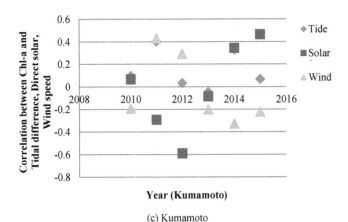

Year (Kumamoto)

(c) Kumamoto

Fig. 12. Correlations between chlorophyll-a concentration and the meteorological conditions

IV. CONCLUSION

Through experiments with Terra and AQUA MODIS data derived chlorophyll-a concentration and meteorological data and tidal data which are acquired for 6 years (winter 2010 to winter 2015), it is found that strong relation between the chlorophyll-a concentration and tidal height changes. Also it is found that the relations between ocean wind speed and chlorophyll-a concentration. Meanwhile, there is a relatively high correlation between sunshine duration a day and chlorophyll-a concentration.

It is found that strong correlation between the truth data of chlorophyll-a and MODIS derived chlorophyll-a concentrations with R square value ranges from 0.677 to 0.791. Also, there is relatively low chlorophyll-a concentration sea area between Isahaya Bay and Ariake Bay. Therefore, chlorophyll-a concentration variation are isolated between both sea areas. Meantime, chlorophyll-a concentrations of Isahaya Bay, Ariake Bay and Kumamoto Offshore are different each other. It seems that chlorophyll-a concentrations at Isahaya Bay, Kumamoto Offshore and Ariake Bay are originated from the mouth of rivers while that of Kumamoto Offshore is migrated from the south.

ACKNOWLEDGMENT

The authors would like to thank Dr. Toshiya Katano of Tokyo University of Marine Science and Technology, Dr. Yuichi Hayami, Dr. Kei Kimura, Dr. Kenji Yoshino, Dr. Naoki Fujii and Dr. Takaharu Hamada of Institute of Lowland and Marine Research, Saga University for their great supports through the experiments.

REFERENCES

[1] Yuji Ito, Toshiya Katano, Naoki Fujii, **Masumi Koriyama,** Kenji Yoshino, and Yuichi Hayami, Decreases in turbidity during neap tides initiate late winter large diatom blooms in a macrotidal embayment, Journal of Oceanography,69: 467-479. 2013.

[2] Nishikawa T (2002) Effects of temperature, salinity and irradiance on the growth of the diatom *Eucampia zodiacus* caused bleaching seaweed *Porphyra* isolated from Harima-Nada, Seto Inland Sea, Japan. Nippon Suisan Gakk 68: 356-361. (in Japanese with English abstract)

[3] Nishikawa T (2007) Occurrence of diatom blooms and damage tocultured *Porphyra* thalli by bleaching. Aquabiology 172: 405-410. (in Japanese with English abstract)

[4] Nishikawa T, Hori Y (2004) Effects of nitrogen, phosphorus and silicon on the growth of the diatom *Eucampia zodiacus* caused bleaching of seaweed *Porphyra* isolated from Harima-Nada, Seto Inland Sea, Japan. Nippon Suisan Gakk 70: 31-38. (in Japanese with English abstract)

[5] Nishikawa T, Hori Y, Nagai S, Miyahara K, Nakamura Y, Harada K, Tanda M, Manabe T, Tada K (2010) Nutrient and phytoplankton dynamics in Harima-Nada, eastern Seto Inland Sea, Japan during a 35-year period from 1973 to 2007. Estuaries Coasts 33: 417-427.

[6] Nishikawa T, Hori Y, Tanida K, Imai I (2007) Population dynamics of the harmful diatom *Eucampia zodiacus* Ehrenberg causing bleachings of *Porphyra* thalli in aquaculture in Harima- Nada, the Seto Inland Sea, Japan. Harmful algae 6: 763-773.

[7] Nishikawa T, Miyahara K, Nagai S (2000) Effects of temperature and salinity on the growth of the giant diatom *Coscinodiscus wailesii* isolated from Harima-Nada, Seto Inland Sea, Japan. Nippon Suisan Gakk 66: 993-998. (in Japanese with English abstract)

[8] Nishikawa T, Tarutani K, Yamamoto T (2009) Nitrate and phosphate uptake kinetics of the harmful diatom *Eucampia zodiacus* Ehrenberg, a causative organism in the bleaching of aquacultured *Porphyra* thalii. Harmful algae 8: 513-517.

[9] Nishikawa T, Yamaguchi M (2006) Effect of temperature on lightlimited growth of the harmful diatom *Eucampia zodiacus* Ehrenberg, a causative organism in the discoloration of *Porphyra* thalli. Harmful Algae 5: 141-147.

[10] Nishikawa T, Yamaguchi M (2008) Effect of temperature on lightlimited growth of the harmful diatom *Coscinodiscus wailesii,* a causative organism in the bleaching of aquacultured *Porphyra* thalli. Harmful Algae 7: 561-566.

[11] Syutou T, Matsubara T, Kuno K (2009) Nutrient state and nori aquaculture in Ariake Bay. Aquabiology 181: 168-170. (in Japanese with English abstract)

[12] Harada K, Hori Y, Nishikawa T, Fujiwara T (2009) Relationship between cultured *Porphyra* and nutrients in Harima-Nada, eastern part of the Seto Inland Sea. Aquabiology 181: 146-149. (in Japanese with English abstract)

[13] Arai K., T. Katano, Trend analysis of relatively large diatoms which appear in the intensive study area of the ARIAKE Sea, Japan, in winter (2011-2015) based on remote sensing satellite data, Internationa Journal of Advanced Research in Artificial Intelligence (IJARAI), 4, 7, to appear, 2015.

25

Attribute Reduction for Generalized Decision Systems*

Bi-Jun REN, Yan-Ling FU

Department of Information Engineering
Henan College of Finance and Taxation
Zhengzhou, Henan 451464, China

Ke-Yun QIN

College of Mathematics
Southwest Jiaotong University
Chengdu, Sichuan 610031, China

Abstract—Attribute reduction of information system is one of the most important applications of rough set theory. This paper focuses on generalized decision system and aims at studying positive region reduction and distribution reduction based on generalized indiscernibility relation. The judgment theorems for attribute reductions and attribute reduction approaches are presented. Our approaches improved the existed discernibility matrix and discernibility conditions. Furthermore, the reduction algorithms based on discernible degree are proposed.

Keywords—Rough set; generalized indiscernibility relation; positive region reduction; distribution reduction

I. Introduction

The theory of rough sets, proposed by Pawlak[6], is an extension of the set theory. Rough set theory has been conceived as a tool to conceptualize, organize, and analyze various types of data, in particular, to deal with inexact, uncertain or vague knowledge in applications related to artificial intelligence.

Information systems (sometimes called data tables, attribute-value systems, decision system etc.) are used for representing knowledge. A basic problem related to many practical applications of information systems is whether the whole set of attributes is always necessary to define a given partition of a universe. This problem is referred to as knowledge reduction, i.e., removing superfluous attributes from the information systems in such a way that the remaining attributes are the most informative. A large variety of approaches have been proposed in the literatures for effective and efficient reduction of knowledge. Of all paradigms, rough set theory is perhaps the most recent one making significant contribution to the field. Based on this theory and discernibility functions, some approaches for attribute reduction in complete and discrete decision systems are proposed[5,9,11,14,16].

In many practical situations, it may happen that the precise values of some of the attributes in an information system are not known, i.e. are missing or known partially. Such a system is called an incomplete information system. In order to deal with incomplete information systems, classical rough sets have been extended to several general models by using other binary relations or covers on the universe[1,2,7,8,10,15,18,19]. Based on these extended rough set models, the researchers have put forward several meaningful indiscernibility relations in incomplete information system to characterize the similarity of objects. For instance, Kryszkiewicz[3,4] introduced a kind of indiscernibility relation, called tolerance relation, to handle incomplete information tables. Stefanowski[12] introduced two generalizations of the rough sets theory to handle the missing value. The first generalization introduces the use of a non symmetric similarity relation in order to formalize the idea of absent value semantics. The second proposal is based on the use of valued tolerance relations. The tolerance relation has also been generalized to constrained similarity relation and constrained dissymmetrical similarity relation[2,13,17]. Accordingly, some attribute reduction approached for incomplete decision systems have been proposed. In this paper, an approach to attribute reduction for incomplete decision systems based on generalized indiscernibility relation is presented. Specifically, this study is not limited to a particular indiscernibility relation, but focus on the indiscernibility relation that satisfies reflexivity and symmetry. A general theory frame of attribute reduction for incomplete decision system will be presented. The paper is organized as follows: In Section 2, we recall some notions and properties of rough sets and decision systems. In Section 3, we propose an approach for positive region reduction. The reduction algorithm based on discernible degree is also presented. Section 4 is devoted to distribution reduction. The paper is completed with some concluding remarks.

II. Generalized Decision Systems

An information system is a triplet (U, A, F), where U is a nonempty finite set of objects called the universe of discourse, $A = \{a_1, \cdots, a_m\}$ is a nonempty finite set of attributes, $F = \{f_j; j \le m\}$ is a set of information functions such that $f_j(x) \in V_j$ for all $x \in U$, where V_j is the domain of attribute a_j. A decision system $(U, C \cup \{d\}, F)$ is a special case of an information system, where d is a special attribute called decision. The elements of C are called conditional attributes.

In a generalized decision system, we do not care about the information function, but focus on the indiscernibility relations generated by attributes. Concretely, a generalized decision system is a triple $S = (U, A, d)$, where U is a nonempty universe of objects, A is a set of conditional attributes, and d is a distinguished decision attribute. Each conditional attribute a determines an indiscernibility relation which is denoted by R_a.

*This work has been supported by the National Natural Science Foundation of China (Grant No. 61473239), The Key Scientific and Technological Funds (Grant No. 142102310096) of Henan, China and Soft-scientific Item (Grant No. 142400410671) of Henan, China.

In what follows we suppose that R_a is reflexive. Additionally, the decision attribute d determine a partition $U/d = \{D_1, \cdots, D_r\}$ of U. If $x \in D_i$, then we take i as the decision value of x and denoted by $d(x) = i$.

Let $S = (U, A, d)$ be a generalized decision system. For any $B \subseteq A$, the indiscernibility relation generated by B is defined as $R_B = \underset{a \in B}{\cap} R_a$. For $x \in U$, the neighborhood of x related to R_B is denoted as $R_B(x) = \{y \in U; (x,y) \in R_B\}$. Obviously, $R_B(x) = \underset{a \in B}{\cap} R_a(x)$. Additionally, because of the reflexivity of R_B, $\{R_B(x); x \in U\}$ forms a cover of U.

Definition 2.1[2,15] Let $S = (U, A, d)$ be a generalized decision system. For any $B \subseteq A$, $X \subseteq U$, the lower approximation and upper approximation of X with respect to R_B are defined as

$$\underline{R_B}(X) = \{x \in U; R_B(x) \subseteq X\} \tag{1}$$

$$\overline{R_B}(X) = \{x \in U; R_B(x) \cap X \neq \varnothing\} \tag{2}$$

Theorem 2.1[2,15] Let $S = (U, A, d)$ be a generalized decision system, $B \subseteq A$, $X, Y \subseteq U$. Then

(1) $\underline{R_B}(X) \subseteq X \subseteq \overline{R_B}(X)$.

(2) If $X \subseteq Y$, then $\underline{R_B}(X) \subseteq \underline{R_B}(Y)$, $\overline{R_B}(X) \subseteq \overline{R_B}(Y)$.

(3) $\underline{R_B}(X \cap Y) = \underline{R_B}(X) \cap \underline{R_B}(Y)$, $\overline{R_B}(X \cup Y) \subseteq \overline{R_B}(X) \cup \overline{R_B}(Y)$.

(4) $\overline{R_B}(X) = \sim \underline{R_B}(\sim X)$, $\underline{R_B}(X) = \sim \overline{R_B}(\sim X)$.

III. ATTRIBUTE REDUCTION BASED ON POSITIVE REGION

The section is devoted to the discussion of positive region reduction of generalized decision systems.

Definition 3.1[11] Let $S = (U, A, d)$ be a generalized decision system, $B \subseteq A$, $U/d = \{D_1, \cdots, D_r\}$. The positive region of d with respect to B is defined as

$$Pos_B(d) = \underset{X \in U/d}{\cup} \underline{R_B}(X) = \underset{i \leq r}{\cup} \underline{R_B}(D_i) \tag{3}$$

The above definition shows that $x \in Pos_B(d)$ if and only if the objects in $R_B(x)$ have the same decision values. Thus, $Pos_B(d)$ is the set of all elements of U that can be uniquely classified to blocks of the partition U/d by means of B. If we take B as the set of conditional attributes, then $x \in Pos_B(d)$ means the decision rule with respect to x is definite.

Theorem 3.1[9] Let $S = (U, A, d)$ be a generalized decision system, $B \subseteq A$, $U/d = \{D_1, \cdots, D_r\}$. Then

(1) $Pos_B(d) \subseteq Pos_A(d)$.

(2) $Pos_B(d) = Pos_A(d)$ if and only if $\underline{R_B}(D_i) = \underline{R_A}(D_i)$ for each $i \leq r$.

(3) $x \in Pos_B(d)$ if and only if $x \in \underline{R_B}([x]_d)$.

Definition 3.2 Let $S = (U, A, d)$ be a generalized decision system. If $B \subseteq A$ such that $Pos_B(d) = Pos_A(d)$, then B is called a positive region consistent set of S. The minimal positive region consistent set of S (with respect to set inclusion relation) is called as positive region reduction of S.

Let $S = (U, A, d)$ be a generalized decision system, $x, y \in U$. We consider the following condition $\omega(x, y)$:

$$\omega(x, y) : x \in Pos_A(d) \wedge d(x) \neq d(y).$$

We note that $\omega(x, y)$ is not symmetric to x and y.

Theorem 3.2[9] Let $S = (U, A, d)$ be a generalized decision system. If $x, y \in U$ satisfy $\omega(x, y)$, then $\alpha_A(x, y) \neq \varnothing$, where $\alpha_A(x, y) = \{a \in A; (x, y) \notin R_a\}$.

Theorem 3.3 Let $S = (U, A, d)$ be a generalized decision system, $B \subseteq A$. B is a positive region consistent set of S if and only if $B \cap \alpha_A(x, y) \neq \varnothing$ for $x, y \in U$ satisfy $\omega(x, y)$.

Proof: Suppose that B is a positive region consistent set of S and $x, y \in U$ satisfy $\omega(x, y)$. Then $d(x) \neq d(y)$ and $x \in Pos_A(d)$. By $x \in Pos_A(d) = Pos_B(d)$, we have $R_B(x) \subseteq [x]_d$. Because of $[x]_d \cap [y]_d = \varnothing$, it follows that $R_B(x) \cap [y]_d = \varnothing$, and consequently $y \notin R_B(x)$. Thus there exists $a \in B$ such that $(x, y) \notin R_a$, namely $a \in \alpha_A(x, y)$, and thus $B \cap \alpha_A(x, y) \neq \varnothing$.

Conversely, we suppose that $B \cap \alpha_A(x, y) \neq \varnothing$ for $x, y \in U$ satisfy $\omega(x, y)$. It only need to prove $Pos_A(d) \subseteq Pos_B(d)$. For any $x \in U$, if $x \notin Pos_B(d)$, then $R_B(x) \not\subset [x]_d$. Thus there exists $y \in R_B(x)$ such that $y \notin [x]_d$. By $B \cap \alpha_A(x, y) = \varnothing$, we know that x, y do not satisfy $\omega(x, y)$. It follows that $x \notin Pos_A(d)$ by $d(x) \neq d(y)$. Thus $Pos_A(d) \subseteq Pos_B(d)$ as required.

This theorem shows that, with respect to positive region reduction, x and y need to be discerned if x, y satisfy $\omega(x, y)$. In this case, we let $\vee \alpha_A(x, y) = \underset{a \in \alpha_A(x, y)}{\vee} a$ denote the disjunction of all attributes in $\alpha_A(x, y)$, where each attribute is looked upon as a Boolean variable. In what follows, $\Delta^* = \underset{(x, y) \in D^*}{\wedge} \vee \alpha_A(x, y)$ is called the positive discernibility function of S, where $D^* = \{(x, y) \in U \times U; \omega(x, y)\}$. It is noted that R_A is reflexive, therefore, D^* need not to be symmetry in general.

Theorem 3.4 Let $S = (U, A, d)$ be a generalized decision system and Δ^* be the positive discernibility function of S. If

$$\Delta^{*\prime} = (a_{11} \wedge \cdots \wedge a_{1m_1}) \vee \cdots \vee (a_{k1} \wedge \cdots \wedge a_{km_k})$$

is the reduced disjunctive form of Δ^*, then $\text{Re}d = \{T_1, \cdots, T_k\}$ is the set of all positive region reductions of S, where $T_i = \{a_{i1}, \cdots, a_{im_i}\}$ for each $i \leq k$.

Proof: (1) For any $i \leq k$, T_i is a positive region reduction of S. In fact, if there exist $x, y \in U$ such that x, y satisfy $\omega(x, y)$ and $T_i \cap \alpha_A(x, y) = \varnothing$, then we let all Boolean variable in T_i be assigned 1 and the other Boolean variables be assigned 0. It follows that $\Delta^* = 0$ because $\vee \alpha_A(x, y) = 0$ and $\Delta^{*'} = 1$ because $a_{i_1} \wedge \cdots \wedge a_{i_{m_i}} = 1$. This contradicts the fact that $\Delta^{*'}$ is the disjunctive form of Δ^*. Thus T_i is a positive region consistent set.

We suppose that there exists a proper subset $T \subset T_i$ such that $T \cap \alpha_A(x, y) \neq \varnothing$ for any $(x, y) \in D^*$. By the property of Boolean function, there exist $j \leq k$ such that $T_j \subseteq T$. It follows that $T_j \subset T_i$. This contradicts the fact that $\Delta^{*'}$ is the reduced disjunctive form of Δ^*. Thus T_i is a positive region reduction of S.

(2) We suppose that B is a reduction of S. It follows that $B \cap \alpha_A(x, y) \neq \varnothing$ for $(x, y) \in D^*$. It follows that there exist $i \leq k$ such that $T_i \subseteq B$. Because T_i is a positive region consistent set, we have $T_i \subseteq B$. Thus, $\{T_1, \cdots, T_k\}$ is just the set of all positive region reductions of S.

If R_A is reflexive and symmetric, then $\alpha_A(x, y) = \alpha_A(y, x)$ for any $x, y \in U$. Hence we have the following corollary.

Corollary 3.1 Let $S = (U, A, d)$ be a generalized decision system and $U = \{x_1, x_2, \cdots, x_n\}$. If R_a is reflexive and symmetric for any $a \in A$, then the positive discernibility function of S is

$$\Delta^* = \underset{(x,y) \in D_1^*}{\wedge} \vee \alpha_A(x, y)$$

where $D_1^* = \{(x_i, x_j); 1 \leq j < i \leq n, \omega_1(x_i, x_j)\}$, $\omega_1(x, y)$ represents the condition: $(x \in Pos_A(d) \vee y \in Pos_A(d)) \wedge d(x) \neq d(y)$.

Theorem 3.5 Let $S = (U, A, d)$ be a generalized decision system and R_a an equivalence relation for any $a \in A$. If $x \in Pos_A(d)$, $y \notin Pos_A(d)$ and $d(x) = d(y)$, then there exists $z \notin Pos_A(d)$ such that $d(x) \neq d(z)$ and $\alpha_A(x, y) = \alpha_A(x, z)$.

Proof: It is trivial that R_A is an equivalence relation on U. We use $[y]_A$ to denote $R_A(y)$. By $x \in Pos_A(d)$, $y \notin Pos_A(d)$ we have $[x]_A \subseteq [x]_d$, $[y]_A \not\subset [y]_d$. It follows that there exists $z \in [y]_A$ such that $z \notin [y]_d$. Thus $d(y) \neq d(z)$, and hence $d(x) \neq d(z)$. By $z \in [y]_A$, we have $(y, z) \in R_a$ for any $a \in A$. In consequence,

$$\alpha_A(x, y) = \{a \in A; a(x) \neq a(y)\} = \{a \in A; a(x) \neq a(z)\} = \alpha_A(x, z).$$

Furthermore, by $z \in [y]_A$, it follows that $[z]_A = [y]_A$. Thus we have $z \notin Pos_A(d)$ by $y \notin Pos_A(d)$.

Remark: Let $S = (U, A, d)$ be a decision system and R_a an equivalence relation for any $a \in A$.

Skowron[11] proposed the discernibility conditions for object pairs that need to discern with respect to positive region reduction. The discernibility conditions are

$$\omega_S(x, y): x \in Pos_A(d) \wedge y \notin Pos_A(d);$$

$$\text{or } x \notin Pos_A(d) \wedge y \in Pos_A(d);$$

$$\text{or } x \in Pos_A(d) \wedge y \in Pos_A(d) \wedge d(x) \neq d(y).$$

According to above theorem, the object pair (x, y) that satisfies $d(x) = d(y)$ do not need to discern in the criterion of positive region reduction. To be specific, Skowrons' discernibility conditions can be simplified as following:

$$\omega_1(x, y): (x \in Pos_A(d) \vee y \in Pos_A(d)) \wedge d(x) \neq d(y).$$

In essence, based on Corollary 3.1, the discernibility condition is $\omega_1(x, y)$ when the indiscernibility relation satisfies reflexivity and symmetry.

Theorem 3.4 presents an approach to calculate the positive region reductions by discernibility function. Similarly as pointed out in [11], the approach is NP hard. In the following of this section, we present a heuristic algorithm based on discernibility matrix to calculate positive region reduction.

Let $S = (U, A, d)$ be a generalized decision system, $B \subseteq A$. By Theorem 3.3, B is a positive region consistent set of S if and only if $B \cap \alpha_A(x, y) \neq \varnothing$ for $x, y \in U$ satisfy $\omega(x, y)$. It follows that D^* is the set of element pairs that needs to be discerned with respect to positive region reduction. For an attribute $a \in A$, $\{(x, y) \in D^*; a \in \alpha_A(x, y)\}$ is the set of object pairs that a can discern. Thus, the bigger the set $\{(x, y) \in D^*; a \in \alpha_A(x, y)\}$, the more possible that a is an element of a reduction. Based on this observation, we propose the notion of discernible degree.

Definition 3.3 Let $S = (U, A, d)$ be a generalized decision system, $E = \underset{(x,y) \in D^*}{\cup} \alpha_A(x, y)$. For any $a \in E$, the positive region discernible degree $\lambda(a)$ of a is defined as

$$\lambda(a) = \frac{\left| \{(x, y) \in D^*; a \in \alpha_A(x, y)\} \right|}{\left| D^* \right|},$$

where $\left| \{(x, y) \in D^*; a \in \alpha_A(x, y)\} \right|$ and $\left| D^* \right|$ are cardinalities of $\{(x, y) \in D^*; a \in \alpha_A(x, y)\}$ and D^* respectively.

Intuitively speaking, the bigger the $\lambda(a)$, the more important the attribute a. We propose the following algorithm.

Algorithm 1

1) Input the generalized decision system $S = (U, A, d)$.
2) Compute the positive region $Pos_A(d)$ of d and $\alpha_A(x, y)$ for every $(x, y) \in D^*$.
3) Place $\alpha_A(x, y)$ in discernibility matrix DM_1.
4) Compute the positive region discernible degree $\lambda(a)$ for each $a \in \underset{\alpha_A(x,y) \in DM_1}{\cup} \alpha_A(x, y)$, where

$$\lambda(a) = \frac{\left|\{\alpha_A(x,y) \in DM_1; a \in \alpha_A(x,y)\}\right|}{\left|DM_1\right|}.$$

5) *Choose a_1 such that $\lambda(a_1) = \max_{b \in E} \lambda(b)$ (If there are more than one attributes with this property, then any one of the attribute may be chosen), delete $\alpha_A(x,y)$ which contain a from discernibility matrix DM_1 to obtain DM_2.*

6) *Go back to step 3 till $DM_{i+1} = \varnothing$. Then $T = \{a_1, \cdots, a_i\}$ is a positive region reduction.*

Example 3.1 We consider the generalized decision system $S = (U,A,d)$, where $U = \{x_1, x_2, x_3, x_4\}$, $A = \{a,b,c\}$, the neighborhoods are given by:

$$R_a(x_1) = \{x_1, x_2\}, \; R_a(x_2) = \{x_2, x_3, x_4\}, \; R_a(x_3) = \{x_2, x_3\},$$

$$R_a(x_4) = \{x_3, x_4\}, \; R_b(x_1) = \{x_1, x_2, x_4\}, \; R_b(x_2) = \{x_2, x_3\},$$

$$R_b(x_3) = \{x_1, x_3, x_4\}, \; R_b(x_4) = \{x_4\}, \; R_c(x_1) = \{x_1, x_2\},$$

$$R_c(x_2) = \{x_2, x_3, x_4\}, \; R_c(x_3) = \{x_2, x_3, x_4\}, \; R_c(x_4) = \{x_4\}.$$

Furthermore, $U/d = \{D_1, D_2\}$, $D_1 = \{x_1, x_2\}$, $D_2 = \{x_3, x_4\}$. It follows that $R_A(x_1) = \{x_1, x_2\}$, $R_A(x_2) = \{x_2, x_3\}$, $R_A(x_3) = \{x_3\}$, $R_A(x_4) = \{x_4\}$. We note that R_A is reflexive, but not symmetric and transitive. By routine computation, $Pos_A(d) = \{x_1, x_3, x_4\}$,

$$DM_1 = \begin{pmatrix} & x_1 & x_2 & x_3 & x_4 \\ x_1 & & & \{a,b,c\} & \{a,c\} \\ x_2 & & & & \\ x_3 & \{a,c\} & & \{b\} & \\ x_4 & \{a,b,c\} & & \{a,b,c\} & \end{pmatrix}.$$

Thus $\lambda(a) = \frac{5}{6}$, $\lambda(b) = \frac{4}{6}$, $\lambda(c) = \frac{5}{6}$. Choose a, then

$$DM_2 = \begin{pmatrix} & x_1 & x_2 & x_3 & x_4 \\ x_1 & & & & \\ x_2 & & & & \\ x_3 & & \{b\} & & \\ x_4 & & & & \end{pmatrix},$$

and choose b, then $DM_3 = \varnothing$. Thus $T = \{a,b\}$ is a positive region reduction.

Note: If we firstly choose c, then we obtain another positive region reduction $T = \{b,c\}$.

IV. DISTRIBUTION REDUCTIONS FOR GENERALIZED DECISION SYSTEMS

Kryszkiewicz[3] proposed an rough set approach to incomplete information systems where the indiscernibility relation is a tolerance relation (reflexive and symmetric relation). In this section, we generalized the approach to generalized decision systems.

Let $S = (U,A,d)$ be a generalized decision system, $B \subseteq A$. We define $d_B : U \to P(V_d)$ as

$$d_B(x) = d(R_B(x)) = \{d(y); y \in R_B(x)\}.$$

Namely, $d_B(x)$ is the set of d attribute values of objects in $R_B(x)$. The mapping d_B is called decision function determined by B.

Definition 4.1 Let $S = (U,A,d)$ be a generalized decision system, $B \subseteq A$. If $d_B = d_A$, then B is called a distribution consistent set of S, and the minimal distribution consistent set of S (with respect to set inclusion relation) is called a distribution reduction of S.

Theorem 4.1[11] Let $S = (U,A,d)$ be a generalized decision system, $B \subseteq A$, $U/d = \{D_1, \cdots, D_r\}$. Then B is a distribution consistent set if and only if $\overline{R_B}(D_i) = \overline{R_A}(D_i)$ for each $i \le r$.

Theorem 4.2 Let $S = (U,A,d)$ be a generalized decision system, $B \subseteq A$. Then B is a distribution consistent set if and only if $B \cap \alpha_A(x,y) \neq \varnothing$ for any $(x,y) \in D_2^*$, where $D_2^* = \{(x,y); d(y) \notin d_A(x)\}$.

Proof: Necessity: Notice that when $(x,y) \in D_2^*$, we have $d(y) \notin d_A(x)$ and hence $y \notin R_A(x)$, $\alpha_A(x,y) \neq \varnothing$. Let $d_B = d_A$ and $(x,y) \in D_2^*$. By $d(y) \notin d_A(x)$ it follows that $d(y) \notin d_B(x)$. Thus $R_B(x) \cap [y]_d = \varnothing$, and $y \notin R_B(x)$. It follows that there exists $b \in B$ such that $(x,y) \notin R_b$, namely, $B \cap \alpha_A(x,y) \neq \varnothing$.

Sufficiency: For any $x \in U$, we have $d_A(x) \subseteq d_B(x)$. Suppose that u is a decision value of d, $u \notin d_A(x)$ and $d(y) = u$. For any $z \in [y]_d$, it follows that $d(z) = d(y) \notin d_A(x)$, and hence $(x,z) \in D_2^*$. Consequently, we have $B \cap \alpha_A(x,z) \neq \varnothing$, namely, there exists $b \in B$ such that $(x,z) \notin R_b$, and thus $z \notin R_B(x)$. It follows that $R_B(x) \cap [y]_d = \varnothing$, and in consequence $u = d(y) \notin d(R_B(x)) = d_B(x)$. Thus $d_B(x) \subseteq d_A(x)$. It follows that $d_B = d_A$ and B is a distribution consistent set of S as required.

Let $S = (U,A,d)$ be a generalized decision system. In what follows, $\Delta^0 = \underset{(x,y) \in D_2^*}{\wedge} \vee \alpha_A(x,y)$ is called the distribution discernibility function of S.

Corollary 4.1 Let $S = (U,A,d)$ be a generalized decision system. If $\Delta^{0*} = (a_{11} \wedge \cdots \wedge a_{1m_1}) \vee \cdots \vee (a_{k1} \wedge \cdots \wedge a_{km_k})$ is the reduced disjunctive form of Δ^0, then $\mathrm{Re}d = \{T_1, \cdots, T_k\}$ is the set of all distribution reductions of S, where $T_i = \{a_{i1}, \cdots, a_{im_i}\}$ for each $i \le k$.

Theorem 4.2 and Corollary 4.1 show the method of distribution reduction based on generalized indiscernibility relation, which only satisfies reflexivity. Obviously, the methods improve the conclusion of literature. Similarly, We propose the following algorithm to compute distribution reduction.

Algorithm 2

1) *Input the generalized decision system $S = (U,A,d)$.*
2) *Compute D_2^* and $\alpha_A(x,y)$ for every $(x,y) \in D_2^*$.*
3) *Place $\alpha_A(x,y)$ in discernibility matrix DM_1^*.*

4) *Compute the distribution discernible degree* $\lambda(a)$ *for each* $a \in \underset{\alpha_A(x,y) \in DM_1^*}{\cup} \alpha_A(x,y)$, *where*

$$\lambda(a) = \frac{\left|\{\alpha_A(x,y) \in DM_1^*; a \in \alpha_A(x,y)\}\right|}{\left|DM_1^*\right|}.$$

5) *Choose* a_1 *such that* $\lambda(a_1) = \underset{b \in E}{\max} \lambda(b)$ *(If there are more than one attributes with this property, then any one of the attribute may be chosen), delete* $\alpha_A(x,y)$ *which contain a from discernibility matrix* DM_1^* *to obtain* DM_2^*.

6) *Go back to step 3 till* $DM_{i+1}^* = \varnothing$. *Then* $T = \{a_1, \cdots, a_i\}$ *is a distribution reduction.*

The following theorem shows the connection between the concepts of distribution reduction and positive region reduction.

Theorem 4.3 Let $S = (U, A, d)$ be a generalized decision system, $B \subseteq A$, $U/d = \{D_1, \cdots, D_r\}$. If B is a distribution consistent set, then B is a positive region consistent set.

Proof: We suppose that B is a distribution consistent set. It follows that $\overline{R_B}(D_i) = \overline{R_A}(D_i)$ for each $i \le r$. Thus

$$\underline{R_B}(D_i) = \sim \overline{R_B}(\sim D_i) = \sim \overline{R_B}(\underset{j \neq i}{\cup} D_j) = \sim \underset{j \neq i}{\cup} \overline{R_B}(D_j)$$

$$= \sim \underset{j \neq i}{\cup} \overline{R_A}(D_j) = \sim \overline{R_A}(\underset{j \neq i}{\cup} D_j) = \sim \overline{R_A}(\sim D_i) = \underline{R_A}(D_i).$$

Consequently, B is a positive region consistent set.

V. CONCLUSIONS

Rough set under incomplete information has been extensively studied. Researchers have put forward several similarity relations on objects and some attribute reduction approaches for incomplete information systems. This paper is devoted to the study of positive region reduction and distribution reduction based on generalized indiscernibility relation.

The judgment theorems for positive region reduction and distribution reduction of generalized decision systems and attribute reduction approaches are presented. Furthermore, the reduction algorithms based on discernible degree are proposed. Based on this work, we can further probe the rough set model under incomplete information and its application in knowledge discovery.

REFERENCES

[1] Z.Bonikowski, E.Bryniarski, U.Wybraniec, "Extensions and intentions in the rough set theory," Information Sciences, vol. 107, pp. 149-167, 1998.

[2] L.H.Guan, G.Y.Wang, Generalized approximations defined by non-equivalence relations, Information Sciences, vol. 193, pp. 163-179, 2012.

[3] M.Kryszkiewicz, Rough set approach to incomplete information system, Information Sciences, vol. 112, pp. 39-49, 1998.

[4] M.Kryszkiewicz, Properties of incomplete information systems in the framework of rough sets, Rough Sets in Data Mining and Knowledge Discovery, Physica-Verlag, 1998, pp. 422-450.

[5] K.Marzena K, Comparative study of alternative types of knowledge reduction in inconsistent systems, International Journal of Intelligent Systems, vol. 16, pp. 105-120, 2001.

[6] Z.Pawlak, Rough sets, Int. J. Computer and Information Sci., vol. 11, pp. 341-356, 1982.

[7] Z.Pawlak, A.Skowron, Rough sets: Some extensions, Information Sciences, vol. 177, pp. 28-40, 2007.

[8] K.Qin, Z.Pei, J.Yang, Y.Xu, Approximation operators on complete completely distributive lattices, Information Sciences, vol. 247, pp. 123-130, 2013

[9] K.Qin, H.Zhao, Z.Pei, The reduction of decision table based on generalized indiscernibility relation, Journal of Xihua University, vol.32(4), pp.1-4, 2013

[10] A.M.Radzikowska, E.E.Kerre, A comparative study of fuzzy rough sets, Fuzzy Sets and Systems, vol. 126, pp.137-155, 2002

[11] A.Skowron, C.Rauszer, The discernibility matrices and functions in information systems, In: R. Slowinski (Ed.), Intelligent Decision Support-Handbook of Applications and Advances of the Rough Sets Theory, Kluwer Academic Publishers, London, pp.331-362, 1992.

[12] J.Stefanowski, A.Tsoukias, Incomplete information tables and rough classification, Computational Intelligence, 17(3)2001, pp. 545-566

[13] G.Y.Wang, Extension of rough set under incomplete information systems, Journal of Computer Research and Development (in Chinese), vol. 39, pp. 1238-1243, 2002.

[14] G.Y.Wang, H.Yu, D.Yang, Decision table reduction based on conditional information entropy, Chinese Journal of Computers(in Chinese), vol. 25, pp. 759-766, 2002.

[15] Y.Y.Yao, Relational interpretation of neighborhood operators and rough set approximation operator, Information Sciences, 111, pp.239-259, 1998.

[16] Y.Y.Yao, Y.Zhao, Discernibility matrix simplification for constructing attribute reducts, Information Sciences, 179, pp.867-882, 2009.

[17] X.Yin, X.Jia, L.Shang, A new extension model of rough sets under incomplete information, Lecture Notes in Artificial Intelligence, 4062, pp. 141-146, 2006.

[18] X.H.Zhang, B.Zhou, P.Li, A general frame for intuitionistic fuzzy rough sets, Information Sciences, 216, pp.34-49, 2012.

[19] X.H.Zhang, J.H.Dai, Y.C.Yu, On the union and intersection operations of rough sets based on various approximation spaces, Information Sciences, 292, pp.214-229, 2015.

26

A Trust-based Mechanism for Avoiding Liars in Referring of Reputation in Multiagent System

Manh Hung Nguyen

Posts and Telecommunications Institute of Technology (PTIT)
Hanoi, Vietnam
UMI UMMISCO 209 (IRD/UPMC), Hanoi, Vietnam

Dinh Que Tran

Posts and Telecommunications Institute of Technology (PTIT)
Hanoi, Vietnam

Abstract—**Trust is considered as the crucial factor for agents in decision making to choose the most trustworthy partner during their interaction in open distributed multiagent systems. Most current trust models are the combination of experience trust and reference trust, in which the reference trust is estimated from the judgements of agents in the community about a given partner. These models are based on the assumption that all agents are reliable when they share their judgements about a given partner to the others. However, these models are no more longer appropriate to applications of multiagent systems, where several concurrent agents may not be ready to share their private judgement about others or may share the wrong data by lying to their partners.**

In this paper, we introduce a combination model of experience trust and experience trust with a mechanism to enable agents take into account the trustworthiness of referees when they refer their judgement about a given partner. We conduct experiments to evaluate the proposed model in the context of the e-commerce environment. Our research results suggest that it is better to take into account the trustworthiness of referees when they share their judgement about partners. The experimental results also indicate that although there are liars in the multiagent systems, combination trust computation is better than the trust computation based only on the experience trust of agents.

Keywords—Multiagent system, Trust, Reputation, Liar.

I. INTRODUCTION

Many software applications are open distributed systems whose components are decentralized, constantly changed, and spread throughout network. For example, peer-to-peer networks, semantic web, social network, recommender systems in e-business, autonomic and pervasive computing are among such systems. These systems may be modeled as open distributed multiagents in which autonomous agents often interact with each other according to some communication mechanisms and protocols. The problem of how agents decide with whom and when to interact has become the active research topic in the recent years. It means that they need to deal with degrees of uncertainty in making decisions during their interaction. Trust among agents is considered as one of the most important foundations based on which agents decide to interact with each other. Thus, the problem of how do agents decide to interact may reduce to the one of how do agents estimate their trust on their partners. The more trust an agent commits on a partner, the more possibility with such partner he decides to interact.

Trust has been defined in many different ways by researchers from various points of view [7], [15]. It has been being an active research topic in various areas of computer science, such as security and access control in computer networks, reliability in distributed systems, game theory and multiagent systems, and policies for decision making under uncertainty. From the computational point of view, trust is defined as a quantified belief by a truster with respect to the competence, honesty, security and dependability of a trustee within a specified context [8].

These current models utilize the combination of experience trust (confidence) and reference trust (reputation) in some way. However, most of them are based on the assumption that all agents are reliable when they share their private trust about a given partner to others. This constraint limits the application scale of these models in multiagent systems including concurrent agents, in which many agents may not be ready to share with each other about their private trust about partners or even share the wrong data by lying to their opponents.

Considering a scenario of the following e-commerce application. There are two concurrent sellers S_1 and S_2 who sell the same product x. An independent third party site w is to collect the consumer's opinions. All clients could submit their opinions about sellers. In this case, the site w could be considered as a reputation channel for clients. It means that a client could refer the given opinions on the site w to select the best seller. However, since the site w is a public reputation and all clients could submit their opinions. Imagining that S_1 is really trustworthy, but S_2 is not fair, some of its employments intentionally submit some negative opinions about the seller S_1 in order to attract more clients to them. In this case, how a client could trust on the reputation given by the site w? These proposed models of trust may not be applicable to such a situation.

In order to get over this limitation, our work proposes a novel computational model of trust that is a weighted combination of experience trust and reference trust. This model offers a mechanism to enable agents take into account the trustworthiness of referees when they refer the the judgement about a given partner from these referees. The model is evaluated experimentally on two issues in the context of the e-commerce environment: (i) It is whether necessary to take into account the trust of referees (in sharing their private trust about partners) or not; (ii) Combination of experience trust and reputation is more useful than the trust based only on the experience trust of agents in multiagent systems with liars.

The rest of paper is organized as follows. Section II presents some related works in literature. Section III describes the model of weighted combination trust of experience trust, reference trust with and without lying referees. Section IV describes the experimental evaluation of the model. Section V is offered to some discussion. Section VI is the conclusion and the future works.

II. RELATED WORKS

By basing on the contribution factors of each model, we try to divide the proposed models into three groups. Firstly, The models are based on *personal experiences* that a truster has on some trustee after their transactions performed in the past. For instance, Manchala [19] and Nefti et al. [20] proposed models for the trust measure in e-commerce based on fuzzy computation with parameters such as cost of a transaction, transaction history, customer loyalty, indemnity and spending patterns. The probability theory-based model of Schillo et al. [28] is intended for scenarios where the result of an interaction between two agents is a boolean impression such as good or bad but without degrees of satisfaction. Shibata et al. [30] used a mechanism for determining the confidence level based on agent's experience with Sugarscape model, which is artificially intelligent agent-based social simulation. Alam et al. [1] calculated trust based on the relationship of stake holders with objects in security management. Li and Gui [18] proposed a reputation model based on human cognitive psychology and the concept of direct trust tree (DTT).

Secondly, the models combine both personal experience and reference trusts. In the trust model proposed by Esfandiari and Chandrasekharan [4], two one-on-one trust acquisition mechanisms are proposed. In Sen and Sajja's [29] reputation model, both types of direct experiences are considered: direct interaction and observed interaction. The main idea behind the reputation model presented by Carter et al. [3] is that "the reputation of an agent is based on the degree of fulfillment of roles ascribed to it by the society". Sabater and Sierra [26], [27] introduced ReGreT, a modular trust and reputation system oriented to complex small/mid-size e-commerce environments where social relations among individuals play an important role. In the model proposed by Singh and colleagues [36], [37] the information stored by an agent about direct interactions is a set of values that reflect the quality of these interactions. Ramchurn et al. [24] developed a trust model, based on confidence and reputation, and show how it can be concretely applied, using fuzzy sets, to guide agents in evaluating past interactions and in establishing new contracts with one another. Jennings et colleagues [12], [13], [25] presented FIRE, a trust and reputation model that integrates a number of information sources to produce a comprehensive assessment of an agent's likely performance in open systems. Nguyen and Tran [22], [23] introduced a computational model of trust, which is also combination of experience and reference trust by using fuzzy computational techniques and weighted aggregation operators. Victor et al. [33] advocate the use of a trust model in which trust scores are (trust, distrust)-couples, drawn from a bilattice that preserves valuable trust provenance information including gradual trust, distrust, ignorance, and inconsistency. Katz and

Golbeck [16] introduces a definition of trust suitable for use in Web-based social networks with a discussion of the properties that will influence its use in computation. Hang et al. [10] describes a new algebraic approach, shows some theoretical properties of it, and empirically evaluates it on two social network datasets. Guha et al. [9] develop a framework of trust propagation schemes, each of which may be appropriate in certain circumstances, and evaluate the schemes on a large trust network. Vogiatzis et al. [34] propose a probabilistic framework that models agent interactions as a Hidden Markov Model. Burnett et al. [2] describes a new approach, inspired by theories of human organisational behaviour, whereby agents generalise their experiences with known partners as stereotypes and apply these when evaluating new and unknown partners. Hermoso et al. [11] present a coordination artifact which can be used by agents in an open multi-agent system to take more informed decisions regarding partner selection, and thus to improve their individual utilities.

Thirdly, the models also compute trust by means of combination of the experience and reputation, but consider unfair agents in sharing their trust in the system as well. For instances, Whitby et al. [35] described a statistical filtering technique for excluding unfair ratings based on the idea that unfair ratings have some statistical pattern being different from fair ratings. Teacy et al. [31], [32] developed TRAVOS (Trust and Reputation model for Agent-based Virtual OrganisationS) which models an agent's trust in an interaction partner, using probability theory taking account of past interactions between agents, and the reputation information gathered from third parties. And HABIT, a Hierarchical And Bayesian Inferred Trust model for assessing how much an agent should trust its peers based on direct and third party information. Zhang, Robin and collegues [39], [14], [5], [6] proposed an approach for handling unfair ratings in an enhanced centralized reputation system.

The models in the third group are closed to our model. However, most of them used Bayes network and statistical method to detect the unfairs in the system. This approach may result in difficulty when the number of unfair agents become major.

This paper is a continuation of our previous work [21] in order to update our approach and perform experimental evaluation of this model.

III. COMPUTATIONAL MODEL OF TRUST

Let $A = \{1, 2, ...n\}$ be a set of agents in the system. Assume that agent i is considering the trust about agent j. We call j is a *partner* of agent i. This consideration includes: (i) the direct trust betwwen agent i and agent j, called *experiment trust* E_{ij}; and (ii) the trust about j refered from community called *reference trust (or reputation)* R_{ij}. Each agent l in the community that agent i refers for the trust of partner j is called a *referee*. This model enables agent i to take into account the trustworthiness of referee l when agent l shares its private trust (judgement) about agent j. The trustworthiness of agent l on the point of view of agent i, in sharing its private trust about partners, is called a *referee trust* S_{il}. We also denote T_{ij} to be the overall trust that agent i obtains on agent j. The following sections will describe a computational model to estimate the values of E_{ij}, S_{il}, R_{ij} and T_{ij}.

TABLE I: Summary of recent proposed models regarding the fact of avoiding liar in calulation of reputation

Models	Experience Trust	Reputation	Liar Judger
Alam et al. [1]	✓		
Burnett et al. [2]	✓	✓	
Esfandiari and Chandrasekharan [4]	✓	✓	
Guha et al. [9]	✓	✓	
Hang et al. [10]	✓	✓	
Hermoso et al. [11]	✓	✓	
Jennings et al. [12], [13]	✓	✓	
Katz and Golbeck [16]	✓	✓	
Lashkari et al.[17]	✓	✓	
Li and Gui [18]		✓	
Manchala [19]	✓		
Nefti et al. [20]	✓		
Nguyen and Tran [22], [23]	✓	✓	
Ramchurn et al. [24]	✓	✓	
Sabater and Sierra [26], [27]	✓	✓	
Schillo et al. [28]	✓		
Sen and Sajja's [29]	✓	✓	
Shibata et al. [30]	✓		
Singh and colleagues [36], [37]	✓	✓	
Teacy et al. [31], [32]	✓	✓	✓
Victor et al. [33]	✓	✓	
Vogiatzis et al. [34]	✓	✓	
Whitby et al. [35]	✓	✓	✓
Zhang, Robin and collegues [39], [14], [5], [6]	✓	✓	✓
Our model	✓	✓	✓

A. Experience trust

Intuitively, experience trust of agent i in agent j is the trustworthiness of j that agent i collects from all transactions between i and j in the past.

Experience trust of agent i in agent j is defined by the formula:

$$E_{ij} = \sum_{k=1}^{n} t_{ij}^k * w_k \qquad (1)$$

where:

- t_{ij}^k is the transaction trust of agent i in its partner j at the kth latest transaction.

- w_k is the weight of the kth latest transaction such that

$$\begin{cases} w_{k_1} \geqslant w_{k_2} \text{ if } k_1 < k_2 \\ \sum_{k=1}^{n} w_k = 1 \end{cases}$$

- n is the number of transactions taken between agent i and agent j in the past.

The weight vector $\vec{w} = \{w_1, w_2, ..w_n\}$ is decreasing from head to tail because the aggregation focuses more on the later transactions and less on the older transactions. It means that the later the transaction is, the more its trust is important to estimate the experience trust of the correspondent partner. This vector may be computed by means of Regular Decreasing Monotone (RDM) linguistic quantifier Q (Zadeh [38]).

B. Trust of referees

Suppose that an agent can refer all agents he knows (referee agents) in the system about their experience trust (private judgement) on a given partner. This is called *reference trust* (this will be defined in the next section). However, some referee agents may be liar. In order to avoid the case of lying

referee, this model proposes a mechanism which enables an agent to evaluate its referees on sharing their private trust about partners.

Let $X_{il} \subseteq A$ be a set of partners that agent i refers their trust via referee l, and that agent i has already at least one transaction with each of them. Since the model supposes that agent always trusts in itself, the trust of referee l from the point of view of agent i is determined based on the difference between experience trust E_{ij} and the trust r_{ij}^l of agent i about partner j referred via referee l (for all $j \in X_{il}$).

Trust of referee (sharing trust) S_{il} of agent i on the referee l is defined by the formula:

$$S_{il} = \frac{1}{\mid X_{il} \mid} * \sum_{j \in X_{il}} h(E_{ij}, r_{ij}^l) \qquad (2)$$

where:

- h is a *referee-trust-function* $h : [0,1] \times [0,1] \to [0,1]$, which satisfies the following conditions:

$$h(e_1, r_1) \leqslant h(e_2, r_2) \text{ if } \mid e_1 - r_1 \mid \geqslant \mid e_2 - r_2 \mid .$$

These constraints are based on the following intuitions:

- The more the difference between E_{ij} and r_{ij}^l is large, the less agent i trust on the referee l, and conversely;
- The more the difference between E_{ij} and r_{ij}^l is small, the more agent i trusts on the referee l.

- E_{ij} is the experience trust of i on j

- r_{ij}^l is the reference trust of agent i on partner j that is referred via referee l:

$$r_{ij}^l = E_{lj} \qquad (3)$$

C. Reference trust

Reference trust (also called reputation trust) of agent i on partner j is the trustworthiness of agent j given by other referees in the system. In order to take into account the trust of referee, the reference trust R_{ij} is a combination between the single reference trust r_{ij}^l and the trust of referee S_{il} of referee l.

Reference trust R_{ij} of agent i on agent j is a non-weighted average:

$$R_{ij} = \begin{cases} \dfrac{\sum\limits_{l \in X_{ij}} g(S_{il}, r_{ij}^l)}{|X_{ij}|} & \text{if } X_{ij} \neq \varnothing \\ 0 & \text{otherwise} \end{cases} \quad (4)$$

where:

- g is a *reference-function* $g : [0,1] \times [0,1] \to [0,1]$, which satisfies the following conditions:

    (i) $g(x_1, y) \leqslant g(x_2, y)$ if $x_1 \leqslant x_2$
    (ii) $g(x, y_1) \leqslant g(x, y_2)$ if $y_1 \leqslant y_2$

    These constraints are based on the intuitions:
    - The more the trust of referee l is high in the point of view of agent i, the more the reference trust R_{ij} is high;
    - The more the single reference trust r_{ij}^l is high, the more the final reference trust R_{ij} is high

- S_{il} is the trust of i on the referee l

- r_{ij}^l is the single reference trust of agent i about partner j referred via referee l

D. Overall trust

Overall trust T_{ij} of agent i in agent j is defined by the formula:

$$T_{ij} = t(E_{ij}, R_{ij}) \quad (5)$$

where:

- t is a *overall-trust-function*, $t : [0,1] \times [0,1] \to [0,1]$, which satisfies the following conditions:

    (i) $min(e, r) \leqslant t(e, r) \leqslant max(e, r)$;
    (ii) $t(e_1, r) \leqslant t(e_2, r)$ if $e_1 \leqslant e_2$;
    (iii) $t(e, r_1) \leqslant t(e, r_2)$ if $r_1 \leqslant r_2$.

    This combination satisfies these intuitions:
    - It must neither lower than the minimal and nor higher the maximal of experience trust and reference trust;
    - The more the experience trust is high, the more the *overall trust* is high;
    - The more the reference trust is high, the more the *overall trust* is high.

- E_{ij} is the experience trust of agent i about partner j.

- R_{ij} is the reference trust of agent i about partner j.

E. Updating trust

Agent i's trust in agent j can be changed in the whole its life-time whenever there is at least one of these conditions occurs (as showed in Algorithm 1, line 2):

- There is a new transaction between i and j occurring (line 3), so the experience trust of i on j changed.

- There is a referee l who shares to i his new experience trust about partner j (line 10). Thus the reference trust of i on j is updated.

```
1:  for all agent i in the system do
2:    if (there is a new transaction k−th with agent j) or
      (there is a new reference trust E_lj from agent l about
      agent j) then
3:      if there is a new transaction k with agent j then
4:        t_ij^k ← a value in interval [0,1]
5:        t_ij ← t_ij ∪ t_ij^k
6:        t_ij ← Sort(t_ij)
7:        w ← GenerateW(k)
                k
8:        E_ij ← ∑ t_ij^h * w_h
               h=1
9:      end if
10:     if there is a new reference trust E_lj from agent l
        about agent j then
11:       r_ij^l ← E_lj
12:       X_il ← X_il ∪ {j}
                    1
13:       S_il ← ───── * ∑ h(E_ij, r_ij^l)
                 |X_il|   j∈X_il
                  ∑_{l∈X_ij} g(S_il, r_ij^l)
14:       R_ij ← ──────────────────────────
                         |X_ij|
15:     end if
16:     T_ij ← t(E_ij, R_ij)
17:   end if
18: end for
```

Algorithm 1: Trust Updating Algorithm

E_{ij} is updated after the occur of each new transaction between i and j as follows (lines 3 - 9):

- The new transaction's trust value t_{ij}^k is placed at the first position of vector t_{ij} (lines 4 - 6). Function $Sort(t_{ij})$ sorts the vector t_{ij} in ordered in time.

- Vector w is also generated again (line 7) in function $GenerateW(k)$.

- E_{ij} is updated by applying formulas 1 with the new vector t_{ij} and w (line 8).

Once E_{ij} is updated, agent i sends E_{ij} to its friend agents. Therefore, all i's friends will update their reference trust when they receive E_{ij} from i. We suppose that all friend relations in system are bilateral, this means that if agent i is a friend of agent j then j is also a friend of i. After having received E_{lj} from agent l, agent i then updates her/his reference trust R_{ij} on j as follows (lines 10 - 15):

- In order to update the individual reference trust r_{ij}^l, the value of E_{lj} is placed at the position of the old one (line 11).

- Agent j will be also added into X_{il} to recalculate the referee trust S_{il} and recalculate the reference trust R_{ij} (lines 12 - 14).

Finally, T_{ij} is updated by applying the formulas 5 from new E_{ij} and R_{ij} (line 16).

IV. EXPERIMENTAL EVALUATION

This section presents the evaluation of the proposed model by taking emperimental data. Section IV-A presents the setting up our experiment application. Section IV-B evaluates the need of avoiding liars in refering of reputation. Section IV-C evaluates the need of combination of experience trust and reputation even if there are liars in refering reputation.

A. Experiment Setup

1) An E-market: An e-market system is composed of a set of seller agents, a set of buyer agents, and a set of transactions. Each transaction is performed by a buyer agent and a seller agent. A seller agent plays the role of a seller who owns a set of products and it could sell many products to many buyer agents. A buyer agent plays the role of a buyer who could buy many products from many seller agents.

- Each seller agent has a set of products to sell. Each product has a quality value in the interval $[0, 1]$. The quality of product will be assigned as the transaction trust of the transaction in which the product is sold.

- Each buyer agent has a transaction history for each of its sellers to calculate the experience trust for the corresponding seller. It has also a set of reference trusts referred from its friends. The buyer agent will update its trust on its sellers once it finishes a transaction or receives a reference trust from one of its friends. The buyer chooses the seller with the highest final trust when it want to buy a new product. The calculation to estimate the highest final trust of sellers is based on the proposed model in this paper.

2) Objectives: The purpose of these experiments is to answer two following questions:

- First, is it better if buyer agent judges the sharing trust of its referees than does not judge it? In order to answer to this question, the proposed model will be compared with the model of Jennings et al.'s model [12], [13] (Section IV-B).

- Second, what is better if buyer agent uses only its experience trust in stead of combination of experience and reference trust? In order to answer this question, the proposed model will be compared with the model of Manchala's model [19] (Section IV-C).

3) Initial Parameters: In order to make the results comparable, and in order to avoid the effect of random aspect in value initiation of simulation parameters, the same values for input parameters of all simulation scenarios will be used: number of sellers; number of products; number of simulations. These values are presented in the Table.II.

TABLE II: Value of parameters in simulations

Parameters	Values
Number of runs for each scenario	100 (times)
Number of sellers	100
Number of buyers	500
Number of products	500000
Average number of bought products/buyer	100
Average number of friends/buyer	300 (60% of buyers)

4) Analysis and evaluation criteria: Each simulation scenario will be ran at least 100 times. At the output, the following parameter will be calculated:

- The average quality (in %) of brought products for all buyers. A model (strategy) is considered better if it brings the higher average quality of brought products for all buyers in the system.

B. The need of avoiding liar in reputation

1) Scenarios: The question need to be answerd is: is it better if buyer agent uses reputation with trust of referees (agent judges the sharing trust of its referees) or uses reputation without trust of referees (agent does not judge the sharing trust of its referees)? In order to answer this question, there are two strategies will be simulated:

- *Strategy A - using proposed model*: Buyer agent refers the reference trust (about sellers) from other buyers *with* taking into account the trust of referee.

- *Strategy B - using model of Jennings et al. [12], [13]*: Buyer agent refers the reference trust (about sellers) from other buyers *without* taking into account the trust of referee.

The simulations are launched in various values of the percentage of lying buyers in the system (0%, 30%, 50%, 80%, and 100%).

2) Results: The results indicate that the average quality of bought products of all buyers in the case of using reputation with considering of trust of referees is always significantly higher than those in the case using reputation without considering of trust of referees.

When there is no lying buyer (Fig.1.a). The average quality of bought products for all buyers in the case using strategy A is not significantly different from that in the case using strategy B ($M(A) = 85.24\%$, $M(B) = 85.20\%$, significant difference with $p\text{-}value > 0.7$)[1].

When there is 30% of buyers is liar (Fig.1.b). The average quality of bought products for all buyers in the case using strategy A is significantly higher than in the case using strategy B ($M(A) = 84.64\%$, $M(B) = 82.76\%$, significant difference with $p\text{-}value < 0.001$).

When there is 50% of buyers is liar (Fig.1.c). The average quality of bought products for all buyers in the case using strategy A is significantly higher than in the case using strategy

---

[1]We use the *t-test* to test the difference between two sets of average quality of bought products of two scenarios, therefore if the probability value $p\text{-}value < 0.05$ we could conclude that the two sets are *significantly different*.

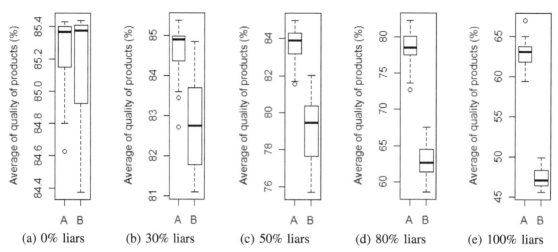

(a) 0% liars (b) 30% liars (c) 50% liars (d) 80% liars (e) 100% liars

Fig. 1: Significant difference of average quality of bought products of all buyers from the case using proposed model (strategy A) and the case using Jennings et al.'s model (strategy B)

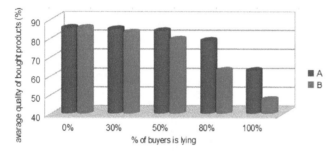

Fig. 2: Summary of difference of average quality of bought products of all buyers between the case using our model (A) and the case using Jennings et al.'s model (B)

B ($M(A) = 83.68\%$, $M(B) = 79.11\%$, significant difference with $p\text{-}value < 0.001$).

When there is 80% of buyers is liar (Fig.1.d). The average quality of bought products for all buyers in the case using strategy A is significantly higher than in the case using strategy B ($M(A) = 78.55\%$, $M(B) = 62.76\%$, significant difference with $p\text{-}value < 0.001$).

When all buyers are liar (Fig.1.e). The average quality of bought products for all buyers in the case using strategy A is significantly higher than in the case using strategy B ($M(A) = 62.78\%$, $M(B) = 47.31\%$, significant difference with $p\text{-}value < 0.001$).

In summary, as being depicted in the Fig.2, the more the percentage of liar in buyers is high, the more the average quality of bought products of all buyers in the case using our model (strategy A) is significantly higher than those in the case using Jennings et al.'s model [12], [13] (strategy B).

C. The need of combination of experience with reputation

1) Scenarios: The results of the first evaluation suggest that using reputation with considering of trust of referees is better than using reputation without considering of trust of

referees, especially in the case there are some liars in sharing their private trust about partners to others. And in turn, another question arises: in the case there are some liars in sharing data to their friends, is it better if buyer agent use reputation with considering of trust of referees or use only experience trust to avoid liar reputation? In order to answer this question, there are two strategies also simulated:

- *Strategy A - using proposed model*: Buyer agent refers the reference trust (reputation) from other buyers by taking into account their considering of trust of referees.

- *Strategy C - using Manchala's model [19]*: Buyer agent does not refer any reference trust from other buyers. It bases only on its experience trust.

The simulations are also launched in various values of the percentage of lying buyers in the system (0%, 30%, 50%, 80%, and 100%).

2) Results: The results indicate that the average quality of bought products of all buyers in the case with considering of trust of referees is almost significantly higher than those in the case using only the experience trust.

When there is no lying buyer (Fig.3.a). The average quality of bought products for all buyers in the case using strategy A is significantly higher than in the case using strategy C ($M(A) = 85.24\%$, $M(C) = 62.75\%$, significant difference with $p\text{-}value < 0.001$).

When there is 30% of buyers is liar (Fig.3.b). The average quality of bought products for all buyers in the case using strategy A is significantly higher than the in case using strategy C ($M(A) = 84.64\%$, $M(C) = 62.74\%$, significant difference with $p\text{-}value < 0.001$).

When there is 50% of buyers is liar (Fig.3.c). The average quality of bought products for all buyers in the case using strategy A is significantly higher than in the case using C ($M(A) = 83.68\%$, $M(C) = 62.76\%$, significant difference with $p\text{-}value < 0.001$).

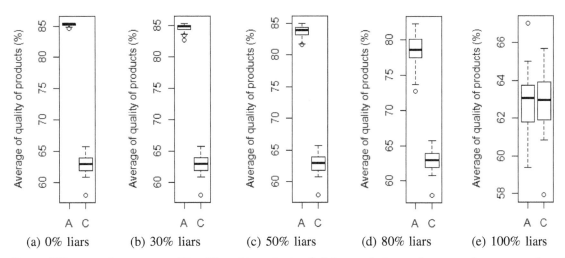

Fig. 3: Significant difference of average quality of bought products of all buyers between the case using proposed model (strategy A) and the case using Manchala's model (strategy C)

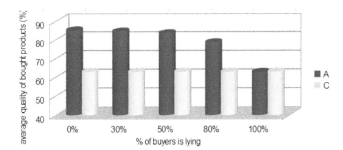

Fig. 4: Summary of difference of average quality of bought products of all buyers between the case using our model (A), and the case using Manchala's model (C)

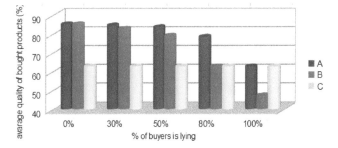

Fig. 5: Summary of difference of average quality of bought products of all buyers among the case using our model (A), the case using Jennings et al.'s model (B), and the case using Manchala's model (C)

When there is 80% of buyers is liar (Fig.3.d). The average quality of bought products for all buyers in the case using strategy A is significantly higher than in the case using strategy C ($M(A) = 78.55\%$, $M(C) = 62.78\%$, significant difference with $p\text{-}value < 0.001$).

When all buyers are liar (Fig.3.e). There is no significant difference between the case using strategy A and the case using strategy C ($M(A) = 62.78\%, M(C) = 62.75\%$, significant difference with $p\text{-}value > 0.6$). It is intuitive because in our model (strategy A), when almost referees are not trustworthy, the trustor tends to trust in himself instead of other. In other word, the trustor has the tendency to base on its won experience rather than others.

The overall result is depicted in the Fig.4. In almost cases, the average quality of bought products of all buyers in the case of using our model is always significantly higher than those in the case of using Manchala's model [19]. In the case that all buyers are liar, there is no significant difference of the average quality of bought products from all buyers between two strategies.

In summary, Fig.5 illustrates the value of average quality of bought products of all buyers in three scenarios. In the case there is no lying buyer, this value is the highest in the case

using our model and Jennings et al.'s model [12], [13] (there is no significant difference between two mosels in this situation). Using Manchala's model [19] is the worst case in this situation. In the case there are 30%, 50% and 80% buyers to be lying, the value is always highest in the case of using our model. In the case that all buyers are liar, there is no significant difference between agents using our model and agents using Manchala's model [19]. Both of these two strategies win a much more higher value compared with the case using Jennings et al.'s model [12], [13].

V. DISCUSSION

Let us consider a scenario of an e-commerce application. There are two concurrent sellers S_1 and S_2 who sell the same product x, there is an independent third party site w which collects the consumer's opinions. All clients could submit its opinions about sellers. In this case, the site w could be considered as a reputation channel for client: a client could refer the given opinions on the site w to choose the best seller. However, because the site w is a public reputation: all clients could submit their opinions. Imagining that S_1 is really trustworthy, but S_2 is not fair, some of its employments intentionally submit some negative opinions about the seller S_1 in order to attract more clients from S_1 to S_2.

Let consider this application in two cases. Firstly, the case without mechanism to avoid liars in the applied trust model. If an user i is considering to buy a product x that both S_1 and S_2 are selling. User i refers the reputation of S_1 and S_2 on the site w. Since there is not any mechanism to avoid liars in the trust model, the more negative opinions from S_2's employments are given about S_1, the lower the reputation of S_1 is. Therefore, the lower the possibility that user i chooses buying the product x from S_1.

Secondly, the case of our proposed model with lying against mechanism. User i will refer the reputation of S_1 and S_2 on the site w with considering the sharing trust of the owner of each opinion. Therefore, the ones from S_2 who gave negative opinions about S_1 will be detected as liars. Their opinion weights thus will be decreased (considered as unimportant ones) when calculating the reputation of S_1. Consequently, the reputation of S_1 will stay high no matter how many people from S_2 intentionally lie about S_2. In other word, our model helps agent to avoid some liars in calculating the reputation of a given partner in multiagent systems.

VI. Conclusion

This paper presented a model of trust which enables agents to calculate, estimate and update trust's degree on their partners based not only on their own experiences, but also based on the reputation of partners. The partner reputation is estimated from the judgements from referees in the community. In which, the model taken into account the trustworthiness of the referee in judging a partner.

The experimental evaluation of the model has been set up for multiagent system in the e-commerce environment. The research results indicate, firstly, that it is better to take into account the trust of referees to estimate the reputation of partners. Seconly, it is better to combine the experience trust and the reputation than using only the experience trust in estimating the trust of a partner in the multiagent system.

Constructing and selecting a strategy, which is appropriate to the context of some application of a multiagent system, need to be investigated furthermore. These research issues will be presented in our future work.

References

[1] Masoom Alam, Shahbaz Khan, Quratulain Alam, Tamleek Ali, Sajid Anwar, Amir Hayat, Arfan Jaffar, Muhammad Ali, and Awais Adnan. Model-driven security for trusted systems. *International Journal of Innovative Computing, Information and Control*, 8(2):1221–1235, 2012.

[2] Chris Burnett, Timothy J. Norman, and Katia Sycara. Bootstrapping trust evaluations through stereotypes. In *Proceedings of the 9th International Conference on Autonomous Agents and Multiagent Systems: volume 1 - Volume 1*, AAMAS '10, pages 241–248, Richland, SC, 2010. International Foundation for Autonomous Agents and Multiagent Systems.

[3] J. Carter, E. Bitting, and A. Ghorbani. Reputation formalization for an information-sharing multi-agent sytem. *Computational Intelligence*, 18(2):515–534, 2002.

[4] B. Esfandiari and S. Chandrasekharan. On how agents make friends: Mechanisms for trust acquisition. In *Proceedings of the Fourth Workshop on Deception, Fraud and Trust in Agent Societies*, pages 27–34, Montreal, Canada, 2001.

[5] Hui Fang, Yang Bao, and Jie Zhang. Misleading opinions provided by advisors: Dishonesty or subjectivity. IJCAI/AAAI, 2013.

[6] Hui Fang, Jie Zhang, and Nadia Magnenat Thalmann. A trust model stemmed from the diffusion theory for opinion evaluation. In *Proceedings of the 2013 International Conference on Autonomous Agents and Multi-agent Systems*, AAMAS '13, pages 805–812, Richland, SC, 2013. International Foundation for Autonomous Agents and Multiagent Systems.

[7] D. Gambetta. Can we trust trust? In D. Gambetta, editor, *Trust: Making and Breaking Cooperative Relations*, pages 213–237. Basil Blackwell, New York, 1990.

[8] Tyrone Grandison and Morris Sloman. Specifying and analysing trust for internet applications. In *Proceedings of the 2nd IFIP Conference on e-Commerce, e-Business, e-Government*, Lisbon, Portugal, October 2002.

[9] R. Guha, Ravi Kumar, Prabhakar Raghavan, and Andrew Tomkins. Propagation of trust and distrust. In *Proceedings of the 13th international conference on World Wide Web*, WWW '04, pages 403–412, New York, NY, USA, 2004. ACM.

[10] Chung-Wei Hang, Yonghong Wang, and Munindar P. Singh. Operators for propagating trust and their evaluation in social networks. In *Proceedings of The 8th International Conference on Autonomous Agents and Multiagent Systems - Volume 2*, AAMAS '09, pages 1025–1032, Richland, SC, 2009. International Foundation for Autonomous Agents and Multiagent Systems.

[11] Ramón Hermoso, Holger Billhardt, and Sascha Ossowski. Role evolution in open multi-agent systems as an information source for trust. In *Proceedings of the 9th International Conference on Autonomous Agents and Multiagent Systems: volume 1 - Volume 1*, AAMAS '10, pages 217–224, Richland, SC, 2010. International Foundation for Autonomous Agents and Multiagent Systems.

[12] Dong Huynh, Nicholas R. Jennings, and Nigel R. Shadbolt. Developing an integrated trust and reputation model for open multi-agent systems. In *Proceedings of the 7th Int Workshop on Trust in Agent Societies*, pages 65–74, New York, USA, 2004.

[13] Trung Dong Huynh, Nicholas R. Jennings, and Nigel R. Shadbolt. An integrated trust and reputation model for open multi-agent systems. *Autonomous Agents and Multi-Agent Systems*, 13(2):119–154, 2006.

[14] Siwei Jiang, Jie Zhang, and Yew-Soon Ong. An evolutionary model for constructing robust trust networks. In *Proceedings of the 2013 International Conference on Autonomous Agents and Multi-agent Systems*, AAMAS '13, pages 813–820, Richland, SC, 2013. International Foundation for Autonomous Agents and Multiagent Systems.

[15] Audun Josang, Claudia Keser, and Theo Dimitrakos. Can we manage trust? In *Proceedings of the 3rd International Conference on Trust Management, (iTrust)*, Paris, 2005.

[16] Yarden Katz and Jennifer Golbeck. Social network-based trust in prioritized default logic. In *Proceedings of the 21st National Conference on Artificial Intelligence (AAAI-06)*, volume 21, pages 1345–1350, Boston, Massachusetts, USA, jul 2006. AAAI Press.

[17] Y. Lashkari, M. Metral, and P. Maes. Collaborative interface agents. In *Proceedings of the Twelfth National Conference on Artificial Intelligence*. AAAIPress, 1994.

[18] Xiaoyong Li and Xiaolin Gui. Tree-trust: A novel and scalable P2P reputation model based on human cognitive psychology. *International Journal of Innovative Computing, Information and Control*, 5(11(A)):3797–3807, 2009.

[19] D. W. Manchala. E-commerce trust metrics and models. *IEEE Internet Comp.*, pages 36–44, 2000.

[20] Samia Nefti, Farid Meziane, and Khairudin Kasiran. A fuzzy trust model for e-commerce. In *Proceedings of the Seventh IEEE International Conference on E-Commerce Technology (CECÓ5)*, pages 401–404, 2005.

[21] Manh Hung Nguyen and Dinh Que Tran. A computational trust model with trustworthiness against liars in multiagent systems. In Ngoc Thanh Nguyen et al., editor, *Proceedings of The 4th International Conference on Computational Collective Intelligence Technologies and Applications (ICCCI), Ho Chi Minh City, Vietnam, 28-30 November 2012*, pages 446–455. Springer-Verlag Berlin Heidelberg, 2012.

[22] Manh Hung Nguyen and Dinh Que Tran. A multi-issue trust model

in multiagent systems: A mathematical approach. *South-East Asian Journal of Sciences*, 1(1):46–56, 2012.

[23] Manh Hung Nguyen and Dinh Que Tran. A combination trust model for multi-agent systems. *International Journal of Innovative Computing, Information and Control (IJICIC)*, 9(6):2405–2421, June 2013.

[24] S. D. Ramchurn, C. Sierra, L. Godo, and N. R. Jennings. Devising a trust model for multi-agent interactions using confidence and reputation. *International Journal of Applied Artificial Intelligence*, 18(9–10):833–852, 2004.

[25] Steven Reece, Alex Rogers, Stephen Roberts, and Nicholas R. Jennings. Rumours and reputation: Evaluating multi-dimensional trust within a decentralised reputation system. In *Proceedings of the 6th International Joint Conference on Autonomous Agents and Multiagent Systems*, AAMAS '07, pages 165:1–165:8, New York, NY, USA, 2007. ACM.

[26] Jordi Sabater and Carles Sierra. Regret: A reputation model for gregarious societies. In *Proceedings of the Fourth Workshop on Deception, Fraud and Trust in Agent Societies*, pages 61–69, Montreal, Canada, 2001.

[27] Jordi Sabater and Carles Sierra. Reputation and social network analysis in multi-agent systems. In *Proceedings of the First International Joint Conference on Autonomous Agents and Multiagent Systems (AAMAS-02)*, pages 475–482, Bologna, Italy, July 15–19 2002.

[28] M. Schillo, P. Funk, and M. Rovatsos. Using trust for detecting deceitful agents in artificial societites. *Applied Artificial Intelligence (Special Issue on Trust, Deception and Fraud in Agent Societies)*, 2000.

[29] S. Sen and N. Sajja. Robustness of reputation-based trust: Booblean case. In *Proceedings of the First International Joint Conference on Autonomous Agents and Multiagent Systems (AAMAS-02)*, pages 288–293, Bologna, Italy, 2002.

[30] Junko Shibata, Koji Okuhara, Shogo Shiode, and Hiroaki Ishii. Application of confidence level based on agents experience to improve internal model. *International Journal of Innovative Computing, Information and Control*, 4(5):1161–1168, 2008.

[31] W. T. Luke Teacy, Jigar Patel, Nicholas R. Jennings, and Michael Luck. Travos: Trust and reputation in the context of inaccurate information sources. *Journal of Autonomous Agents and Multi-Agent Systems*, 12(2):183–198, 2006.

[32] W.T. Luke Teacy, Michael Luck, Alex Rogers, and Nicholas R. Jennings. An efficient and versatile approach to trust and reputation using hierarchical bayesian modelling. *Artif. Intell.*, 193:149–185, December 2012.

[33] Patricia Victor, Chris Cornelis, Martine De Cock, and Paulo Pinheiro da Silva. Gradual trust and distrust in recommender systems. *Fuzzy Sets and Systems*, 160(10):1367–1382, 2009. Special Issue: Fuzzy Sets in Interdisciplinary Perception and Intelligence.

[34] George Vogiatzis, Ian Macgillivray, and Maria Chli. A probabilistic model for trust and reputation. AAMAS, 225-232 (2010)., 2010.

[35] Andrew Whitby, Audun Josang, and Jadwiga Indulska. Filtering out unfair ratings in bayesian reputation systems. In *Proceedings of the 3rd International Joint Conference on Autonomous Agenst Systems Workshop on Trust in Agent Societies (AAMAS)*, 2005.

[36] B. Yu and M. P. Singh. Distributed reputation management for electronic commerce. *Computational Intelligence*, 18(4):535–549, 2002.

[37] B. Yu and M. P. Singh. An evidential model of distributed reputation management. In *Proceedings of the First International Joint Conference on Autonomous Agents and Multiagent Systems (AAMAS-02)*, pages 294–301, Bologna, Italy, 2002.

[38] L. A. Zadeh. A computational approach to fuzzy quantifiers in natural languages. pages 149–184, 1983.

[39] Jie Zhang and Robin Cohen. A framework for trust modeling in multiagent electronic marketplaces with buying advisors to consider varying seller behavior and the limiting of seller bids. *ACM Trans. Intell. Syst. Technol.*, 4(2):24:1–24:22, April 2013.

27

A New Trust Evaluation for Trust-based RS

Sajjawat Charoenrien
Department of Mathematics
Chulalongkorn University
Bangkok, Thailand

Saranya Maneeroj
Department of Mathematics
Chulalongkorn University
Bangkok, Thailand

Abstract—Trust-based recommender systems provide the recommendations on the most suitable items for the individual users by using the trust values from their trusted friends. Usually, the trust values are obtained directly from the users, or by calculated using the similarity values between the pair of users. However, the current trust value evaluation can cause the following three problems. First, it is difficult to identify the co-rated items for calculating the similarity values between the users. Second, the current trust value evaluation still has symmetry property which makes the same trust value on both directions (trustor and trustee). Finally, the current trust value evaluation does not focus on how to adjust the trust values for the remote user. To eliminate all of these problems, our purposed method consists of three new factors. First, the similarity values between the users are calculated using a latent factor model instead of the co-rated items. Second, in order to identify the trustworthiness for every user in trust network, the degrees of reliability are calculated. Finally, we use the number of hops for adjusting the trust value for the remote users who are expected to be low trust as shown in the real-world application concept. This trust evaluation leads to better predicted rating and getting more predictable ratings. Consequently, from our experiment, the more efficiency trust-based recommender system is obtained, comparing with the classical method on both accuracy and coverage.

Keywords—trust-based recommender systems; trust values; similarity values

I. INTRODUCTION

Recommender systems (RS) act as the tools that help selecting the most relevant items to the target users. First, the users' preference ratings on items are collected. After that, the similarities between users are calculated by using the ratings on their co-rated items. These similarities are then applied to select the nearest neighbors for each user. Finally, recommender systems predict the ratings for a target user by using the ratings from his neighbors. However, usually most users tend to provide the small fraction of all possible ratings. This leads to a sparsity problem which the system cannot provide the accurate prediction and, for some items, unpredictable.

In order to solve this problem, Trust-Awared Recommender systems (TARS) [3] have been implemented. The system uses not only the users' ratings data, but also the trust information for prediction. These trust values are usually collects directly from users or calculated as the same way as similarity values. However, calculating trust values this way has cause many problems. First, the similarity value has symmetry property which makes the trust value on both directions of the pair of users to be the same.

Actually, the trust value between two persons might not be the same. For example, user A might trust user B but user B might not trust user A. Therefore, the trust between these two should not be the same. Second, the systems may not be able to effectively find the trust values due to the sparsity problem. For the last, it cannot use the transitive property of graph to calculate the trust value between the friends of friends for a user. TidalTrust [1][4] and MoleTrust [5] were proposed to solve this problem. Both of them use the propagation technique to propagate the trust values for the raters (who rated score on the target item). The propagated trust values are calculated every time when the friends of friends of the target users are visited. However, finding the propagation in a very large trust network is time comsuming. Thus, another model has been proposed by Y. Guo. [9] This model tries to find the trust factor of the raters toward the ratings.

The trust factor is used to calculate the trust value as the important factor. The trust factor calculates from the number of friends and number of evaluated items belonging to that rater and the experience of rater on the past rated item, with the rater. That means the trust values are calculated from the relations between the raters and target users. However, this method does not concern the number of the hops, which might reduce the trust values of the remoteness users. While exploring the trust network, a target user has to visit the friends of his friends until he retrieves the wanted rating. Every time of the visiting, the number of hops is increased. If the number of hops is large, the two users are far from each other and they are less related. Therefore, the number of hops should be included in the model to suit the real world applications.

In this work, a new trust evaluation method is proposed. This method has three factors for calculating the trust value of each rater, e.g. similarity value between a pair of user, trustworthiness and number of hops. By using the latent feature model for calculating similarity value for each pair of users, the sparsity and symmetry problems can be solved. The second factor, trust worthiness is the extended idea from Y. GUO, which tries to find the degree of reliability of each rater in the trust.

Finally, the number of hops is used to adjust the trust values of the raters based on the distance from the target users in the trust network. These three factors are combined in this work to make a new trust value calculation. After that, the new trust value is used in rating prediction. It improves the quality of the prediction than the classical trust-based recommender systems on both accuracy and coverage as shown in the experiment in the fourth section of this paper.

II. RELATED WORKS

Recommender system involves three major steps. First, the system collects rating data representing the user's preference towards the different items. Second, it generates the user's pattern based on his past experience towards those items. Finally, RS makes the prediction for the new items based on the user's preference pattern. However, usually, the numbers of ratings are not large enough to give an effective prediction. This problem is called sparsity problem, which leads to inaccurate rating prediction or unpredictable for some items. After that, the study of Swearingen & Sinha [7] on the usability of three book RS and three movie RS has found that, by integrating trust information into the RS, the prediction is improved. Moreover, Ziegler & Golbeck [8] investigated the correlations between trust and similarity definition. They found that the trust values between each pair of user in the RS were be calculated by using their similarity values on the co-rated items (the overlap items that have been rated from both users). However, it still leads to two problems. First, the trust value from similarity value is symmetry on the both side of the users, such as A → B equals to B → A. In fact, two users who trust each other might not have the same trust value because it opposes to the fact which two persons not necessary to trust each other. One person can trust the other by one side. Another problem is that sometimes, the system cannot calculate the similarity values for some pairs of users because they do not have the co-rated items. To solve this, J. O' Donovan [6] proposed the work called Trust in Recommender Systems. In their work, the trust values can be found from the reliabilities of the raters which are indicated the amount of corrected ratings they have made. The trust value is the ratio between the number of correctly predicted ratings and the whole number of the predicted ratings as (1).

$$Trust^P = \frac{|CorrectSet(p)|}{|RecSet(p)|} \tag{1}$$

Where p is rater, $CorrectSet(p)$ is the set of correctly predicted items , and $RecSet(p)$ is the set of all predicted items. Although the trust values calculated by J. O' Donovan can solved the symmetry problem, it did not use information in trust network which contains the relationships among the users to make more accurate prediction. Trust network can be represented as a graph that consists of nodes (as the user) and edges (as the relationship between the users).

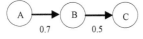

Fig. 1.   The sample of Trust network

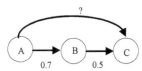

Fig. 2.   Example of a figure caption. (*figure caption*)

The weight on each edge shows the trust value from source user to destination users in as Fig. 1. Sometimes, when the friends of the target user cannot provide the prediction for him, the system might use the opinions of the friends of that user's friends instead. The opinions from the target user's friends of friends can be transitively obtained from their directed friends. However, the target user might not have directed trust to these friends of friends. The example of this is shown in Fig. 2, if user A is a friend with user B and user B is a friend with user C, then how much user A trusts user C? Messa [3] proposed TAR architecture to provide trust value calculation for friends of friends.

From the TAR architecture, the predicted ratings are generated by rating predictor module using output from both User Similarity and Estimated Trust modules. The User Similarity is the process which calculates the similarity value between a pair of users.

While the Estimated Trust finds the trust values on friends of friends, which can solve the previous problem. TidalTrust and MoleTrust are the traditional methods that used to find the trust values from friends of friends for the target users. Both of them use transitive property of the graph to propagate the trust value from friends to friends. They use depth first search to propagate trust value from the target user to the rater. In the propagation process, the trust values are calculated by using the trust values of the previous friends as shown in (2).

$$trust(u) = \frac{\sum_{i \in predecessors}(trust(i) * trust\_edge(i, u))}{\sum_{i \in predecessors}(trust(i))} \tag{2}$$

Where u is target user, i is the user whom the target user trusted, $trust(i)$ is the trust value of user i, $trust\_edge(i, u)$ is the trust between user i and user u and predecessors are the previous friends of the user.

The different between their methods is that, in MoleTrust the cycle from the trust network is removed to reduce the distance of the trust propagation, in order to improve the performance. In contrast, TidalTrust propagates to every node in the trust network that is connected.

On the very large trust network, the propagation technique is not good because the exploration on every node on the graph takes a large amount of time, depending on the complexity of that the trust network.

Y. Guo [9] proposed trust value calculation without propagation. This method finds the trust factor which is the main factor of the trust factor. It calculates from the combination of the number of friends and number of evaluated items.   If a user has the number of friends more than the others, he is more reliable.

$$T_i = \frac{2(1 - 1/lnf_i) * (1 - 1/ln(q_i + 3))}{2 - 1/lnf_i - 1/ln(q_i + 3)} \tag{3}$$

Where i is the user whom the target user trusted, T_i is the trust factor of user i, f_i is the number of user i's evaluation on each item and q_i is the number of recommendations user i has made for the others.

In the real world application, when considering the transitive property, a friend who is far away from target user should have lesser trust value than close friend. While exploring the trust network, the target user will visit the friend of friend until retrieve the rating. Every time of the visiting, the number of hops is increased every time. If the number of hops is large, the two users are far from each other and they are less related. Therefore, the number of hops should be included into the model to suit the real world applications.

In this work, a new trust value calculation that solves the problems mentioned above is proposed. First, this method uses the latent features of users for finding similarity instead of using co-rated items. Second, the symmetry problem is eliminated by exploiting the degree of reliability of each user. And, finally, this method considers the distance among friends by using the number of hops.

Therefore, we can summarize pros and cons of related trust methods comparing with our proposed method on three attributes: symmetry or asymmetry of initial trust value, transitive opinion from the directed friends to the friend of the directed friends and concerning the number of hops of remoteness users as shown in the following Table I.

TABLE I.   PROS AND CONS COMPARISON OF EACH METHOD

Method	Initial Trust Value	Transitive Opinion	Concern the number of hops
J. O'Donovan's Method	Asymmetry	No	No
TidalTrust	Symmetry	Yes	No
MoleTrust	Symmetry	Yes	No
Y Guo's Method	Asymmetry	Yes	No
Proposed Method	Asymmetry	Yes	Yes

III.   PROPOSE METHOD

The new trust value calculation proposed in this work consists of three main steps. First, the similarity values for all pairs of users are calculated. Then, then reliability concept is combined into trust value calculation in order to get rid of symmetry problem. Finally, the number of hops is included in the calculation to reduce the trust of remote friends.

To guarantee that the trust values can be calculated for every pair of users, the singular value decomposition (SVD) [2] is applied in this work instead of relying on the co-rated items. First, the latent features of the user are extracted by the following.

$$R = USV^t \quad (4)$$

Where R is the user-rating matrix, U is the user matrix, S is the reduced matrix and V^t is the transpose matrix of the item matrix. From matrix U, each row is represented as the user feature vector. Each feature vector of user contains the latent features representing the user's characteristic. The cosine similarity is then applied on these feature vectors to find the similarity for every pair of users by the following.

$$sim_{A,B} = \frac{\sum_{i=1}^{n} A_i \times B_i}{\sqrt{\sum_{i=1}^{n}(A_i)^2} \times \sqrt{\sum_{i=1}^{n}(B_i)^2}} \quad (5)$$

Where A_i is the ith latent feature of user A (target user), B_i is the ith latent feature of user B and n is the number of latent features.

To prevent the symmetry problem of the trust values on both directions of the pair of users, the system uses not only the similarity value between a pair of users ($sim(A,B)$) but also concerns the reliability of rater(B) in the network. That is, reliability concept by Y. Guo [9] is applied in this work by using the number of in-degree which is represented the number of friends who trust rater (B).

$$trustworthiness_B =$$

$$\frac{ln(n_{in,B}+e)}{ln(\max(\{n_{in,C}|C \in \{user\ in\ Trust\ Network\}\})+e)} \quad (6)$$

Where e is a natural number, $n_{in,B}$ is the number of the in-degree edges of the user B and

$\max(\{n_{in,C}|C \in \{user\ in\ Trust\ Network\}\})$ is the maximum number of the in-degree edge of the user in the trust network. Also, the confidence of B towards A is calculated by merging $sim(A,B)$ and $trustworthiness_B$ using harmonic mean as shown in (7).

$$confidence_B = \frac{2 \times trustworthiness_B \times sim_{A,B}}{trustworthiness_B + sim_{A,B}} \quad (7)$$

Moreover, to make the method suitable for the real world situation, the confidence level of user who is far from target user should be adjusted by using number of hops as shown in (8).

$$trust_{A \to B} = confidence_B \times \frac{1}{d_B} \quad (8)$$

Where d is the number of hops from user A to B

After gathering the trust values of all raters from previous steps, the predicted rating can be calculated by using trust values of target user to all the raters who have rated the target item as weights.

$$predict\ rating_t = \frac{\sum_{i \in Rater} rating(i)*trust_{t \to i}}{\sum_{i \in Rater} trust_{t \to i}} \quad (9)$$

Where t is target user, $rating(i)$ is an actual rating of rater(i) on the target item and $trust_{t \to i}$ is trust value of target user(t) toward rater (i).

IV.   EXPERIMENT

In order to evaluate the performance of our proposed method, we compares our work with 2 classical trusted based RS methods: TidalTrust and MoleTrust

A. Dataset

The Epinions dataset is used in this work. This dataset consists of user-rating data containing 664,824 reviews from 49,290 users on 139,738 items, and also the trust network data containing 487,181 issued trust statements.

B. Evaluation Metric

To evaluate the model, we use two measurements: RMSE and coverage.

1) RMSE (Root Mean Square Error): It represents the accuracy of the prediction. The lower the value means the method is more accurate that the method with higher RMSE. This value can be calculated by the following

$$RMSE = \sqrt{\frac{\sum_{(u,i)|R_{u,i}} (r_{u,i} - \hat{r}_{u,i})^2}{|\{(u,i)|R_{u,i}\}|}} \qquad (10)$$

Where $r_{u,i}$ is an actual rating of user(u) on target item(i), $\hat{r}_{u,i}$ is a predicted rating of user(u) on target item(i) and $\{(u,i)|R_{u,i}\}$ is the set of user who rated on item(i).

2) Prediction Coverage: This value indicates the fraction of the ratings that can be predicted from all of the rating available.

$$coverage = \frac{|\{(u,i)|\hat{r}_{u,i}\}|}{|\{(u,i)|r_{u,i}\}|} \qquad (11)$$

Where $|\{(u,i)|\hat{r}_{u,i}\}|$ is the number of ratings that can be predictable and $|\{(u,i)|r_{u,i}\}|$ is number of all ratings. The higher coverage means the system provides more predictable rating.

C. Experimental Results

In the experiment, we compare TidalTrust and MoleTrust with the proposed method by using the dataset mentioned above. To predict the rating for the target item, we use the leave one out technique (hide only the actual rating on the target item of the target user in the dataset, and uses the rest for prediction) . Instead of using all ratings in the dataset, we randomly select one target item per target user and only the first 5,000 users are use as the test set. To avoid the bias, we use the same random data on TidalTrust, MoleTrust and proposed methods. The results of the experiment are shown as Fig. 3.

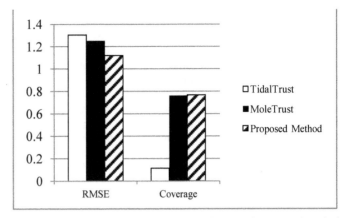

Fig. 3. Comparison of RMSE and Coverage between the proposed method and others

From the results shown above, RMSE of our proposed method is lower than TidalTrust and MoleTrust. It can be concluded that proposed method provides better accuracy. In the aspect of prediction coverage, our proposed method provides higher coverage values than both TidalTrust and MoleTrust. It can be concluded that proposed method provides more predictable items.

V. DISSCUSION

A. Accuracy

The reasons that proposed method has better accuracy than both TidalTrust and MoleTrust are following by this:

Both current trust-based RSs calculate the trust values by using the co-rated items which is hard to be identified because of the sparsity problem. The small number of co-rated items leads to the low quality neighbor which causes the low accuracy in prediction. On the other hand, the purpose method uses latent features of user instead of co-rated items. About 5,000 user features are extracted from user-rating matrix, so, it can represent user characteristic more correctly.

The current trust-based RSs do not consider the degree of reliability for each user in trust network. In the proposed method, the ratio between the numbers of friends on each specified user to the maximum number of friends from all users in the trust network is used to determine the reliability.

The current trust based RSs do not concern the remoteness of the friends of the target users. This type of friends should have the lesser trust value comparing to the close friends. The proposed method deals with this by adjusting the trust values using the number of hops.

B. Coverage

The reason that proposed method can provide more prediction coverage than the other two methods is that the proposed uses the latent features of user for calculating the similarity values without using co-rated items. Therefore, it can calculate similarity values for all pairs of users. However, it cannot provide 100% coverage because the target item obtains the rating only from the target user, not from other users. Therefore, there is no rating from rater for the prediction step. However, this case occurs not only in our proposed method, but also in current trust based RS such as MoleTrust and TidalTrust as well.

VI. CONCLUSION

A new trust evaluation calculation is proposed in this work. It consists of three new factors. First, the similarity values between the users are calculated using a latent factor model instead of the co-rated items. Second, in order to identify the trustworthiness for every user in trust network, the degrees of reliability are calculated. Finally, we use the number of hops for adjusting the trust value for the remote users who are expected to be low trust as shown in the real-world application concept. From the experiment results, our proposed method is more efficiency than the classical trust based RS (MoleTrust and TidalTrust) on both accuracy and coverage.

VII. FUTURE WORK

Usually, Recommender System calculates a prediction by using opinion of the directed friends who rated the target item. However, in the trust-based recommender system, the rater who rated the target item may not be the directed friend of the target user. Therefore, there should be the way to translate the rater's opinion into directed friend's opinion. This is an objective of our near-future work.

REFERENCES

[1] J. Golbeck, "Computing and Applying Trust in Web-based Social Networks", PhD thesis, University of Maryland College Park, 2005.

[2] B. M. Sarwar, G. Karypis, J. A. Konstan, and J. Riedl, "Application of Dimensionality Reduction in Recommender Systems: a Case Study", ACM WebKDD 2000 Workshop, 2000.

[3] P. Massa and P. Avesani, "Trust-aware recommender systems", In ACM Recommender Systems Conference (RecSys), USA, 2007.

[4] J. Golbeck, "Personalizing Applications through Integration of Inferred Trust Values in Semantic Web-Based Social Networks", Proceedings of Semantic Network Analysis Workshop, Galway, Ireland, 2005.

[5] P. Massa and P. Avesani, "Trust metrics on controversial users: balancing between tyranny of the majority and echo chambers", International Journal on Semantic Web and Information Systems (IJSWIS), 3(1), 2007.

[6] J. O'Donovan , B. Smyth, "Trust in recommender systems", Proceedings of the 10th international conference on Intelligent user interfaces, San Diego, California, USA , January 10-13, 2005.

[7] R. Sinha and K. Swearingen, "Comparing recommendations made by online systems and friends.", In Proceedings of the DELOS-NSF Workshop on Personalization and Recommender Systems in Digital Libraries. Dublin, Ireland, 2001

[8] C. Ziegler and J. Golbeck, "Investigating interactions of trust and interest similarity", Decision Support Systems, v.43 n.2, p.460-475, March, 2007

[9] Y. Guo, X. Cheng, D. Dong and C. L. Rishuang Wang, "An Improved Collaborative Filtering Algorithm Based on Trust in E-Commerce Recommendation Systems," research supported by a grant from the Chinese National Science Foundation Key Project, IEEE 2010, pp 1-4

28

Comparative Study of Optimization Methods for Estimation of Sea Surface Temperature and Ocean Wind with Microwave Radiometer Data

Kohei Arai[1]

Graduate School of Science and Engineering
Saga University
Saga City, Japan

Abstract—Comparative study of optimization methods for estimation sea surface temperature and ocean wind with microwave radiometer data is conducted. The well known mesh method (Grid Search Method: GSM), regressive method, and simulated annealing method are compared. Surface emissivity is estimated with the simulated annealing and compared to the well known Thomas T. Wilheit model based emissivity. On the other hand, brightness temperature of microwave radiometer as a function of observation angle is estimated by the simulated annealing method and compares it to the actual microwave radiometer data. Also, simultaneous estimation of sea surface temperature and ocean wind speed is carried out by the simulated annealing and compared it to the estimated those by the GSM method. The experimental results show the simulated annealing which allows estimation of global optimum is superior to the other method in some extent.

Keywords—Microwave radiometer; remote sensing; sea surface temperature; nonlinear optimization theory; simulated annealing

I. Introduction

Microwave scanning radiometer allows estimation of geophysical parameters such as soil moisture, salinity, ocean wind, sea surface temperature, water vapor, cloud liquid, and so on with all weather conditions and in day and night basis [1]-[24]. Several microwave radiometers are carried on the several satellites and used for weather prediction and climate change research. One of the major concerns on the microwave radiometer is estimation accuracy of the geophysical parameters. Most of the methods for estimation of geophysical parameters are based on statistical models, regressive analysis. The estimation accuracy is not good enough because the regressive coefficients are determined with some observation conditions, areas of concerns, specific seasons. Therefore, the estimation accuracy is not good enough when the actual conditions are not matched to the conditions for the determination of regressive coefficients. Other than this, there is physical model based approaches. Through minimization processes between the actual acquired brightness temperature and the estimated brightness temperature derived from the model based method.

Microwave radiometer allows estimation of geophysical parameters such as water vapor, rainfall rate, ocean wind speed, salinity, soil moisture, air-temperature, sea surface temperature, cloud liquid, etc. based on least square method. Due to the fact that relation between microwave radiometer data (at sensor brightness temperature at the specified frequency) and geophysical parameters is non-linear, non-linear least square method is required for the estimations. Although there are some methods which allow estimation optimum solutions, Simulated Annealing: SA method [25] is just one method for finding global optimum solution.

Other methods, such as steepest descending method, conjugate gradient method, etc. gives one of local minima, not the global optimum solution. SA, on the other hand, requires huge computer resources for convergence. In order to accelerate the convergence process, not the conventional exponential function with the temperature control, but osculated decreasing function is employed for cool down function. Geophysical parameter estimation based on simulated annealing is proposed previously [6]. It takes relatively long computational time for convergence. Moreover, optimization with constraints makes much accurate estimation of geophysical parameters. Some of the constraints is relation among the geophysical parameters.

Geophysical parameters have relations each other. For instance, sea surface temperature and water vapor has a positive relation, in general. Therefore, it is better to estimate several geophysical parameters simultaneously rather than the estimation for single parameter. The proposed method is based on modified SA algorithm and is for simultaneous estimation for several geophysical parameters at once. Some experiments are conducted with Advanced Microwave Scanning Radiometer: AMSR [2] onboard AQUA satellite. Then it is confirmed that the proposed method surely works for improvement of estimation accuracy for all the geophysical parameters.

The related research works is described the following section. Then the proposed method is described followed by experiments. The experimental results are validated in the following section followed by conclusion with some discussions.

II. Related Research Works

A. Geophysical Parameter Estimation by Regressive Analysis

There are some atmospheric and ocean surface models in the microwave wavelength region. Therefore, it is possible to estimate at sensor brightness temperature (microwave radiometer) with the geophysical parameters. The real and the imaginary part of dielectric constant of the calm ocean surface is modeled with the SST, salinity (conductivity). From the dielectric constant, reflectance of the ocean surface is estimated together with the emissivity (Debue, 1929 [26]; Cole and Cole,1941 [27]). There are some geometric optics ocean surface models (Cox and Munk, 1954 [28]; Wilheit and Chang, 1980 [29]). According to the Wilheit model, the slant angle against the averaged ocean surface is expressed by Gaussian distribution function.

There is a relation between ocean wind speed and the variance of the Gaussian distribution function as a function of the observation frequency. Meanwhile the influence due to foams, white caps on the emissivity estimation is expressed with the wind speed and the observation frequency so that the emissivity of the ocean surface and wind speed is estimated with the observation frequency simultaneously. Meanwhile, the atmospheric absorptions due to oxygen, water vapor and liquid water were well modeled (Waters, 1976 [30]). Then atmospheric attenuation and the radiation from the atmosphere can be estimated using the models. Thus the at-sensor-brightness temperature is estimated with the assumed geophysical parameters.

Sea surface temperature estimation methods with AMSR data are proposed and published [31] while ocean wind retrieval methods with AMSR data are also proposed and investigated [32]. Furthermore, water vapor and cloud liquid estimation methods with AMSR data are proposed and studied [33]. The conventional geophysical parameter estimation method is based on regressive analysis with a plenty of truth data and the corresponding microwave radiometer data [34].

The brightness temperature which acquired with microwave radiometer depends on geophysical parameters, (1) Sea Surface Temperature: SST, (2) ocean Wind Speed: WS, (3) Cloud Liquid: CL, (4) Water Vapor: WV in the atmosphere, (5) Salinity: SAL, etc. Also, the brightness temperature depends on observation frequency and abservation angle.

There are physical model based approach and statistical model based approach. The most typical statistical model is proposed by Frank Wentz [33]. His model is expressed with the following second order of equation,

$$\text{Geophysical}(x) = c_0 + \Sigma\, a_i\, T_{Bi} + \Sigma\, b_i\, T_{Bi}^2 \qquad (1)$$

where Geophysical(x) denotes geophysical parameter of (x) while a_i, b_i denotes regressive coefficients while T_{Bi} denotes observed brightness temperature with microwave radiometer, respectively. When truth data of the geophysical parameter are given, then regressive coefficients are derived through regressive analysis.

Once the regressive coefficients, geophysical parameter can be estimated with the regressive equation and the observed brightness temperature. Example of the regressive coefficients for geophysical parameter of SST for Advanced Microwave Scanning Radiometer: AMSR of the 10GHz frequency band which is carried by AQUA, etc. is shown in Table 1.

TABLE I.   EXAMPLE OF THE REGRESSIVE COEFFICIENTS FOR GEOPHYSICAL PARAMETER OF SEA SURFACE TEMPERATURE

	Coefficient
c_0	122.317
a_1	2.1117
a_2	0.9079
a_3	0.4618
a_4	-0.6192
a_5	-1.0579
a_6	0.6242
a_7	-8.915
a_8	25.6123
a_9	-0.4318
a_{10}	0.2244
b_1	0.0335
b_2	0.00468
b_3	-0.0293
b_4	0.003914
b_5	-0.4718
b_6	0.000753
b_7	-5.9235
b_8	5.4932
b_9	0.001703
b_{10}	0.0001107

Although this regressive approach is convenient and ensures a marginal accuracy, it is not enough SST estimation accuracy. It depends on the ocean areas, seasons, etc. Therefore, the regressive equation with only one set of coefficients cannot cover these dependencies which results in not so good estimation accuracy.

B. Physical Model Based Approach

Minimizing the difference between a geophysical model based Brightness Temperature: T_m and an acquired actual Brightness Temperature: T_a, input parameter of geophysical parameter can be estimated. T_a, depends on the observation frequency, observation angle, and the geophysical parameters as mentioned above. The observation frequency and angle is known. Therefore, the geophysical parameters can be estimated through minimization of the difference between both of T_m and T_a,. The important thing for this approach is accurate geophysical model. There is the well known sea surface model which is proposed by Thomas T. Wilheit [28].

III.   PROPOSED MODEL

A. Basic Idea

The brightness temperatures of the several observation frequency bands can be acquired in both horizontal and vertical polarizations. If the users focus water vapor and cloud liquid, then 23 GHz and 31 GHz of observation frequency bands are needed. It is totally up to frequency dependency of brightness temperature of frequency. There is strong absorption of water vapor at the 23.235 GHz while dual frequency channels allow simultaneous estimation of water vapor and cloud liquid. Therefore, 23 GHz and 31 GHz of frequency bands are effective for water vapor and cloud liquid estimations. And if we focus SST and wind speed, only 6.925 and 10.69 GHz of observation frequency bands are taken into account. In this paper, targeted geophysical parameters are SST and Wind Speed.

The observed brightness temperature at the certain frequency band in horizontal and vertical polarizations are expressed as follows,

$$T_{bh}=\varepsilon_h(T,W)T + n_h \quad (2)$$
$$T_{bv}=\varepsilon_v(T,W)T + n_v \quad (3)$$

where T_{bh}, ε_h, T, W, n_h denotes brightness temperature, emissivity of the sea surface, Planck function of surface temperature, ocean wind speed, and observation noise for horizontal polarization while these for suffix of v denotes those for vertical polarization. Cost function of optimization processes is defined as follows,

$$\| T_{bh} - \varepsilon_h(T,W)T\|^2 + \| T_{bv} - \varepsilon_v(T,W)T\|^2 \quad (4)$$

Minimizing the cost function of equation (4) with the changing the input parameter of T and W, T and W can be estimated by using the observed brightness temperature. The most important thing for this method is how to estimate sea surface emissivity. In accordance with the Wilheit model, emissivity in horizontal and vertical polarizations is estimated. Fig.1 shows the example of the calculated emissivity.

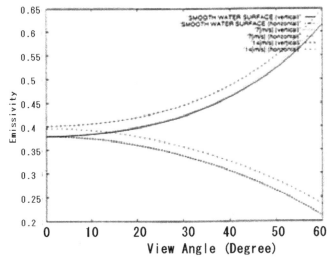

Fig. 1.   Emissivity model originated from the Wilheit model

B. Simulated Annealing

The proposed geophysical parameter estimation here is based on the physical model based approach. Minimization of the difference between T_m and T_a, is total identical to optimization model. The problem situated here is how to find the global optimum. Only the solution for that is Simulated Annealing: SA. It, however, takes huge computational resources. Therefore, the proposed model here is modified SA model which has a limitation of iteration. Namely, iterations is stopped at the previously designated upper limit. Therefore, the proposed modified SA is not real SA essentially because the solution does not reach to a global optimum. In the case of the estimation of geophysical parameter with microwave radiometer data, residual error is gradually reduced when the current solution is approaching to a global optimum (the solution does not jump in this stage). Therefore, we may stop the iteration at the certain number of iterations or elapsed computation time.

IV. Experiments

A. Validation of Emissivity Model

As an example of brightness temperature, the brightness temperature of Microwave Imager: TMI onboard Tropical Rainfall Measuring Mission: TRMM satellite of 10.65 GHz for horizontal and vertical polarizations is shown in Fig.2. The actual brightness temperature as a function of observation angle is plotted in Fig.2. The location of intensive study area is the following,

Longitude and latitude: 31.6 North, 109.1 East

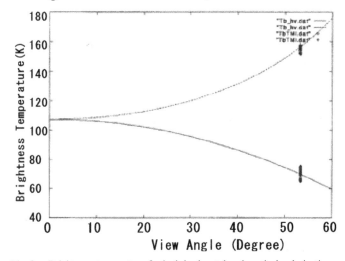

Fig. 2.   Brightness temperature for both horizontal and vertical polarizations derived from the proposed physical model based method and actual received brightness temperature with TRMM/TMI of 10.65 GHz of frequency channel acquired on June 2 1998

The actual brightness temperature data are situated at the observation angle of 53 degree because the brightness temperature for horizontal polarization does not depend on ocean wind speed at the observation angle of 53 degree. The estimated brightness temperature is coincident to the actual brightness temperature. This is the same thing for the different observation frequency and both of horizontal and vertical polarizations. Therefore, emissivity model originated from the Wilheit model is validated.

The actual TMI data of the location (Longitude and latitude: 31.6 North, 109.1 East) which is acquired on June 2 1998 is used for the experiment. From the measured data at the site, it is found that SST=294 K, WS=7 m/s, Salinity=36ppm, respectively. The truth geophysical parameters of SST are set at 292 K, 294 K, and 296 K while that of wind speed is set at 7 m/s. The brightness temperature estimated by the proposed physical model based method. The results are as follows,

1) Theoretical brightness temperature: 70.549
The mean of observed brightness temperature: 100.589

The standard deviation of the actual brightness temperature: 9.634

2) Theoretical brightness temperature: 156.574
The mean of observed brightness temperature: 173.814

The standard deviation of the actual brightness temperature: 2.906

   3) Theoretical brightness temperature: 70.3
The mean of observed brightness temperature: 100.589

The standard deviation of the actual brightness temperature: 9.635

   4) Theoretical brightness temperature: 155.905
The mean of observed brightness temperature: 173.814

The standard deviation of the actual brightness temperature: 2.906

   5) Theoretical brightness temperature: 70.081
The mean of observed brightness temperature: 100.589

The standard deviation of the actual brightness temperature: 9.635

   6) Theoretical brightness temperature: 155.284
The mean of observed brightness temperature: 173.814

The standard deviation of the actual brightness temperature: 2.906

Thus the proposed model is validated with some extent of estimation errors.

B. Comparison of Estimated Sea Surface Temperature

In order to show the advantage of the proposed method, the estimated SST and WS with the proposed method is compared to those with the statistical model based method, conventional GSM method. Fig.3 shows the results from the comparative study. In the experiment, observation frequency channels are set at 6.925 GHz and 10.69 GHz. Fig.3 shows RMS error of SST and WS with the designated biases of plus minus 1(K), 3(K) for SST and plus minus 1(m/s), 3(m/s) for WS as well as without any bias for the proposed SA based method and the conventional GSM method.

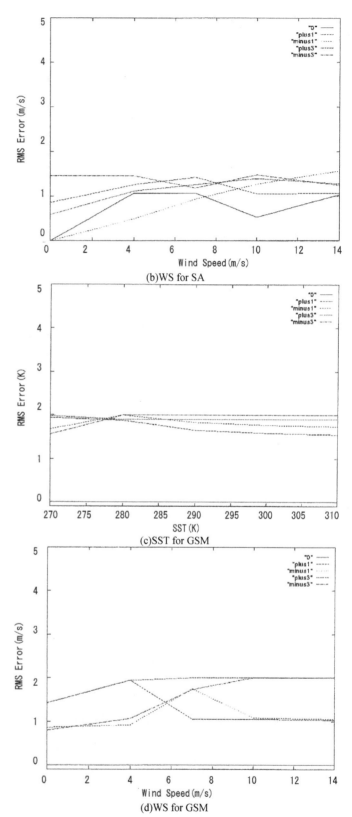

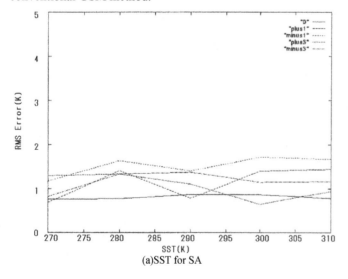

Fig. 3. RMS error of SA and GSM for the estimation of SST and WS with the designated bias of plus minus 1(K), 5(K) for SST and plus minus 1(m/s), 3(m/s) for WS as well as without any bias

As the results, it is found that RMS error of the proposed SA based method is superior to the conventional GSM method by approximately 50 (%) for both of SST and WS. Also, it is found that the RMS error is getting large in accordance with increasing of additive biases.

Root Mean Square: RMS error is evaluated and compared. Table 2 shows the results of RMS errors for the statistical model based method, GSM method and the proposed SA method.

TABLE II. RMS ERROR COMPARISONS AMONG THE STATISTICAL MODEL BASED METHOD, GSM METHOD AND THE PROPOSED SA METHOD

Method	SST(K)	WS(m/s)
Statistical Approach	0.46	0.66
GSM	0.274	0.327
SA	0.492	0.435

If the biases are added to the theoretical SST and WS intentionally, then the RMS errors are varied as shown in Table 3 for GSM method while those for SA method is shown in Table 4.

TABLE III. RMS ERRORS OF SST AND WS FOR GSM METHOD AS A FUNCTION OF DEVIATIONS

Biases	SST(K)	WS(m/s)
0	0	0
+1	1.722	1.302
-1	1.805	1.135
+3	1.916	1.874
-3	1.912	1.520

TABLE IV. RMS ERRORS OF SST AND WS FOR SA METHOD AS A FUNCTION OF DEVIATIONS

Deviation	SST(K)	WS(m/s)
0	0.809	0.739
+1	1.146	1.132
-1	1.520	0.853
+3	1.064	1.127
-3	1.169	1.363

By using the actual brightness temperature data of TMI, SST and WS estimation errors are evaluated. Table 5 shows the estimated SST and WS as well as RMS errors for the cases of SST are set at 292, 294 and 296(K). In these cases, the estimated SST and WS are compared to the actual TMI data derived SST and WS. RMS error of SST shows around 4.5(K) while that of WS is approximately 3.7(m/s) respectively.

TABLE V. ESTIMATED SST AND WS AS WELL AS RMS ERRORS FOR THE CASES OF SST ARE SET AT 292, 294 AND 296(K)

Case	SST(K)	RMSE(SST)	WS(m/s)	RMSE(WS)
296(K)	291.8	4.297(K)	6.708	4.036(m/s)
294(K)	289.498	4.619(K)	7.313	2.997(m/s)
292(K)	287.348	4.753(K)	6.604	4.156(m/s)

As the results from the experiments, it is found that the proposed SA based method is superior to the statistical model based method and the GSM method.

V. CONCLUSION

Comparative study of optimization methods for estimation sea surface temperature and ocean wind with microwave radiometer data is conducted. The well known mesh method (Grid Search Method: GSM), regressive method, and simulated annealing method are compared. Surface emissivity is estimated with the simulated annealing and compared to the well known Thomas T. Wilheit model based emissivity. On the other hand, brightness temperature of microwave radiometer as a function of observation angle is estimated by the simulated annealing method and compares it to the actual microwave radiometer data. Also, simultaneous estimation of sea surface temperature and ocean wind speed is carried out by the simulated annealing and compared it to the estimated those by the GSM method. The experimental results show the simulated annealing which allows estimation of global optimum is superior to the other method in some extent.

As the results, it is confirmed that the well known Wilheit sea surface model is appropriate for estimation of geophysical parameters. Also, it is confirmed that the statistical model based method for geophysical parameter estimation shows marginal estimation accuracies of SST and WS (0.46(K) and 0.66(m/s), respectively). It is found that the estimated SST and WS are compared to the actual TMI data derived SST and WS. RMS error of SST for the proposed SA based method shows around 4.5(K) while that of WS is approximately 3.7(m/s) respectively.

ACKNOWLEDGMENT

The author would like to thank Ms. Emi Shimomura of Saga University for her effort to conduct the experiments.

REFERENCES

[1] Kohei Arai, Preliminary Assessment of Radiometric Accuracy for MOS-1 Sensors, International Journal of Remote Sensing, Vol.9, No.1, pp.5-12, Apr.1988.

[2] K.Tachi, Kohei Arai and Y.Satoh, Advanced Microwave Scanning Radiometer -Requirements and Preliminary Design Study-, IEEE Trans. on Geoscience and Remote Sensing, Vol.27, No.2, pp.177-183, Jan.1989.

[3] Kenbu Teramoto, Kohei Arai, Toshio Imatani, Antenna Pattern Correction of Microwave Radiometer Based on 凸 Projection Method, Journal of Remote Sensing Society of Japan, Vol.15, No.4, pp.38-49, Sep.1994.

[4] Kohei Arai, E.Ishiyama and Y.Terayama, A method for ice concentration estimation with microwave radiometer data by means of inversion techniques, Advances in Space Research, Vol.16, No.10, pp.129-132, A31-32, Jul.1994.

[5] Kohei Arai, Kenbu Teramoto, Algorithm of Multi-Frequency Microwave Radiometer Data Correction, Notes of Meteorological Research, No.187, pp.169-176, Aug.1996.

[6] Kohei Arai and J.Sakakibara, Estimation of SST, wind speed and water vapor with microwave radiometer data based on simulated annealing, Advances in Space Research, 37, 12, 2202-2207, 2006.

[7] Kohei Arai, Kenta Azuma, Ocean wind speed estimation accuracy improvement using microwave radiometer data taking into account wind direction dependency of the observed brightness temperature, Journal of Remote Sensing Society of Japan, 27, 5, 465-473, 2007

[8] Kohei Arai, Kenta Azuma, Rainfall rate estimation with AMSR data on board AQUA satellite taking into account geometric relation between the microwave radiometer and observation targets, Journal of Japan Society of Photogrammetry and Remote Sensing, 49, 1, 32-40, 2010

[9] K.Arai, Simultaneous estimation of geophysical parameters with microwave radiometer data based on accelerated Simulated Annealing: SA, International Journal of Advanced Computer Science and Applications, 3, 7, 90-95, 2012.

[10] K.Arai, Nonlinear Optimization Based Sea Surface Temperature: SST Estimation Methods with Remote Sensing Satellite Based Microwave

Scanning Radiometer: MSR Data, International Journal of Research and Reviews in Computer Science (IJRRCS) Vol. 3, No. 6, 1881-1886, December 2012, ISSN: 2079-2557

[11] Kohei Arai, Data fusion between microwave and thermal infrared radiometer data and its application to skin sea surface temperature, wind speed and salinity retrievals, International Journal of Advanced Computer Science and Applications, 4, 2, 239-244, 2013.

[12] K.Arai, T.Igarashi and C.Ishida, Evaluation of MOS-1 Microwave Scanning Radiometer (MSR) data in field experiments, Proc. of the 18th International Symposium on Remote Sensing of Environment, 1-8, 1984.

[13] K.Arai, T.Igarashi and Y.Takagi, Emissivity model of snowpack for passive microwave observations, Proc. of the 36th International Astronautics Federation (IAF) Congress, IAF-85-98, 1-8, 1985.

[14] K.Arai and T.Suzuki, Beam compressed microwave scanning radiometer, Proc. of the IGARSS'89, I1-2, 268-270, 1989.

[15] Y.Itoh, K.Tachi, Y.Sato and K.Arai, Advanced Microwave Scanning Radiometer: AMSR, Preliminary study, Proc. of the IGARSS'89, I1-4, 273-276, 1989.

[16] K.Arai, K.Teramoto and T.Imatani, Influence due to antenna pattern changes in brightness temperature estimation for a space based microwave radiometer, Proc. of the European ISY Conference, 311-316, 1992.

[17] K.Arai, E.Ishiyama and Y.Terayama, Method for ice concentration estimation with microwave scanning radiometer data by means of inversion, Proc. of the 30th COSPAR Congress, A3.1-032, 1993.

[18] Arai,K., E.Ishiyama and Y.Terayama, A method for ice concentration estimation with microwave radiometer data by means of inversion techniques, Proceedings of the 30th COSPAR Symposium,., A31-032, 1994

[19] Arai,K., New algorithms for ice concentration estimation with passive microwave data, Proceedings of the 1st ADEOS-II Science Symposium, Nov., 1994.

[20] K.Teramoto, K.Arai and T.Imatani, Antenna Pattern Correction for Microwave Radiometry Using A Prior Knowledge Based on Projection Convex Sets Method, Proceeding of the International Geoscience and Remote Sensing Symposium, IGARSS'95, Florence,July 1995.

[21] K.Teramoto and K.Arai, POCS Based Array Processing in Incoherent Microwave Radiometric Image Reconstruction, Proceedings of the ICASSP'96, SSAP#615, Atlanta, May 1996.

[22] Kohei Arai, Sea Surface Temperature (SST) estimation with microwave radiometers by means of simulated annealing based on an ocean surface model, Proceedings of the NASA Oceanography Scientific Conference, Florida, USA, 2001.

[23] Kohei Arai, Sea Surface Temperature (SST) retrieval with microwave radiometer data based on simulated annealing, Proc. of the Kyushu Brunch Symposium of the electronics related Society of Japan, Asian Session, 2001.

[24] Kohei Arai and Jun Sakakibara, Estimation of SST, wind speed and water vapor with microwave radiometer data based on simulated

annealing, Abstracts of the 35th Congress of the Committee on Space Research of the ICSU, A1.1-0130-04, (2004)

[25] S. Kirkpatrick, C.D. Gelett, M.P. Cecchi, Optimization by simulated annealing, Science, 220, 621-630, 1983.

[26] Debue, R. Polar Molecules, Chemical Catalog, New York, 1929.

[27] Cole, K.S., Cole, R.H. Dispersion and absorption in dielectrics. J. Chem. Phys. 9, 341–351, 1941.

[28] Cox, C.S., Munk, W.H. Measurement of the roughness of the sea surface from photographs of the sun_s glitter. J. Opt. Sci. Am. 44, 838–850, 1954.

[29] Wilheit, T.T., Chang, A.T.C. An algorithm for retrieval of ocean surface and atmospheric parameters from the observations of the Scanning Multichannel Microwave Radiometer (SMMR). Radio Sci. 15, 525–544, 1980.

[30] Waters, J.R. Absorption and emission by atmospheric gasses. in: Meeks, M.L. (Ed.), Methods of Experimental Physics, vol. 12B.Academic, Orland, 1976 (Chapter 2.3).

[31] Dong, SF; Sprintall, J; Gille, ST, Location of the antarctic polar front from AMSR-E satellite sea surface temperature measurements, *JOURNAL OF PHYSICAL OCEANOGRAPHY*, Nov 2006, 2075-2089.

[32] Konda, M., A. Shibata, N. Ebuchi, and K. Arai, An evaluation of the effect of the relative wind direction on the measurement of the wind and the instantaneous latent heat flux by Advanced Microwave Scanning Radiometer, *J. Oceanogr.*, vol. 62, no. 3, pp. 395-404, 2006.

[33] Cosh, M. H., T. J. Jackson, R. Bindlish, J. Famiglietti, and D. Ryu, A comparison of an impedance probe for estimation of surface soil water content over large region, *Journal of Hydrology*, vol. 311, pp. 49-58, 2005.

[34] Wentz, F. AMSR Ocean Algorithm, second version of ATBD, NASA/GSFC, 2000.

29

Digital Library of Expert System Based at Indonesia Technology University

Dewa Gede Hendra Divayana[1]
Chair of Information Technology Department
Indonesia Technology University
Bali, Indonesia

I Made Sugiarta[3]
Lecture of Mathematics Education
Ganesha University of Education
Bali, Indonesia

I Putu Wisna Ariawan[2]
Lecture of Mathematics Education
Ganesha University of Education
Bali, Indonesia

I Wayan Artanayasa[4]
Chair of Sport & Health Education Department
Ganesha University of Education
Bali, Indonesia

Abstract—**Digital library is a very interesting phenomenon in the world of libraries. In this era of globalization, the digital library is needed by students, faculty, and the community in the search for quick reference through internet access, so that students, faculty, and the community does not have to come directly to the library. Accessing collections of digital libraries can also be done anytime and anywhere. Digital Library development also occurred at Indonesia Technology University. That University offers a digital library based of expert system. The concept of digital library is utilizing science expert system in the process of cataloging and searching digital collections. By using this digital library based of expert system, users can search the collection, reading collection, and download the desired collection by online system. The digital library based of expert system at Indonesia Technology University is built using the PHP programming language, MySQL database as a data base management system, and developed the method of forward chaining and backward chaining as inference engine.**

Keywords—*Digital Library; Expert System; Forward Chaining; Backward Chaining*

I. INTRODUCTION

In the current era of globalization, information technology has a very important role in supporting the activities carried out by the community. The development of information technology very rapidly leads us toward the use of documents in digital form.

Library as one of the sources of knowledge needs to be organized and presented for the system services can be accessed from anywhere and anytime with the involvement of information technology in the form of an integrated system, so the user does not have to come directly to the library. This phenomenon is supported by the use of the Internet that facilitate processing, dissemination, and accessing information in digital form from anywhere and anytime. The development of increasingly sophisticated technology also affects the formation of a library in a digital form.

Digital library is a very interesting phenomenon in the world of libraries. In this era of globalization, the digital library is needed by students, faculty, and the community in the search for quick reference through internet access, so that students, faculty, and the community do not have to come directly to the library. Accessing collections of digital libraries can also be done anytime and anywhere. Digital Library development also occurred at Indonesia Technology University. That University offers a digital library based of expert system. The concept of digital library is utilizing science expert system in the process of cataloging and searching digital collections. By using this digital library based of expert system, users can search the collection, reading collection, and download the desired collection by online system.

II. LITERATURE REVIEW

A. Digital Library

In [1], The library has become digital: processes such as mass digitization, web archiving, and to a smaller extent digital preservation, are no longer isolated but disseminated among relevant production teams within the library.

In [2], A digital library (DL) is a library in which collections are stored in digital formats (as opposed to print, microform, or other media) and accessible by computers.

In [3], A digital library is a particular kind of information system which consists of a set of components, typically a collection (or collections) of computer system offering diverse services on a technical infrastructure, people, and the environment or usage.

In [4], Digital libraries are set of library activities and services which facilitate by electronic means the processing, transmission and display of information.

B. Expert System

In [5], an expert system is a computer program designed to simulate the problem-solving behaviour of a human who is an expert in a narrow domain or discipline. An expert system is normally composed of a knowledge base (information, heuristics, etc.), inference engine (analyses the knowledge base), and the end user interface (accepting inputs, generating outputs). The concepts for expert system development come from the subject domain of artificial intelligence (AI), and require a departure from conventional computing practices and programming techniques.

In [6], an Expert system is software that simulates the performance of human experts in a specific field. Today's expert systems have been used in many areas where require decision making or predicting with expertise.

In [7], an expert system is a set of programs that manipulate encoded knowledge to solve problems in a specialized domain that normally requires human expertise.

In [8], Expert System is a branch of Artificial Intelligence that makes extensive use of specialized knowledge to solve problems at the human expert level.

In [9], the Expert System (ES) is one of the well-known reasoning techniques that is utilized in diagnosis applications domain. In ES, human knowledge about a particular expertise to accomplish a particular task is represented as facts and rules in its knowledge base.

In [10], an expert system is the computer system that emulates the behaviour of human experts in a well-specified manner, and narrowly defines the domain of knowledge. It captures the knowledge and heuristics that an expert employs in a specific task. An overview of current technologies applied with an expert system that is developed for Database Management System, Decision Support System, and the other Intelligent Systems such as Neural Networks System, Genetic Algorithm, etc.

In [11], expert system is an artificial intelligence system that combines knowledge base with inference engine so that it can adopt the ability of the experts into a computer, so the computer can solve problems such as the often performed by experts.

C. Forward Chaining

The inference engine contains the methodology used to perform reasoning on the information in the knowledge base and used to formulate conclusions. Inference engine is the part that contains the mechanism and function of thought patterns of reasoning systems that are used by an expert. The mechanism will analyze a specific problem and will seek answers, conclusions or decisions are best. Because the inference engine is the most important part of an expert system, that plays a role in determining the effectiveness and efficiency of the system.

There are several ways that can be done in performing inference, including the Forward Chaining. In [12], an inference engine using forward chaining searches the inference rules until it finds one where the IF clause is known to be true.

When found it can conclude, or infer, the THEN clause, resulting in the addition of new information to its dataset. In other words, it starts with some facts and applies rules to find all possible conclusions. Therefore, it is also known as Data Driven Approach.

In [13], forward chaining is matching facts or statements starting from the left (first IF).

D. Backward Chaining

Also in [13], backward chaining is matching facts or statements starting from the right (first THEN). In other words, the reasoning starts from the first hypothesis, and to test the truth of this hypothesis to look for the facts that exist in the knowledge base.

In [14], an inference engine using backward chaining would search the inference rules until it finds one which has a THEN clause that matches a desired goal. If the IF clause of that inference rule is not known to be true, then it is added to the list of goals (in order for goal to be confirmed it must also provide data that confirms this new rule). In other words, this approach starts with the desired conclusion and works backward to find supporting facts. Therefore, it is also known as Goal-Driven Approach.

In [15], backward chaining systems are good for diagnostic and classification tasks, but they are not good for planning, design, process monitoring, and quite a few other tasks. In backward chaining, the search is goal directed, so rules can be applied that are necessary to achieve the goal.

E. Digital Library Based of Expert System

In [16], Digital Library Based of Expert System is a digital library that implements the basic concept of expert system includes a knowledge base and the inference engine in helping service mechanism. The knowledge base is used for storage and cataloging process of making collections that exist in digital library, while the inference engine is used to search the detail collection that available in digital libraries such as the ability to work like an expert.

III. METHODOLOGY

A. Object dan Research Site

1) Research Object is Digital Library Based of Expert System
2) Research Site at Indonesia Technology University.

B. Data Type

In this research, the authors use primary data, secondary data, quantitative data and qualitative data.

C. Data Collection Techniques

In this research, the authors use data collection techniques such as interviews, observation, and documentation.

D. Analysis Techniques

Analysis techniques used in this research is descriptive statistical.

IV. RESULT AND DISCUSSION

A. Result

1) Early Trial
At this early trial, the authors conducted a limited scale testing of the digital library based of expert system that have been made previously by involving five staff at Indonesia Technology University to perform *white box* and *black box* testing. This test can be done by giving 10 questionnaires

early trials digital library based of expert system to staff at Indonesia Technology University.

Diagram form of answers score percentage given by the respondents in early trial can be described as follows:

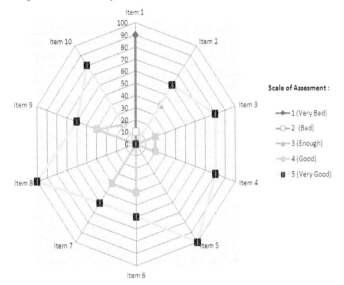

Fig. 1.   Percentage Diagram of Respondents Answer Score In Early Trial

Based on the diagram above, it can be seen that the results of early trials of the digital library based of expert system, find a constraint that is the answer to a very bad score by 90% of the questions on the questionnaire 1st initial trials. This is due to the unavailability of the form for the create of a new username and password for administrator in the future if there is a mutation of the staff who operate the digital library based of expert system. Given these constraints, then the system needs to be revised again.

2) Field Trial

At this field trial, the authors tested in a larger scale, involving an expert is understood about the digital library and ten staff at Indonesia Technology University. This test can be done by giving 15 questionnaires field trials for digital library based of expert system to the librarian and staff at Indonesia Technology University.

Diagram form of answers score percentage given by the respondents in field trial can be described as follows:

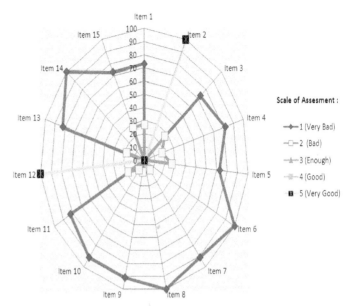

Fig. 2.   Percentage Diagram of Respondents Answer Score In Field Trial

Based on the diagram above, it can be seen that the results of a field trial of the digital library based of expert system, the presence of obstacles that scores are very bad answer the score of 73% to the question 1st, 3rd, 5th, and 15th, 82% of the questions 4th, 11th and 13th, at 91% of the questions 7th, 9th, 10th, and at 100% of the questions 6th, 8th and 14th on field trial questionnaire.

This is due to the unavailability of the collection input or edits form if in the future there is a new journal and article. Of the constraints are found, then the system needs to be revised to obtain collection more interactive and dynamic.

3) Usage Test

At this usage test, the authors conducted a trial involving with the use of 50 people (users). The test is performed to test the operation of the overall form available on digital library based of expert system that has undergone revisions to field trials. This test can be done by giving the user satisfaction questionnaire to users who visited Indonesia Technology University.

Diagram form of answers score percentage given by the respondents in usage test can be described as follows:

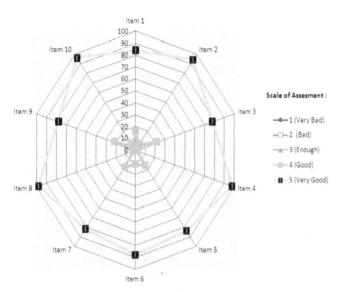

Fig. 3. Percentage Diagram of Respondents Answer Score In Usage Test

Based on the diagram above, it can be seen that the results of testing the use of the digital library based of expert system outline already looks very good and not found again the constraints in terms of technical operation (inputing and editing a new collection) as well as the principle method of expertise (forward chaining and bacward chaining method). This is evidenced by the percentage scoring very good response by 78% of questions 3rd, and 9th. Percentage scoring very good response by 82% of questions 7th. Percentage scoring very good response by 84% of questions 1st and 5th. Percentage scoring very good response by 88% of statement 6th. Percentage scoring very good response by 94% of question 2nd. Percentage scoring very good response by 96% of questions 10th.

As well as scoring 98% of the questions 4th and 8th trial usage. And it would be even better if the digital library based of expert system added help fasility form for written in accordance with the suggestions of the respondents to the improvement of the system, so as to explain the performance of the expert system and the function of the buttons in the design of digital library based of expert system overall with easy to understand and simple language.

B. Discussion

1) Knowledge Base

Knowledge base is used to build the expert system obtained from multiple sources of knowledge, containing data on digital library based of expert system at Indonesia Technology University. The knowledge base contained in a digital library based of expert system at Indonesia Technology University can be described by the following table.

TABLE I. KNOWLEDGE BASE ON DIGITAL LIBRARY BASED OF EXPERT SYSTEM AT INDONESIA TECHNOLOGY UNIVERSITY

No	Properties/Identity	Journal	Book	Magazine	Article
1.	Collection Code	√	√	√	√
2.	Collection Name	√	√	√	√
3.	Author	√	√	√	√
4.	Publisher	√	√	√	√
5.	Year of Issue	√	√	√	√
6.	Pages	√	√	√	√
7.	Place of Issue	√	√	√	√
8.	ISSN	√	√	√	√

2) Shows forward chaining concept in digital library based of expert system

Application of forward chaining method in a digital library can be explained by the following chart:

For example : search by collection code

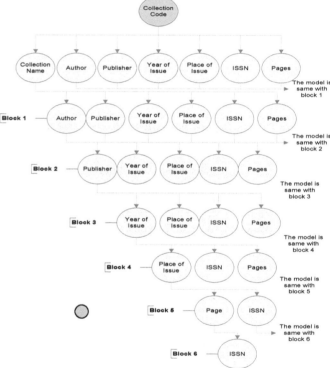

Explanation :   = Search by

Fig. 4. Forward Chaining Concept in Digital Library Based of Expert System

3) Shows backward chaining concept in digital library based of expert system

Application of backward chaining method in a digital library can be explained by the following chart:

For example : search by ISSN

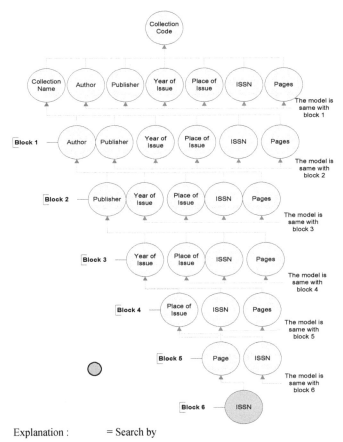

Explanation :   = Search by

Fig. 5. Backward Chaining Concept in Digital Library Based of Expert System

4) Trials forward chaining and backward chaining performed by respondents

Respondents who did this trial was a expert and 19 staff at Indonesia Technology University conducted the field trials. The trial results are shown in table II.

Based on the table results of *trials forward chaining and backward chaining performed by respondents* mentioned above, it can be analyzed that the forward chaining and backward chaining method has been run in accordance with the rule of expert system inference engine.

To view the *forward chaining and backward chaining* method has been run in accordance with the rules can be seen in the percentage diagram of rules conformance testing.

TABLE II.   Trials Forward Chaining and Backward Chaining Method

Respondent	Method	CC	CN	AT	PL	YI	PG	PI	IS	%
RS.01	FC	√	√	√	√	√	√	√	√	100
	BC	√	√	√	√	√	√	√	√	100
RS.02	FC	√	√	√	√	√	√	√	√	100
	BC	√	√	√	√	√	√	√	√	100
RS.03	FC	√	√	√	√	√	√	√	√	100
	BC	√	√	√	√	√	√	√	√	100
RS.04	FC	√	√	√	√	√	√	√	√	100
	BC	√	√	√	√	√	√	√	√	100
RS.05	FC	√	√	√	√	√	√	√	√	100
	BC	√	√	√	√	√	√	√	√	100
RS.06	FC	√	√	√	√	√	√	√	√	100
	BC	√	√	√	√	√	√	√	√	100
RS.07	FC	√	√	√	√	√	√	√	√	100
	BC	√	√	√	√	√	√	√	√	100
RS.08	FC	√	√	√	√	√	√	√	√	100
	BC	√	√	√	√	√	√	√	√	100
RS.10	FC	√	√	√	√	√	√	√	√	100
	BC	√	√	√	√	√	√	√	√	100
RS.11	FC	√	√	√	√	√	√	√	√	100
	BC	√	√	√	√	√	√	√	√	100
RS.12	FC	√	√	√	√	√	√	√	√	100
	BC	√	√	√	√	√	√	√	√	100
RS.13	FC	√	√	√	√	√	√	√	√	100
	BC	√	√	√	√	√	√	√	√	100
RS.14	FC	√	√	√	√	√	√	√	√	100
	BC	√	√	√	√	√	√	√	√	100
RS.15	FC	√	√	√	√	√	√	√	√	100
	BC	√	√	√	√	√	√	√	√	100
RS.16	FC	√	√	√	√	√	√	√	√	100
	BC	√	√	√	√	√	√	√	√	100
RS.17	FC	√	√	√	√	√	√	√	√	100
	BC	√	√	√	√	√	√	√	√	100
RS.18	FC	√	√	√	√	√	√	√	√	100
	BC	√	√	√	√	√	√	√	√	100
RS.19	FC	√	√	√	√	√	√	√	√	100
	BC	√	√	√	√	√	√	√	√	100
RS.20	FC	√	√	√	√	√	√	√	√	100
	BC	√	√	√	√	√	√	√	√	100

Explanation :
FC   : Forward Chaining   BC : Backward Chaining
CC   : Collection Code   CN : Collection Name
AT   : Author   PL : Publisher
YI   : Year of Issue   PG : Pages
PI   : Place of Issue   IS   : ISSN

As for the form of percentage diagram of rules conformance testing given by the respondents can be described as follows:

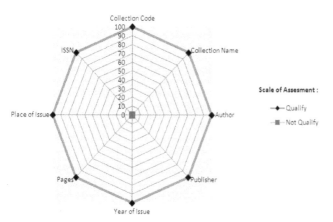

Fig. 6.   Answer Percentage Diagram of Rules Conformance Testing

Based on the diagram above, it can be seen that the results of testing the suitability of digital library based of expert system rules is an outline already looks qualify. This is evidenced by the answer percentage of collection code, collection name, author, publisher, year of issue, pages, place of issue and ISSN according to the rules of backward chaining and forward chaining in the field of testing and each get a percentage of 100%.

5) Implementation of Digital Library Based of Expert System

a) Main Menu Page

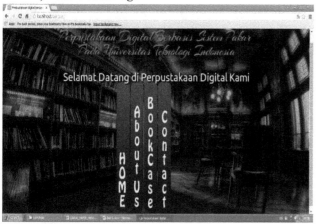

Fig. 7.   Main Page

This main menu page used as link to home menu, about us menu, book case menu, and contact menu.

b) Membership Registration Form

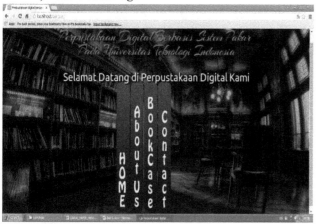

Fig. 8.   Membership Registration Form

This form is used by users who will register to become a member, that all fields must be filled in accordance with the user's identity, and then click the register button to register and click cancel button to cancel the registration.

c) Member Login Form

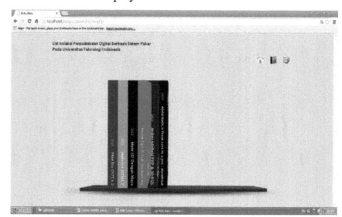

Fig. 9.   Member Login Form

This form contains the username and password for the members, so that members can search and download a collection of all the collections that exist in digital library based of expert system at Indonesia Technology University.

d) Latest Collection Menu Page

After login, the user will go to the latest collection menu, the menu contains the latest collections of Digital Library based of expert system. The following figure is a latest collection menu display.

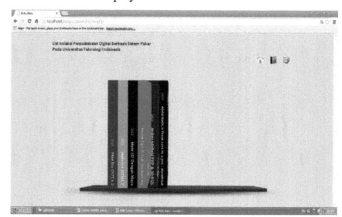

Fig. 10.   Latest Collection Menu Page

e) Collection List Page

Fig. 11.   Collection List Page

This collection list page is a page that is used to display a list of collections that exist in the digital library based of expert system at Indonesia Technology University and displays the complete details of the collection identity.

f) Collection Search Page

This form contains the search facilities of data on existing collections in digital library based of expert system using a keyword based input the desired category.

Fig. 12. Collection Search Page

g) Collection Search Result Page

Collection Search Result Page is the page serves to display the collection search result page. The following figure is a collection search results page display.

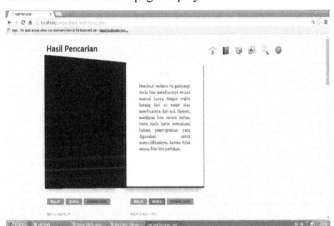

Fig. 13. Collection Search Result Page

h) Administrator Login Form

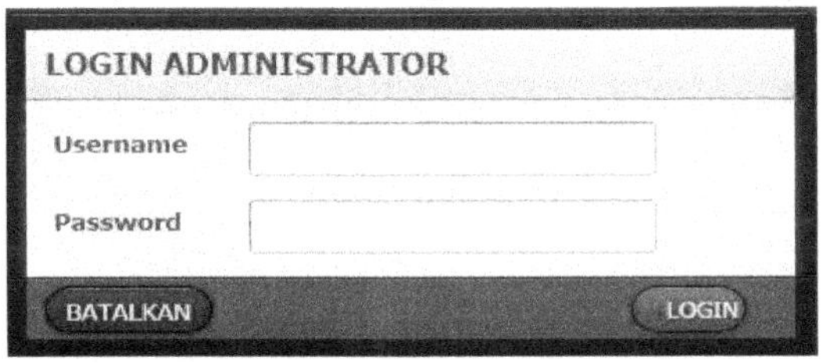

Fig. 14. Administrator Login Form

This form contains the username and password for the Admin, click on login button if you want to login and click cancel button if you want to cancel.

i) Administrator Page

Administrator page is a page that is used by administrators to system perform processing, such as processing of knowledge base and put the rule into an expert system inference engine. The following figure is a administrator page display.

Fig. 15. Administrator Page

V. CONCLUSIONS

Based on the analysis that has been made and the results of the discussion in the previous section, then some conclusions can be drawn as follows:

a) With this digital library based of expert system can help the performance of existing conventional systems towards a computerized system, so as to speed up service.

b) With digital library based of expert system makes it easy for users to search the collection (books, journals, magazines, and articles) in digital form through the facilities of the internet without having to come directly to the campus library.

c) With the digital library digital library based of expert system, the concept of a real expert system can already be applied and useful in solving the existing problems in human life, especially in the field of library.

d) By looking at the optimal use of knowledge base and also the use of forward chaining and backward chaining inference engine implemented as in this digital library based of expert system, it can also be the accuracy result of the system is working at 100%.

ACKNOWLEDGMENT

The authors express their gratefulness to staff at Indonesia Technology University for inspiring words and allowing them to use the examination data. They generously thank Mr. Dayung, President of Indonesia Technology University, and Mr. Ketut Semadi, Dean of Computer Faculty, Indonesia Technology University.

REFERENCES

[1] E.Bermès, and L.Fauduet, "The Human Face of Digital Preservation: Organizational and Staff Challenges, and Initiatives at the Bibliothèque

nationale de France," in The International Journal of Digital Curation vol.6, 2011,pp.226-237.

[2] D.I.Greenstein, Thorin, and S. Elizabeth. The Digital Library: A Biography. Washington: Digital Library Federation, 2002.

[3] N. Fuhr, G. Tsakonas, T. Aalberg, M.Agosti, P. Hansen, and S. Kapidakis,"Evaluation of digital libraries," in International Journal on Digital Libraries vol. 8, 2007, pp.21–38.

[4] K. Towolawi, and Oluwakemi, "School Library Media Specialist's Awareness and Perception of Digital Library Services: A Survey," in Ozean Journal of Social Sciences vol. 6, 2013, pp.77-89.

[5] Y.A. Nada, "Construction of Powerful Online Search Expert System Based on Semantic Web," in International Journal of Advanced Computer Science and Applications vol.4, 2013, pp.181-187.

[6] Y. Qu, F. Tao, and H. Qui, "A Fuzzy Expert System Framework Using Object Oriented Techniques," in IEEE Pacific-Asia Workshop on Computational Intelligence and Industrial Application, 2008, pp. 474-477.

[7] Y. Erdani, "Developing Recursive Forward Chaining Method in Ternary Grid Expert Systems," in International Journal of Computer Science and Network Security, vol.11, No.8, 2011, pp.126-130.

[8] J.C. Giarratano, and G. Riley, Expert Systems : Principles and Programming 4th Edition. USA : PWS Publishing Co, 2004.

[9] A. A. Hopgood, Intelligent Systems for Engineers and Scientists (2nd Edition). USA : CRC Press, 2001.

[10] E. Turban, and J. E. Aronson, Decision Support Systems and Intelligent System. NJ, USA: Prentice-Hall Inc, 2001.

[11] H. Divayana, "Development of Duck Diseases Expert System with Applying Alliance Method at Bali Provincial Livestock Office," in International Journal of Advanced Computer Science and Applications, Vol. 5, No. 8, 2014,pp.48-54.

[12] RC Chakraborty, 2010. Knowledge Representation: AI Course Lecture 15-22.Retrieved December, 2012, from www.myreaders.info/html/aritificial_intelligence.html

[13] S.Kusumadewi, Artificial Intelligence (Technique and Application) 1st Edition. Yogyakarta : Graha Ilmu, 2003.

[14] K. Donna, A Comparison of Forward and Backward Chaining Algorithms For use in a Technical Support Expert System Used for Diagnosing Computer Virus Issues. Chico : Computer Science Department California State University, 2009.

[15] T. Sharma, et.al., "Study of Difference Between Forward and Backward Reasoning," in International Journal of Emerging Technology and Advanced Engineering, vol. 2 No. 10, 2012, pp.271-273.

[16] H. Divayana, et.al., Digital Library Based of Expert System at Indonesia Technology University. Bali : Indonesia Technology University, 2015.

30

New concepts of fuzzy planar graphs

Sovan Samanta

Department of Applied
Mathematics with Oceanology
and Computer Programming,
Vidyasagar University,
Midnapore - 721 102, India.
email: ssamantavu@gmail.com

Anita Pal

Department of Mathematics,
National Institute of
Technology Durgapur,
Durgapur-713209, India.
e-mail: anita.buie@gmail.com

Madhumangal Pal

Department of Applied
Mathematics with Oceanology
and Computer Programming,
Vidyasagar University,
Midnapore - 721 102, India.
email: mmpalvu@gmail.com

Abstract—Fuzzy planar graph is an important subclass of fuzzy graph. Fuzzy planar graphs and its several properties are presented. A very close association of fuzzy planar graph is fuzzy dual graph. This is also defined and several properties of it are studied. Isomorphism on fuzzy graphs are well defined in literature. Isomorphic relation between fuzzy planar graph and its dual graph are established.

Keywords: *Fuzzy graphs, fuzzy planar graphs, fuzzy dual graphs, isomorphism.*

I. INTRODUCTION

Graph theory has vast applications in data mining, image segmentation, clustering, image capturing, networking, communication, planning, scheduling. For example, a data structure can be designed in the form of a tree which utilizes vertices and edges. Similarly, modeling of network topologies can be done using the concept of graph. In the same way, the most important concept of graph colouring is utilized in resource allocation, scheduling, etc. Also, paths, walks and circuits are used to solve many problems, viz. travelling salesman, database design, resource networking. This leads to the development of new algorithms and new theories that can be used in various applications.

There are many practical applications with a graph structure in which crossing between edges is a nuisance such as design problems for circuits, subways, utility lines, etc. Crossing of two connections normally means that the communication lines must be run at different heights. This is not a big issue for electrical wires, but it creates extra expenses for some types of lines, e.g. burying one subway tunnel under another. Circuits, in particular, are easier to manufacture if their connections can be constructed in fewer layers. These applications are designed by the concept of planar graphs. Circuits where crossing of lines is necessary, can not be represented by planar graphs. Numerous computational challenges including image segmentation or shape matching can be solved by means of cuts of planar graph.

After development of fuzzy graph theory by Rosenfeld [23], the fuzzy graph theory is increased with a large number of branches. McAllister [17] characterised the fuzzy intersection graphs. In this paper, fuzzy intersection graphs have been defined from the concept of intersection of fuzzy sets.

Samanta and Pal [25] introduced fuzzy tolerance graphs as the generalisation of fuzzy intersection graphs. They also defined fuzzy threshold graphs [26]. Fuzzy competition graphs [24] are another kind of fuzzy graphs which are the intersection of the fuzzy neighbourhoods of vertices of a fuzzy graph. Many works have been done on fuzzy sets as well as on fuzzy graphs [2], [3], [4], [5], [7], [8], [10], [11], [12], [13], [14], [15], [19]. Abdul-jabbar et al. [1] introduced the concept of fuzzy planar graph. In this paper, the crisp planar graph is considered and the membership values are assigned on vertices and edges. They also defined fuzzy dual graph as a straight forward way as crisp dual graph. Again, Nirmala and Dhanabal [22] defined special fuzzy planar graphs. The work presented in this paper is similar to the work presented in [1]. In these papers, the crossing of edges in fuzzy planar graph is not allowed. But, in our work, we define fuzzy planar graph in such a way that the crossing of edges is allowed. Also, we define the fuzzy planarity value which measures the amount of planarity of a fuzzy planar graph. These two concepts are new and no work has been done with these ideas. It is also shown that an image can be represented by a fuzzy planar graph and contraction of such image can be made with the help of fuzzy planar graph. The fuzzy multigraphs, fuzzy planar graphs and fuzzy dual graphs are illustrated by examples. Also, lot of results are presented for these graphs. These results have certain applications in subway tunnels, routes, oil/gas pipelines representation, etc.

II. PRELIMINARIES

A finite graph is a graph $G = (V, E)$ such that V and E are finite sets. An infinite graph is one with an infinite set of vertices or edges or both. Most commonly in graph theory, it is implied that the graphs discussed are finite. A *multigraph* [6] is a graph that may contain multiple edges between any two vertices, but it does not contain any self loops. A graph can be drawn in many different ways. A graph may or may not be drawn on a plane without crossing of edges.

A drawing of a geometric representation of a graph on any surface such that no edges intersect is called embedding [6]. A graph G is *planar* if it can be drawn in the plane with its edges only intersecting at vertices of G. So the graph is non-planar if it can not be drawn without crossing. A planar graph with cycles divides the plane into a set of regions, also called *faces*. The length of a face in a plane graph G is the total length of the closed walk(s) in G bounding the face. The portion of the plane lying outside a graph embedded in a plane is infinite region.

In graph theory, the dual graph of a given planar graph G is a graph which has a vertex corresponding to each plane region of G, and the graph has an edge joining two neighboring regions for each edge in G, for a certain embedding of G.

A *fuzzy set* A on an universal set X is characterized by a mapping $m : X \to [0,1]$, which is called the membership function. A fuzzy set is denoted by $A = (X, m)$.

A *fuzzy graph* [23] $\xi = (V, \sigma, \mu)$ is a non-empty set V together with a pair of functions $\sigma : V \to [0,1]$ and $\mu : V \times V \to [0,1]$ such that for all $x, y \in V$, $\mu(x, y) \le \min\{\sigma(x), \sigma(y)\}$, where $\sigma(x)$ and $\mu(x, y)$ represent the membership values of the vertex x and of the edge (x, y) in ξ respectively. A loop at a vertex x in a fuzzy graph is represented by $\mu(x, x) \ne 0$. An edge is non-trivial if $\mu(x, y) \ne 0$.

A fuzzy graph $\xi = (V, \sigma, \mu)$ is complete if $\mu(u, v) = \min\{\sigma(u), \sigma(v)\}$ for all $u, v \in V$, where (u, v) denotes the edge between the vertices u and v.

Several definitions of strong edge are available in literature. Among them the definition of [9] is more suitable for our purpose. The definition is given below. For the fuzzy graph $\xi = (V, \sigma, \mu)$, an edge (x, y) is called *strong* [9] if $\frac{1}{2}\min\{\sigma(x), \sigma(y)\} \le \mu(x, y)$ and weak otherwise. The strength of the fuzzy edge (x, y) is represented by the value $\frac{\mu(x,y)}{\min\{\sigma(x),\sigma(y)\}}$.

If an edge (x, y) of a fuzzy graph satisfies the condition $\mu(x, y) = \min\{\sigma(x), \sigma(y)\}$, then this edge is called effective edge [21]. Two vertices are said to be effective adjacent if they are the end vertices of the same effective edge. Then the effective incident degree of a fuzzy graph is defined as number of effective incident edges on a vertex v. If all the edges of a fuzzy graph are effective, then the fuzzy graph becomes complete fuzzy graph. A pendent vertex in a fuzzy graph is defined as a vertex of an effective incident degree one. A fuzzy edge is called a fuzzy *pendant edge* [24], if one end vertex is fuzzy pendant vertex. The membership value of the pendent edge is the minimum among the membership values of the end vertices.

A *homomorphism* [20] between fuzzy graphs ξ and ξ' is a map $h : S \to S'$ which satisfies $\sigma(x) \le \sigma'(h(x))$ for all $x \in S$ and $\mu(x, y) \le \mu'(h(x), h(y))$ for all $x, y \in S$ where S is set of vertices of ξ and S' is that of ξ'.

A *weak isomorphism* [20] between fuzzy graphs is a bijective homomorphism $h : S \to S'$ which satisfies $\sigma(x) = \sigma'(h(x))$ for all $x \in S$.

A *co-weak isomorphism* [20] between fuzzy graphs is a bijective homomorphism $h : S \to S'$ which satisfies $\mu(x, y) = \mu'(h(x), h(y))$ for all $x, y \in S$.

An *isomorphism* [20] between fuzzy graphs is a bijective homomorphism $h : S \to S'$ which satisfies $\sigma(x) = \sigma'(h(x))$ for all $x \in S$ and $\mu(x, y) = \mu'(h(x), h(y))$ for all $x, y \in S$.

The *underlying crisp graph* of the fuzzy graph $\xi = (V, \sigma, \mu)$ is denoted as $\xi^* = (V, \sigma^*, \mu^*)$ where $\sigma^* = \{u \in V | \sigma(u) > 0\}$ and $\mu^* = \{(u, v) \in V \times V | \mu(u, v) > 0\}$.

A (crisp) *multiset* over a non-empty set V is simply a mapping $d : V \to N$, where N is the set of natural numbers. Yager [31] first discussed fuzzy multisets, although he used the term "fuzzy bag". An element of nonempty set V may occur more than once with possibly the same or different membership values. A natural generalization of this interpretation of multiset leads to the notion of *fuzzy multiset*, or *fuzzy bag*, over a non-empty set V as a mapping $\tilde{C} : V \times [0,1] \to N$. The membership values of $v \in V$ are denoted as $v_{\mu^j}, j = 1, 2, \ldots, p$ where $p = \max\{j : v_{\mu^j} \ne 0\}$. So the fuzzy multiset can be denoted as $M = \{(v, v_{\mu^j}), j = 1, 2, \ldots, p | v \in V\}$.

To define fuzzy planar graph, fuzzy multigraph is essential as planar graphs contain multi-edges. In the next section, fuzzy multigraph is defined.

III. Fuzzy Multigraph

In this section, the fuzzy multigraph is defined.

Definition 1: Let V be a non-empty set and $\sigma : V \to [0,1]$ be a mapping. Also let $E = \{((x, y), (x, y)_{\mu^j}), j = 1, 2, \ldots, p_{xy} | (x, y) \in V \times V\}$ be a fuzzy multiset of $V \times V$ such that $(x, y)_{\mu^j} \le \min\{\sigma(x), \sigma(y)\}$ for all $j = 1, 2, \ldots, p_{xy}$, where $p_{xy} = \max\{j | (x, y)_{\mu^j} \ne 0\}$. Then $\psi = (V, \sigma, E)$ is denoted as fuzzy multigraph where $\sigma(x)$ and $(x, y)_{\mu^j}$ represent the membership value of the vertex x and the membership value of the edge (x, y) in ψ respectively.

It may be noted that there may be more than one edge between the vertices x and y. $(x, y)_{\mu^j}$ denotes the membership value of the j-th edge between the vertices x and y. Note that p_{xy} represents the number of edges between the vertices x and y.

IV. Fuzzy Planar Graphs

Planarity is important in connecting the wire lines, gas lines, water lines, printed circuit design, etc. But, some times little crossing may be accepted to these design of such lines/ circuits. So fuzzy planar graph is an important topic for these connections.

A crisp graph is called non-planar graph if there is at least one crossing between the edges for all possible geometrical representations of the graph. Let a crisp graph G has a crossing for a certain geometrical representation between two edges (a, b) and (c, d). In fuzzy concept, we say that this two edges have membership values 1. If we remove the edge (c, d), the graph becomes planar. In fuzzy sense, we say that the edges (a, b) and (c, d) have membership values 1 and 0 respectively.

Let $\xi = (V, \sigma, \mu)$ be a fuzzy graph and for a certain geometric representation, the graph has only one crossing between two fuzzy edges $((w, x), \mu(w, x))$ and $((y, z), \mu(y, z))$. If $\mu(w, x) = 1$ and $\mu(y, z) = 0$, then we say that the fuzzy graph has no crossing. Similarly, if $\mu(w, x)$ has value near to 1 and $\mu(w, x)$ has value near to 0, the crossing will not be important for the planarity. If $\mu(w, x)$ has value near to 1 and

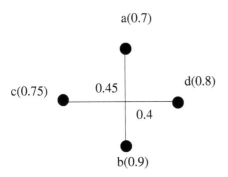

Fig. 1. Intersecting value between two intersecting edges.

$\mu(w, x)$ has value near to 1, then the crossing becomes very important for the planarity.

Before going to the main definition, some co-related terms are discussed below.

A. Intersecting value in fuzzy multigraph

In fuzzy multigraph, when two edges intersect at a point, a value is assigned to that point in the following way. Let in a fuzzy multigraph $\psi = (V, \sigma, E)$, E contains two edges $((a, b), (a, b)_{\mu^k})$ and $((c, d), (c, d)_{\mu^l})$ which are intersected at a point P, where k and l are fixed integers.

Strength of the fuzzy edge (a, b) can be measured by the value $I_{(a,b)} = \frac{(a,b)_{\mu^k}}{\min\{\sigma(a), \sigma(b)\}}$. If $I_{(a,b)} \geq 0.5$, then the fuzzy edge is called strong otherwise weak.

We define the intersecting value at the point P by $\mathcal{I}_P = \frac{I_{(a,b)} + I_{(c,d)}}{2}$. If the number of point of intersections in a fuzzy multigraph increases, planarity decreases. So for fuzzy multigraph, $\mathcal{I}_P$ is inversely proportional to the planarity. Based on this concept, a new terminology is introduced below for a fuzzy planar graph.

Definition 2: Let ψ be a fuzzy multigraph and for a certain geometrical representation $P_1, P_2, \ldots, P_z$ be the points of intersections between the edges. ψ is said to be fuzzy planar graph with fuzzy planarity value f, where

$$f = \frac{1}{1 + \{\mathcal{I}_{P_1} + \mathcal{I}_{P_2} + \ldots + \mathcal{I}_{P_z}\}}.$$

It is obvious that f is bounded and the range of f is $0 < f \leq 1$.

If there is no point of intersection for a certain geometrical representation of a fuzzy planar graph, then its fuzzy planarity value is 1. In this case, the underlying crisp graph of this fuzzy graph is the crisp planar graph. If f decreases, then the number of points of intersection between the edges increases and obviously the nature of planarity decreases. From this analogy, one can say that every fuzzy graph is a fuzzy planar graph with certain fuzzy planarity value.

Example 1: Here an example is given to calculate the intersecting value at the intersecting point between two edges. Two edges (a, b) and (c, d) are intersected where $\sigma(a) = 0.7, \sigma(b) = 0.9, \sigma(c) = 0.75, \sigma(0.8), \mu(a, b) = 0.4, \mu(c, d) = 0.45$ (see Fig. 1). Strength of the edge (a, b) is $\frac{0.4}{0.7} = 0.57$ and that of (c, d) is $\frac{0.45}{0.75} = 0.6$. Thus the intersecting value at the point is $\frac{0.57 + 0.6}{2} = 0.585$.

Fuzzy planarity value for a fuzzy multigraph is calculated from the following theorem.

Theorem 1: Let ψ be a fuzzy multigraph such that edge membership value of each intersecting edge is equal to the minimum of membership values of its end vertices. The fuzzy planarity value f of ψ is given by $f = \frac{1}{1 + N_p}$, where N_p is the number of point of intersections between the edges in ψ.

Proof. Let $\psi = (V, \sigma, E)$ be a fuzzy multigraph such that edge membership values of each intersecting edge is equal to minimum of its vertex membership values. For the fuzzy multigraph, $(x, y)_{\mu^j} = \min\{\sigma(x), \sigma(y)\}$ for each intersecting edge (x, y) and $j = 1, 2, \ldots, p_{xy}$.

Let $P_1, P_2, \ldots, P_k$, be the point of intersections between the edges in ψ, k being an integer. For any intersecting edge (a, b) in ψ, $I_{(a,b)} = \frac{(a,b)_{\mu^j}}{\min\{\sigma(a), \sigma(b)\}} = 1$. Therefore, for P_1, the point of intersection between the edges (a, b) and (c, d), $\mathcal{I}_{P_1}$ is equals to $\frac{1+1}{2} = 1$. Hence $\mathcal{I}_{P_i} = 1$ for $i = 1, 2, \ldots, k$.

Now, $f = \frac{1}{1 + \mathcal{I}_{P_1} + \mathcal{I}_{P_2} + \ldots + \mathcal{I}_{P_k}} = \frac{1}{1 + (1 + 1 + \ldots + 1)} = \frac{1}{1 + N_p}$, where N_p is the number of point of intersections between the edges in ψ. $\qquad \square$

Definition 3: A fuzzy planar graph ψ is called strong fuzzy planar graph if the fuzzy planarity value of the graph is greater than 0.5.

The fuzzy planar graph of Example 5 is not strong fuzzy planar graph as its fuzzy planarity value is less than 0.5.

Thus, depending on the fuzzy planarity value, the fuzzy planar graphs are divided into two groups namely, strong and weak fuzzy planar graphs.

Theorem 2: Let ψ be a strong fuzzy planar graph. The number of point of intersections between strong edges in ψ is at most one.

Proof. Let $\psi = (V, \sigma, E)$ be a strong fuzzy planar graph. Let, if possible, ψ has at least two point of intersections P_1 and P_2 between two strong edges in ψ.

For any strong edge $((a, b), (a, b)_{\mu^j})$, $(a, b)_{\mu^j} \geq \frac{1}{2}\min\{\sigma(a), \sigma(b)\}$. So $I_{(a,b)} \geq 0.5$.

Thus for two intersecting strong edges $((a, b), (a, b)_{\mu^j})$ and $((c, d), (c, d)_{\mu^i})$, $\frac{I_{(a,b)} + I_{(c,d)}}{2} \geq 0.5$, that is, $\mathcal{I}_{P_1} \geq 0.5$. Similarly, $\mathcal{I}_{P_2} \geq 0.5$. Then $1 + \mathcal{I}_{P_1} + \mathcal{I}_{P_2} \geq 2$. Therefore, $f = \frac{1}{1 + \mathcal{I}_{P_1} + \mathcal{I}_{P_2}} \leq 0.5$. It contradicts the fact that the fuzzy graph is a strong fuzzy planar graph.

So number of point of intersections between strong edges can not be two. It is clear that if the number of point of intersections of strong fuzzy edges increases, the fuzzy planarity value decreases. Similarly, if the number of point of intersection of strong edges is one, then the fuzzy planarity value $f > 0.5$. Any fuzzy planar graph without any crossing between edges is a strong fuzzy planar graph. Thus, we conclude that the maximum number of point of intersections between the strong edges in ψ is one. $\qquad \square$

Face of a planar graph is an important feature. We now introduce the fuzzy face of a fuzzy planar graph.

Fuzzy face in a fuzzy graph is a region bounded by fuzzy edges. Every fuzzy face is characterized by fuzzy edges in its boundary. If all the edges in the boundary of a fuzzy face have membership value 1, it becomes crisp face. If one of such edges is removed or has membership value 0, the fuzzy face does not exist. So the existence of a fuzzy face depends on the minimum value of strength of fuzzy edges in its boundary. A fuzzy face and its membership value are defined below.

Definition 4: Let $\psi = (V, \sigma, E)$ be a fuzzy planar graph and
$$E = \{((x,y),(x,y)_{\mu^j}), j = 1, 2, \ldots, p_{xy} | (x,y) \in V \times V\}$$ and $p_{xy} = \max\{j | (x,y)_{\mu^j} \neq 0\}$. A fuzzy face of ψ is a region, bounded by the set of fuzzy edges $E' \subset E$, of a geometric representation of ψ. The membership value of the fuzzy face is

$$\min\left\{ \frac{(x,y)_{\mu^j}}{\min\{\sigma(x), \sigma(y)\}}, j = 1, 2, \ldots, p_{xy} | (x,y) \in E' \right\}.$$

A fuzzy face is called strong fuzzy face if its membership value is greater than 0.5, and weak face otherwise. Every fuzzy planar graph has an infinite region which is called outer fuzzy face. Other faces are called inner fuzzy faces.

Example 2: In Fig. 2, F_1, F_2 and F_3 are three fuzzy faces. F_1 is bounded by the edges $((v_1, v_2), 0.5), ((v_2, v_3), 0.6), ((v_1, v_3), 0.55)$ with membership value 0.833. Similarly, F_2 is a fuzzy bounded face. F_3 is the outer fuzzy face with membership value 0.33. So F_1 is a strong fuzzy face and F_2, F_3 are weak fuzzy faces.

Every strong fuzzy face has membership value greater than 0.5. So every edge of a strong fuzzy face is a strong fuzzy edge.

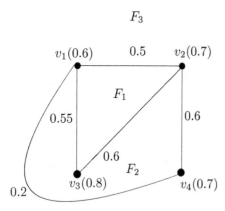

Fig. 2.   Example of faces in fuzzy planar graph.

Theorem 3: If the fuzzy planarity value of a fuzzy planar graph is greater than 0.67, then the graph does not contain any point of intersection between two strong edges.

Proof. Let $\psi = (V, \sigma, E)$ be a fuzzy planar graph with fuzzy planarity value f, where $f \geq 0.67$. Let, if possible, P be a point of intersection between two strong fuzzy edges $((a,b),(a,b)_{\mu^j})$ and $((c,d),(c,d)_{\mu^j})$.

For any strong edge $((a,b),(a,b)_{\mu^j})$, $(a,b)_{\mu^j} \geq \frac{1}{2}\min\{\sigma(a), \sigma(b)\}$. Therefore, $I_{(a,b)} \geq 0.5$. For the minimum value of $I_{(a,b)}$ and $I_{(c,d)}$, $\mathcal{I}_P = 0.5$ and $f = \frac{1}{1+0.5} < 0.67$. A

contradiction arises. Hence, ψ does not contain any point of intersection between strong edges.   □

Motivated from this theorem, we introduce a special type of fuzzy planar graph called 0.67-fuzzy planar graph whose fuzzy planarity value is more than or equal to 0.67. As in mentioned earlier, if the fuzzy planarity value is 1, then the geometrical representation of fuzzy planar graph is similar to the crisp planar graph. It is shown in Theorem 7, if fuzzy planarity value is 0.67, then there is no crossing between strong edges. For this case, if there is any point of intersection between edges, that is the crossing between the weak edge and any other edge. Again, the significance of weak edge is less compared to strong edges. Thus, 0.67-fuzzy planar graph is more significant. If fuzzy planarity value increases, then the geometrical structure of planar graph tends to crisp planar graph.

Any fuzzy planar graph without any point of intersection of fuzzy edges is a fuzzy planar graph with fuzzy planarity value 1. Therefore, it is a 0.67-fuzzy planar graph.

V. FUZZY DUAL GRAPH

We now introduce dual of 0.67-fuzzy planar graph. In fuzzy dual graph, vertices are corresponding to the strong fuzzy faces of the 0.67-fuzzy planar graph and each fuzzy edge between two vertices is corresponding to each edge in the boundary between two faces of 0.67-fuzzy planar graph. The formal definition is given below.

Definition 5: Let $\psi = (V, \sigma, E)$ be a 0.67-fuzzy planar graph and $E = \{((x,y),(x,y)_{\mu^j}), j = 1, 2, \ldots, p_{xy} | (x,y) \in V \times V\}$. Again, let $F_1, F_2, \ldots, F_k$ be the strong fuzzy faces of ψ. The fuzzy dual graph of ψ is a fuzzy planar graph $\psi' = (V', \sigma', E')$, where $V' = \{x_i, i = 1, 2, \ldots, k\}$, and the vertex x_i of ψ' is considered for the face F_i of ψ.

The membership values of vertices are given by the mapping $\sigma' : V' \to [0,1]$ such that $\sigma'(x_i) = \max\{(u,v)_{\mu^j}, j = 1, 2, \ldots, p_{uv} | (u,v)$ is an edge of the boundary of the strong fuzzy face $F_i\}$.

Between two faces F_i and F_j of ψ, there may exist more than one common edge. Thus, between two vertices x_i and x_j in fuzzy dual graph ψ', there may be more than one edge. We denote $(x_i, x_j)_{\nu^l}$ be the membership value of the l-th edge between x_i and x_j. The membership values of the fuzzy edges of the fuzzy dual graph are given by $(x_i, x_j)_{\nu^l} = (u,v)_{\mu^j}^l$ where $(u,v)^l$ is an edge in the boundary between two strong fuzzy faces F_i and F_j and $l = 1, 2, \ldots, s$, where s is the number of common edges in the boundary between F_i and F_j or the number of edges between x_i and x_j.

If there be any strong pendant edge in the 0.67-fuzzy planar graph, then there will be a self loop in ψ' corresponding to this pendant edge. The edge membership value of the self loop is equal to the membership value of the pendant edge.

Fuzzy dual graph of 0.67-fuzzy planar graph does not contain point of intersection of edges for a certain representation, so it is 0.67-fuzzy planar graph with planarity value 1. Thus the fuzzy face of fuzzy dual graph can be similarly described as in 0.67-fuzzy planar graphs.

Example 3: In Fig. 3, a 0.67-fuzzy planar graph $\psi = (V, \sigma, E)$ where $V = \{a, b, c, d\}$ is given. For this graph let $\sigma(a) = 0.6, \sigma(b) = 0.7, \sigma(c) = 0.8, \sigma(d) = 0.9$.

and $E = \{((a, b), 0.5), ((a, c), 0.4), ((a, d), 0.55), ((b, c), 0.45), ((c, d), 0.7)\}$.

Thus, the 0.67-fuzzy planar graph has the following fuzzy faces

F_1 (bounded by $((a, b), 0.5), ((a, c), 0.4), ((b, c), 0.45))$,
F_2 (bounded by $((a, d), 0.55), ((c, d), 0.7), ((a, c), 0.4))$,
and outer fuzzy face
F_3 (surrounded by $((a, b), 0.5), ((b, c), 0.45), ((c, d), 0.7), ((a, d), 0.55))$.

The fuzzy dual graph is constructed as follows. Here all the fuzzy faces are strong fuzzy faces. For each strong fuzzy face, we consider a vertex for the fuzzy dual graph. Thus the vertex set $V' = \{x_1, x_2, x_3, x_4\}$ where the vertex x_i is taken corresponding to the strong fuzzy face F_i, $i = 1, 2, 3, 4$. So $\sigma'(x_1) = \max\{0.5, 0.4, 0.45\} = 0.5$, $\sigma'(x_2) = \max\{0.55, 0.7, 0.4\} = 0.7$, $\sigma'(x_3) = \max\{0.5, 0.45, 0.7, 0.55\} = 0.7$.

There are two common edges (a, d) and (c, d) between the faces F_2 and F_3 in ψ. Hence between the vertices x_2 and x_3, two edges exist in the fuzzy dual graph of ψ. Here membership values of these edges are given by $(x_2, x_4)_{\nu^1} = (c, d)_{\mu^1} = 0.7$, $(x_2, x_4)_{\nu^2} = (a, d)_{\mu^1} = 0.55$.

The membership values of other edges of the fuzzy dual graph are calculated as $(x_1, x_2)_{\nu^1} = (a, c)_{\mu^1} = 0.4$, $(x_1, x_3)_{\nu^1} = (a, b)_{\mu^1} = 0.5, (x_1, x_3)_{\nu^2} = (b, c)_{\mu^1} = 0.45$.

Thus the edge set of fuzzy dual graph is $E' = \{((x_1, x_2), 0.4), ((x_1, x_3), 0.5), ((x_1, x_3), 0.45), ((x_2, x_3), 0.7), ((x_2, x_3), 0.55)\}$.

In Fig. 3, the fuzzy dual graph $\psi' = (V', \sigma', E')$ of ψ is drawn by dotted line.

Theorem 4: Let ψ be a 0.67-fuzzy planar graph without weak edges. The number of vertices, number of fuzzy edges and number of strong faces of ψ are denoted by n, p, m respectively. Also let ψ' be the fuzzy dual graph of ψ. Then
(i) the number of vertices of ψ' is equal to m,
(ii) number of edges of ψ' is equal to p,
(iii) number of fuzzy faces of ψ' is equal to n.
Proof. Proof of (i), (ii) and (iii) are obvious from the definition of fuzzy dual graph. □

Theorem 5: Let ψ' be a fuzzy dual graph of a 0.67-fuzzy planar graph ψ. The number of strong fuzzy faces in ψ' is less than or equal to the number of vertices of ψ.
Proof. Here ψ' is a fuzzy dual graph of a 0.67-fuzzy planar graph ψ. Let ψ has n vertices and ψ' has m strong fuzzy faces. Now, ψ may have weak edges and strong edges. To construct fuzzy dual graph, weak edges are to eliminate. Thus if ψ has some weak edges, some vertices may have all its adjacent edges as weak edges. Let the number of such vertices be t. These vertices are not bounding any strong fuzzy faces. If we remove these vertices and adjacent edges, then the number of vertices is $n - t$. Again, from Theorem 4, $m = n - t$. Hence, in general $m \le n$. This concludes that the number of strong fuzzy faces in ψ' is less than or equal to the number of vertices of ψ. □

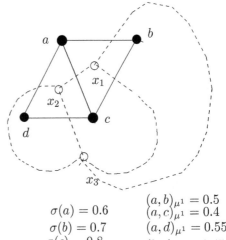

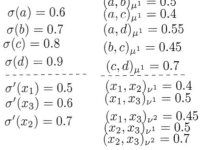

$$\sigma(a) = 0.6 \qquad (a, b)_{\mu^1} = 0.5$$
$$\sigma(b) = 0.7 \qquad (a, c)_{\mu^1} = 0.4$$
$$\sigma(c) = 0.8 \qquad (a, d)_{\mu^1} = 0.55$$
$$\sigma(d) = 0.9 \qquad (b, c)_{\mu^1} = 0.45$$
$$\text{-----} \qquad (c, d)_{\mu^1} = 0.7$$
$$\sigma'(x_1) = 0.5 \qquad (x_1, x_2)_{\nu^1} = 0.4$$
$$\sigma'(x_3) = 0.6 \qquad (x_1, x_3)_{\nu^1} = 0.5$$
$$\sigma'(x_2) = 0.7 \qquad (x_1, x_3)_{\nu^2} = 0.45$$
$$\qquad (x_2, x_3)_{\nu^1} = 0.5$$
$$\qquad (x_2, x_3)_{\nu^2} = 0.7$$

Fig. 3. Example of fuzzy dual graph.

An example is considered to illustrate the statement. Let $\psi = (V, \sigma, E)$ be a 0.67-fuzzy planar graph where $V = \{a, b, c, d\}$. $\sigma(a) = 0.8, \sigma(b) = 0.7, \sigma(c) = 0.9, \sigma(d) = 0.3$. $E = \{((a, b), 0.7), ((b, c), 0.7), ((c, d), 0.2), ((b, d), 0.2), ((a, d), 0.2)\}$. The corresponding fuzzy dual graph is $\psi' = (V', \sigma', E')$ where $V' = \{x_1, x_2, x_3\}$. $\sigma'(x_1) = 0.7, \sigma'(x_2) = 0.7, \sigma'(x_3) = 0.7$. $E' = \{((x_1, x_2), 0.2), ((x_1, x_3), 0.2), ((x_1, x_3), 0.7), ((x_2, x_3), 0.2), ((x_2, x_3), 0.7)\}$. Here number of strong fuzzy face is one while number of fuzzy face is three (see Fig. 4).

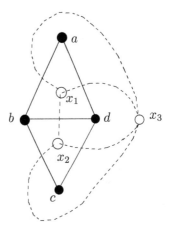

Fig. 4. Example of a fuzzy dual graph with strong face.

Theorem 6: Let $\psi = (V, \sigma, E)$ be a 0.67-fuzzy planar graph without weak edges and the fuzzy dual graph of ψ be $\psi' = (V', \sigma', E')$. The membership values of fuzzy edges of

ψ' are equal to membership values of the fuzzy edges of ψ.
Proof. Let $\psi = (V, \sigma, E)$ be a 0.67-fuzzy planar graph without weak edges. The fuzzy dual graph of ψ is $\psi' = (V', \sigma', E')$ which is a 0.67-fuzzy planar graph as there is no point of intersection between any edges. Let $\{F_1, F_2, \ldots, F_k\}$ be the set of strong fuzzy faces of ψ.

From the definition of fuzzy dual graph we know that $(x_i, x_j)_{\nu^l} = (u, v)^l_{\mu^j}$ where $(u, v)^l$ is an edge in the boundary between two strong fuzzy faces F_i and F_j and $l = 1, 2, \ldots, s$, where s is the number of common edges in the boundary between F_i and F_j.

The numbers of fuzzy edges of two fuzzy graphs ψ and ψ' are same as ψ has no weak edges. For each fuzzy edge of ψ there is a fuzzy edge in ψ' with same membership value. $\square$

VI. ISOMORPHISM ON FUZZY PLANAR GRAPHS

Isomorphism between fuzzy graphs is an equivalence relation. But, if there is an isomorphism between two fuzzy graph and one is fuzzy planar graph, then the other will be fuzzy planar graph. This result is proved in the following theorem.

Theorem 7: Let ψ be a fuzzy planar graph. If there exists an isomorphism $h : \psi \to \xi$ where ξ is a fuzzy graph, ξ can be drawn as fuzzy planar graph with same planarity value of ψ.
Proof. Let ψ be a fuzzy planar graph and there exists an isomorphism $h : \psi \to \xi$ where ξ is a fuzzy graph. Now, isomorphism preserves edge and vertex weights. Also the order and size of fuzzy graphs are preserved in isomorphic fuzzy graphs [20]. So, the order and size of ξ will be equal to ψ. Then, ξ can be drawn similarly as ψ. Hence, the number of intersection between edges and fuzzy planarity value of ξ will be same as ψ. This concludes that ξ can be drawn as fuzzy planar graph with same fuzzy planarity value. $\square$

In crisp graph theory, dual of dual graph of a planar graph is planar graph itself. In fuzzy graph concept, fuzzy dual graph of a fuzzy dual graph is not isomorphic to fuzzy planar graph. The membership values of vertices of fuzzy dual graph are the maximum membership values of its bounding edges of the corresponding fuzzy faces in fuzzy planar graph. Thus vertex weight is not preserved in fuzzy dual graph. But edge weight is preserved in fuzzy dual graph. This result is established in following theorem.

Theorem 8: Let ψ_2 be the fuzzy dual graph of fuzzy dual graph of a 0.67-fuzzy planar graph ψ without weak edges. Then there exists a co-weak isomorphism between ψ and ψ_2.
Proof. Let ψ be a 0.67-fuzzy planar graph which has no weak edges. Also let, ψ_1 be the fuzzy dual graph of ψ and ψ_2 be the fuzzy dual graph of ψ_1. Now we have to establish a co-weak isomorphism between ψ_2 and ψ. As the number of vertices of ψ_2 is equal to that of strong fuzzy faces of ψ_1. Again the number of strong fuzzy faces is equal to the number of vertices of ψ. Thus, the number of vertices of ψ_2 and ψ are same. Also, the numbers of edges of a fuzzy planar graph and its dual graph are same. By the definition of fuzzy dual graph, the edge membership value of an edge in fuzzy dual graph is equal to the edge membership value of an edge in fuzzy

planar graph. Thus we can construct a co-weak isomorphism from ψ_2 to ψ. Hence the result is true. $\square$

The Theorem 8 can be explained by the following example. Here a 0.67-fuzzy planar graph ψ is constructed (See Fig. 5(a)). Then its fuzzy dual graph ψ_1 is drawn in Fig. 5(b). Also the fuzzy dual graph ψ_2 of ψ_1 is drawn in Fig. 5(c). Now, we construct a bijective mapping from vertices of ψ_2 to vertices of ψ as $a_1 \to a, b_1 \to b, c_1 \to c, d_1 \to d$. Similarly, we can extend the mapping from edge set of ψ_2 to the edge set of ψ. It is observed that the vertex membership values of ψ_2 is less than or equal to the vertex membership values of ψ under the mapping and edge membership values are equal under the mapping. Thus the mapping is said to satisfy the co-weak isomorphism property.

Two fuzzy planar graphs with same number of vertices may be isomorphic. But, the relations between fuzzy planarity values of two fuzzy planar graphs may have the following relations.

Theorem 9: Let ξ_1 and ξ_2 be two isomorphic fuzzy graphs with fuzzy planarity values f_1 and f_2 respectively. Then $f_1 = f_2$.
The proof of the theorem is the immediate consequence of Theorem 7.

Theorem 10: Let ξ_1 and ξ_2 be two weak isomorphic fuzzy graphs with fuzzy planarity values f_1 and f_2 respectively. $f_1 = f_2$ if the edge membership values of corresponding intersecting edges are same.
Proof. Here $\xi_1 = (V, \sigma_1, \mu_1)$ and $\xi_2 = (V, \sigma_2, \mu_2)$ are two weak isomorphic fuzzy graphs with fuzzy planarity values f_1 and f_2 respectively. As two fuzzy graphs are weak isomorphic, $\sigma_1(x) = \sigma_2(y)$ for some x in ξ_1 and y in ξ_2. Let the graphs have one point of intersection. Let two intersecting edges be (a_1, b_1) and (c_1, d_1) in ξ_1. Also two corresponding edges in ξ_2 be (a_2, b_2) and (c_2, d_2). Then, intersecting value of the point is given by $\frac{\frac{\mu(a_1, b_1)}{\sigma(a_1) \wedge \sigma(b_1)} + \frac{\mu(c_1, d_1)}{\sigma(c_1) \wedge \sigma(d_1)}}{2}$. The intersecting value of the corresponding point in ξ_2 is given as $\frac{\frac{\mu(a_2, b_2)}{\sigma(a_2) \wedge \sigma(b_2)} + \frac{\mu(c_2, d_2)}{\sigma(c_2) \wedge \sigma(d_2)}}{2}$. Now, $f_1 = f_2$, if $\mu(a_1, b_1) = \mu(a_2, b_2)$. The number of point of intersections may increase. But, if the sum of the intersecting value of ξ_1 is equal to that of ξ_2, fuzzy planarity values of the graphs must be equal. Thus, for equality of f_1 and f_2, the edge membership values of intersecting edges of ξ are equal to the edge membership values of the corresponding edges in ξ_2. $\square$

Theorem 11: Let ξ_1 and ξ_2 be two co-weak isomorphic fuzzy graphs with fuzzy planarity values f_1 and f_2 respectively. $f_1 = f_2$ if the minimum of membership values of the end vertices of corresponding intersecting edges are same.
Proof. Here $\xi_1 = (V, \sigma_1, \mu_1)$ and $\xi_2 = (V, \sigma_2, \mu_2)$ are two co-weak isomorphic fuzzy graphs with fuzzy planarity values f_1 and f_2 respectively. As two fuzzy graphs are co-weak isomorphic, $\mu_1(x, y) = \mu_2(z, t)$ for some edge (x, y) in ξ_1 and (z, t) in ξ_2. Let the graphs have one point of intersection. Let two intersecting edges be (a_1, b_1) and (c_1, d_1) in ξ_1. Also, two corresponding edges in ξ_2 be (a_2, b_2) and (c_2, d_2). Then, inter-

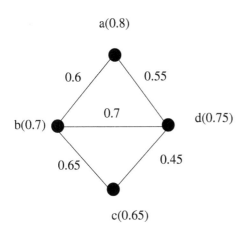

a(0.8)

0.6   0.55

0.7

b(0.7)   d(0.75)

0.65   0.45

c(0.65)

(a): A 0.67-fuzzy planar graph ψ

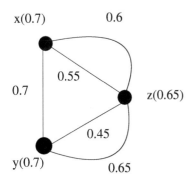

x(0.7)   0.6

0.55

0.7

z(0.65)

0.45

y(0.7)   0.65

(b): Fuzzy dual graph ψ_1 of ψ

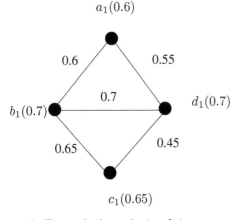

$a_1(0.6)$

0.6   0.55

0.7

$b_1(0.7)$ $d_1(0.7)$

0.65   0.45

$c_1(0.65)$

(c): Fuzzy dual graph ψ_2 of ψ_1

Fig. 5. Dual of dual is co-weak isomorphic to planar graph in fuzzy graph theory.

secting value of the point is given by $\frac{\frac{\mu(a_1,b_1)}{\sigma(a_1)\wedge\sigma(b_1)} + \frac{\mu(c_1,d_1)}{\sigma(c_1)\wedge\sigma(d_1)}}{2}$.
The intersecting value of the corresponding point in ξ_2 is given as $\frac{\frac{\mu(a_2,b_2)}{\sigma(a_2)\wedge\sigma(b_2)} + \frac{\mu(c_2,d_2)}{\sigma(c_2)\wedge\sigma(d_2)}}{2}$. Now, the fuzzy planarity values f_1 = f_2, if $\sigma_1(a_1)\wedge\sigma(b_1) = \sigma_2(a_2)\wedge\sigma_2(b_2)$. The number of point of intersections may increase. But if the sum of the intersecting value of ξ_1 is equal to that of ξ_2, fuzzy planarity values of the graphs must be equal. Thus, for equality of f_1 and f_2, the minimum membership value of end vertices of an edge in ξ_1 is equal to that of a corresponding edge in ξ_2. □

VII. CONCLUSION

This study describes the fuzzy multigraphs, fuzzy planar graphs, and a very important consequence of fuzzy planar graph known as fuzzy dual graphs. In crisp planar graph, no edge intersects other. In fuzzy graph, an edge may be weak or strong. Using the concept of weak edge, we define fuzzy planar graph in such a way that an edge can intersect other edge. But, this facility violates the definition of planarity of graph. Since the role of weak edge is insignificant, the intersection between a weak edge and an other edge is less important. Motivating from this idea, we allow the intersection of edges in fuzzy planar graph. It is well known that if the membership values of all edges become one, the graph becomes crisp graph. Keeping this idea in mind, we define a new term called fuzzy planarity value of a fuzzy graph. If the fuzzy planarity value of a fuzzy graph is one, then no edge crosses other edge. This leads to the crisp planar graph. Thus, the fuzzy planarity value measures the amount of planarity of a fuzzy graph. This is a very interesting concept of fuzzy graph theory. Strong fuzzy planar graphs and a distinguishable subclass of strong fuzzy planar graph namely 0.67-fuzzy planar graphs have been exemplified. From the definitions, it is concluded that 0.67-fuzzy planar graph $\subset$ strong fuzzy planar graph $\subset$ fuzzy planar graph. Another important term of planar graph is "face" which is redefined in fuzzy planar graph. A particular type of fuzzy face called strong fuzzy face is incorporated. Besides, isomorphism properties of fuzzy planar graphs are investigated. It is shown that dual of dual fuzzy graphs are co-weak isomorphism to fuzzy planar graph. Several properties of isomorphism on fuzzy planar graphs are explained. It may be noted that, in this article, fuzzy dual graph is defined for the 0.67-fuzzy planar graph. But, if the planarity value is less than 0.67, then some modification is required to define dual graph. This is to be investigated in near future.

REFERENCES

[1] N. abdul-jabbar, J. H. Naoom and, E. H. Ouda, Fuzzy dual graph, *Journal Of Al-Nahrain University*, 12(4), 168-171, 2009.
[2] M. Akram, Bipolar fuzzy graphs, *Information Science*, 181, 5548-5564, 2011.
[3] M. Akram, Interval-valued fuzzy line graphs, *Neural Computing & Applications*, 21, 145-150, 2012.
[4] M. Akram and W. A. Dudek, Intuitionistic fuzzy hypergraphs with applications, *Information Sciences*, 218, 182-193, 2013.
[5] M. Akram, Bipolar fuzzy graphs with applications, *Knowledge Based Systems*, 39, 1-8, 2013.
[6] V. K. Balakrishnan, Graph Theory, *McGraw-Hill*, 1997.
[7] K. R. Bhutani and A. Battou, On M-strong fuzzy graphs, *Information Sciences*, 155(12), 103-109, 2003.

[8] K. R. Bhutani and A. Rosenfeld, Strong arcs in fuzzy graphs, *Information Sciences*, 152, 319-322, 2003.

[9] C. Eslahchi and B. N. Onaghe, Vertex strength of fuzzy graphs, *International Journal of Mathematics and Mathematical Sciences*, Volume 2006, Article ID 43614, Pages 1-9, DOI 10.1155/IJMMS/2006/43614.

[10] M. Fazzolari and H. Ishibuchi, A review of the application of multiobjective evolutionary fuzzy systems: current status and further directions, *IEEE Transection on Fuzzy Systems*, 21(1), 45-65, 2013.

[11] P. Ghosh, K. Kundu and D. Sarkar, Fuzzy graph representation of a fuzzy concept lattice, *Fuzzy Sets and Systems*, 161(12), 1669-1675, 2010.

[12] L. T. Koczy, Fuzzy graphs in the evaluation and optimization of networks, *Fuzzy Sets and Systems*, 46, 307-319, 1992.

[13] Q. Liang and J. M. Mendel, MPEG VBR video traffic modelling and classification using fuzzy technique, *IEEE Transection on Fuzzy Systems*, 9(1), 183-193, 2001.

[14] K. -C. Lin and M. -S. Chern, The fuzzy shortest path problem and its most vital arcs, *Fuzzy Sets and Systems*, 58, 343-353, 1993.

[15] S. Mathew and M.S. Sunitha, Types of arcs in a fuzzy graph, *Information Sciences*, 179, 1760-1768, 2009.

[16] S. Mathew and M.S. Sunitha, Strongest strong cycles and theta fuzzy graphs, *IEEE Transections on Fuzzy Systems*, 2013.

[17] M. L. N. McAllister, Fuzzy intersection graphs, *Comput. Math. Applic.*, 15 (10) , 871-886, 1988.

[18] J. N. Mordeson and P. S. Nair, Fuzzy Graphs and Hypergraphs, Physica Verlag, 2000.

[19] S. Munoz, M. T. Ortuno, J. Ramirez and J. Yanez, Coloring fuzzy graphs, *Omega*, 33(3), 211-221, 2005.

[20] A. Nagoorgani and J. Malarvizhi, Isomorphism on fuzzy graphs, *World Academy of Science, Engineering and Technology*, 23, 505-511, 2008.

[21] A. Nagoorgani and R. J. Hussain, Fuzzy effective distance K-dominating sets and their applications, *International Journal of Algorithms, Computing and Mathematics*, 2(3), 25-36, 2009.

[22] G. Nirmala and K. Dhanabal, Special planar fuzzy graph configurations, *International Journal of Scientific and Research Publications*, 2(7), 1-4, 2012.

[23] A. Rosenfeld, Fuzzy graphs, in: L.A. Zadeh, K.S. Fu, M. Shimura (Eds.), *Fuzzy Sets and Their Applications*, Academic Press, New York, 77-95, 1975.

[24] S. Samanta and M. Pal, Fuzzy k-competition graphs and p-competition fuzzy graphs, *Fuzzy Engineering and Information*, 5(2), 191-204, 2013.

[25] S. Samanta and M. Pal, Fuzzy tolerance graphs, *International Journal of Latest Trends in Mathematics*, 1(2), 57-67, 2011.

[26] S. Samanta and M. Pal, Fuzzy threshold graphs, *CIIT International Journal of Fuzzy Systems*, 3(12), 360-364, 2011.

[27] S. Samanta and M. Pal, Irregular bipolar fuzzy graphs, *Inernational Journal of Applications of Fuzzy Sets*, 2, 91-102, 2012.

[28] S. samanta and M. Pal, Bipolar fuzzy hypergraphs, *International Journal of Fuzzy Logic Systems*, 2(1), 17 − 28, 2012.

[29] S. Samanta, M. Pal and A. Pal, Some more results on bipolar fuzzy sets and bipolar fuzzy intersection graphs, To appear in *The Journal of Fuzzy Mathematics*.

[30] S. Samanta and M. Pal, A new approach to social networks based on fuzzy graphs, To appear in *Journal of Mass Communication and Journalism*.

[31] R. R. Yager, On the theory of bags, *Int. J. General Systems*, 13, 23-37, 1986.

Driver's Awareness and Lane Changing Maneuver in Traffic Flow based on Cellular Automaton Model

Kohei Arai[1]

1 Graduate School of Science and Engineering
Saga University
Saga City, Japan

Steven Ray Sentinuwo[2]

2 Sam Ratulangi University,
Kampus Unsrat, Manado, Indonesia

Abstract—Effect of driver's awareness (e.g., to estimate the speed and arrival time of another vehicle) on the lane changing maneuver is discussed. "Scope awareness" is defined as the visibility which is required for the driver to make a visual perception about road condition and the speed of vehicle that appears in the target lane for lane changing in the road. Cellular automaton based simulation model is created and applied to simulation studies for driver awareness behavior. This study clarifies relations between the lane changing behavior and the scope awareness parameter that reflects driver behavior. Simulation results show that the proposed model is valid for investigation of the important features of lane changing maneuver.

Keywords—traffic cellular automata; scope awareness; lane changing maneuver; driver perception; speed estimation

I. INTRODUCTION

Recent study on the traffic flow reports that the traffic congestion is influenced not only by the road capacity condition, but also by the driver behavior [1]. The other studies also found the strong relationship between the driver' speed behavior and accidents [2]-[6]. There are two separate components which affect human factors in driving, driving skills and driving style [7]. Driving style has a direct relation to the individual drivers' behavior. The U.S. Department of Transportation recently reported that driver behavior is leading to lane-change crashes and near-crashes [8]. In some countries, the reckless driving behaviors such as sudden-stop by public-buses, tailgating, or vehicles which changing lane too quickly also could give an impact to the traffic flow.

The lane changing maneuver is one of the actions in the highway. Lane changing is defined as a driving maneuver that moves a vehicle laterally from one lane to another lane where both lanes have the same direction of travel. Lane changing maneuvers are occasionally performed in order to avoid hazards, obstacles, vehicle collision, or pass through the slow vehicle ahead. Lane changing requires high attention and visual perceptions compared to normal highway of freeway driving due to the need to continually monitor areas around the subject vehicle [9]. However, in the real traffic situations, there are some reckless drivers that change the lanes at the moment when they signal or who make "last-minutes-decision" on the road. Frequent lane changing in roadway could affect traffic flow and even lead accidents. The lane changing behaviors can be very depended on the characteristic of the driver [10].

There are some crashes of accidents typically referred to as Look-But-Fail-To-See errors because drivers involved in these accidents frequently. These are reported that they failed notification of the conflicting vehicles in spite of looking in the appropriate directions; commonly occur when drivers change lanes [25]. This means that the drivers typically use their visual perception in order to estimate the speed and the arrival time of the other vehicles before making a maneuver, e.g., lane changing maneuver.

A psychology study is also reported that the accuracy level of this visual perception may lead to both failures, detect the collision and judge the crash risk (e.g., time-to-contact). From a certain distance, a short fixation may be enough to identify an approaching vehicle. If gaze duration for stimulus processing is long, then it is complicated processing while it is short for simple processing. Inaccuracy of the gazes' duration is likely to reflect a failure to process these stimuli [26].

The Cellular Automata model of Nagel and Schreckenberg (NaSch) [12] is improved for showing effects of scope awareness that reflect drivers' behavior when they are making a lane changing. This NaSch model has been modified to describe more realistic movement of individual vehicle when make a lane changing maneuver. Moreover, the recent study of spontaneous braking behavior [1] enhances the driver's scope awareness behavior. The proposed model is based on NaSch model which takes into account scope of awareness and spontaneous braking in order to clarify the effects of these drivers' behavior.

This paper is organized as follows. Section 2 presents a theoretical aspect of traffic CA model. Section 3 explains the proposed model. Section 4 describes simulation process and the results in the form of fundamental diagrams and space-time diagrams. Finally, in section 5, a summary and conclusion is described with some discussions.

II. TRAFFIC CELLULAR AUTOMATA MODEL

One of the famous microscopic models for road traffic flow simulation is Cellular Automata (CA) model. CA model is a discrete computability mathematical model. In comparison with another microscopic model, CA model based approach is efficient [11] and is used for dynamic system simulations. CA model consists of two components, a cellular space and a set of state. The state of a cell is completely determined by its nearest neighboring cells.

All neighborhood cells have the same size in the lattice. Each cell can either be empty, or is occupied by exactly one node (car in this simulation model). There is a set of local transition rule that is applied to each cell from one discrete time step to another (i.e., iteration of the system). This parallel updating from local simple interaction leads to the emergence of global complex behavior. Furthermore, the utilization of CA model successfully explains the phenomenon of transportation. These traffic cellular automata (TCA) are dynamical systems that are discrete in nature and powerful to capture all previously mentioned basic phenomena that occur in traffic flows [11].

The one dimensional cellular automata model for single lane freeway traffic is introduced by Nagel and Schreckenberg (NaSch) [12]. This model shows how traffic congestion can be thought of as an emergent or collective phenomenon due to interactions between cars on the road, when the car density is high (cars are close to each on average). According to NaSch model, the randomization rule captures natural speed fluctuations due to human behavior or varying external conditions [14].

In a real traffic situation, most highways have two or more lanes. Regarding this road condition, there are a few analytical models for multi-lane traffic. Nagatani is one of the first researchers who introduced a CA model for two lane traffic [24]. Then, in addition to the Nagatani's model, Rickert et al. [15] considers a model with $v_{max} \geq 1$. Their model introduces the lane changing behavior for two lanes traffic. They proposed a symmetric rule set where the vehicle changes lanes if the following criteria are fulfilled:

$$gap(i) < 1 \tag{1}$$

$$gap_0(i) > l_0 \tag{2}$$

$$gap_{0,back}(i) > l_{0,back} \tag{3}$$

The variable $gap(i)$, $gap_0(i)$, and $gap_{0,back}(i)$ denote the number of empty cells between the vehicle and its predecessor on its current lane, and forward gap on the desired lane, and backward gap on the desired lane, respectively. Rickert also uses the parameters which allow decide how far the vehicle look ahead in the current lane for l, ahead on the desired lane for l_0, and how far the vehicle look back on the desired lane for $l_{0,back}$.

The advanced analysis about lane-changing behavior is reported which includes symmetric and asymmetric rules of lane-changing [16]-[19], [20]-[23].

III. PROPOSED MODEL DEFINITION AND SIMULATION

The proposed model uses two-lane highway with unidirectional traffic character in periodic boundaries condition. Two-lane model is necessary in order to accommodate the lane changing behavior in the real traffic condition. A one-dimensional chain of L cells of length 7.5 m represents each lane. This value is considered as the length of vehicle plus the distance between vehicles in a stopped position. A one-lane consists of 10^3 cells. There are just two possible states of each cell. Each cell can only be empty or containing by just one vehicle. The speed of each vehicle is integer value between $v = 0, 1, \ldots, v_{max}$. In this model, all vehicles are considered as homogeneous and have the same maximum speed $v_{max} = 5$. The speed value number corresponds to the number of cell that the vehicle proceeds at one time step. The state of a road cell at the

next time step, form t to $t + 1$ is dependent on the states of the direct frontal neighborhood cell of the vehicle and the core cell itself of the vehicle.

Rickert et al. [15] discusses criteria of safety by introducing the parameters which decide how far the vehicle looks ahead on current lane, looks ahead on desired lane, and looks back on desired lane. Those criteria have to be fulfilled before a vehicle makes a lane changing. However, in real traffic condition, these criteria of safety rules by Rickert are not sufficient to describe driver's behaviors in highway traffic. This paper introduces a new additional parameter to accommodate the driver behavior when making a lane changing. In addition to considering the gap of cell that consists of vehicle, the speed parameter of the other vehicle that situated in the desired lane is taken into account. The parameter of scope awareness S_a that reflects the various characters of driver is introduced. The scope awareness parameter represents drivers' behavior when lane changing. S_a value reflects the degree of driver aggressiveness and awareness. Fig.1 shows schematic definition diagram of scope awareness S_a from the perspective of the vehicle (the third cell on lane 2) in its current speed and position $v_{(1)}; x_{(1)}$

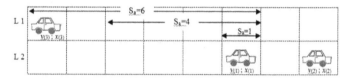

Fig. 1. Schematic definition diagram of scope awareness S_a from the perspective of the vehicle (the third cell on lane 2) in its current speed and position $v_{(1)}; x_{(1)}$

$S_a=1$ implies very dangerous because the driver of vehicle 1 intend to make a lane changing even the gap between the car and the car ahead of the car on the next lane is only one cell. On the other hand, the lane changing can be done safely when $S_a=6$ because the gap between both is six cells.

In contrast to the gap-length model's parameter [15], the scope awareness parameter S_a accommodates the sight distance taken by the driver to make a perception about road-lane condition and the speed of other vehicles that exist in target lane. Therefore, the subject vehicle would consider changing its lane not only due to the comparison value between the number of gap and condition which decide how far the vehicle look ahead in the current lane, but also depending on the current speed of the subject vehicle and the speed of other vehicles that situated along the scope awareness area.

The updating rule for lane changing maneuver is done in accordance with a set of rules. In comparison to the lane changing model of Rickert et al. [15], there are two basic differences rules from the proposed model. The first one, as the result of traffic conditions ahead of subject driver, the subject vehicle would consider changing its lane not only due to the comparison value between the number of gap and condition which decide how far the vehicle look ahead in the current lane, but also depending on the current speed of the subject vehicle that can be varied based on traffic situation. Another difference is the scope awareness value (S_a). The subject vehicle would consider the speed of vehicles that situated along with its scope awareness area then decide whether possible or not to change the lane.

At the beginning of iterations, the subject driver checks desirability of a lane changing. The subject driver looks ahead to check if the existing gap in the current lane can accommodate the current speed. If not, then due to the randomness number of percentage ratio, the subject driver decides whether the driver maintains or decelerates the vehicle speed due to the existing gap number or change the lane. When the subject driver chooses to "lane changing", then driver looks sideways at the other lane to check whether the cell next to the subject vehicle is empty and the forward gap on the other lane is equal or longer than the current lane. If one cell is unoccupied or free-cell then its state is 0. Moreover, the subject driver also looks back at the other lane to check road condition. In real traffic situations, a subject driver also has to look back on the other lane in order to estimate the velocity of the following vehicle to avoid a collision. Equation (8) accommodates the driver behavior that estimate the velocity of vehicle at the moment before making a lane changing.

If there is another vehicle within the area of scope awareness, then the subject driver estimates the speed of the vehicle in order to avoid collision during the lane changing maneuver. The subject driver makes a lane changing maneuver if the speed of the vehicle that is located within the area of scope awareness is less than the existing gap. The lane changing rules can be summarized as follows:

$$gap_{same} < v_{current} \tag{4}$$

$$cell_{next} = 0 \tag{5}$$

$$rand() < p_{change} \tag{6}$$

$$gap_{target} > gap_{same} \tag{7}$$

$$v_{Vehicle,back} \leq gap_{back} \; ; X(vehicle_{back}) \in S_a \tag{8}$$

The lane changing rules are applied to the vehicle that changes from right lane to left lane and conversely. The vehicle only moves sideways and it does not advance. Once all the lane changing maneuvers are made, then the updating rules from a single lane model are applied independently to each lane. Fig.2 shows the schematic diagram of lane changing operation.

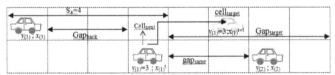

Fig. 2.   Schematic diagram of a lane changing operation

In Fig.2, the subject vehicle $v_{(1)}$; $x_{(1)}$ is assumed that have the current speed $v_{(1)}^t = 3$ cell per time step and the parameter of scope awareness $S_a = 4$ cells. In order to avoid the introduction of any unrealistic artifacts in the simulation then the proposed model uses eq. (7) to express the more realistic lane changing decision. According to eq. (7), the driver must consider that the forward gap in the desired lane is more than the gap in the current lane. This consideration is important because the proposed model uses the different desired velocities for the vehicles.

Once the lane changing maneuvers are made to all possible vehicles, then the updating rules from a single lane model are applied independently to each lane. Together with a set of lane changing rules, the road state is obtained by applying the following rules to all by parallel updated:

$$Acceleration: v(i) \rightarrow min(v(i)+1, v_{max}) \tag{9}$$

$$Deceleration: v(i) \rightarrow min(v(i), gap_{same}(i)) \tag{10}$$

$$Driving: x(i) \rightarrow x(i)+v(i) \tag{11}$$

IV.   SIMULATION RESULTS

The simulation starts with an initial configuration of N vehicles, with fixed distributions of positions on both lanes. The simulation uses the same initial velocity for all vehicle $v_{min} = 0$ and the maximum vehicle speed has been set to $v_{max} = 5$ cell/time-step. The velocity corresponds to the number of cells that a vehicle advances in iteration. Many simulations are done with the different density ρ. The density ρ is defined as the number of vehicles N along with the highway over the number of cells on the highway L.

The traffic model uses close (periodic) boundary conditions. This means that during one simulation, the total number of vehicles on the highway cannot be changed. Vehicles move from left to right. If a vehicle arrives on the right boundary then it moves to the left boundary. Since the proposed model assumes symmetry character of the both lanes, the traffic flow characteristics on both lanes are identical.

A.   Traffic Flow

In order to examine the effect of scope awareness on the traffic flow, then the proposed model is simulated over 1000 iterations on 10^3 cells for all possible density level. The flow indicates the number of moving vehicles per unit of time. Along with the study of the proposed model, a comparative study between with and without using scope awareness parameter.

Fig.3 shows the average flow-density diagram of the proposed model is compared to a two-lane traffic system without using scope awareness parameter.

The following results are made:

1)   The proposed model reproduces a recognizable diagram of flow towards density relationship. Flow is linearly increasing together with the increases in density level. A maximum flow level is achieved at density level $\rho=0.5$ for each value of S_a. After reaches the critical point of flow at $\rho=0.5$, the flow at each level of S_a becomes linearly decreasing in density. In other words, the laminar flow turn into back travelling start-stop waves after density level $\rho=0.5$.

2)   Compared to the model without scope awareness consideration (Fig.3-bottom diagram), the usage of S_a parameter produced a better flow of vehicles, especially above density $\rho=0.4$. The Sa parameter maintained the traffic to keep flowing by carefully calculate the appropriate time to make a lane changing decision, thus the lane changing maneuver does not disturb the traffic in the target lane.

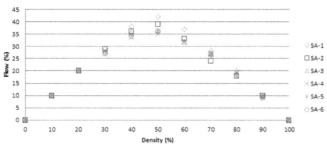

(a) Average flow-density diagram of the proposed model

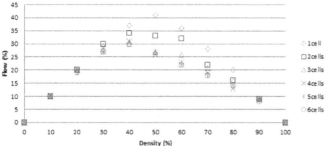

(b) Two-lane traffic system without using scope awareness parameter

Fig. 3. The average flow-density diagram of the proposed model is compared to a two-lane traffic system without using scope awareness parameter

B. Space-Time Diagram

The space-time diagram represents the location of the vehicles at the certain time. The space-time diagram for density $\rho=0.25$, $\rho=0.5$, and $\rho=0.75$ is shown. These three values of density assumed as the light traffic, moderate traffic, and heavy traffic in the real traffic condition, respectively. Fig.4 shows the result for density $\rho=0.25$ at all the values of scope awareness while the lane changing probability is 100%. Fig.4 (a) for Scope awareness Sa=1 ; (b) for Scope awareness Sa=2 ; (c) for Scope awareness Sa=3 ; (d) for Scope awareness Sa=4 ; (e) for Scope awareness Sa=5 ; (f) for Scope awareness Sa=6, respectively. The horizontal axis represents space while the vertical axis represents the time. Vehicles move from left to right in space axis while from top to bottom in time axis.

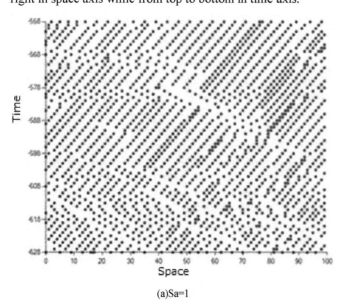

(a)Sa=1

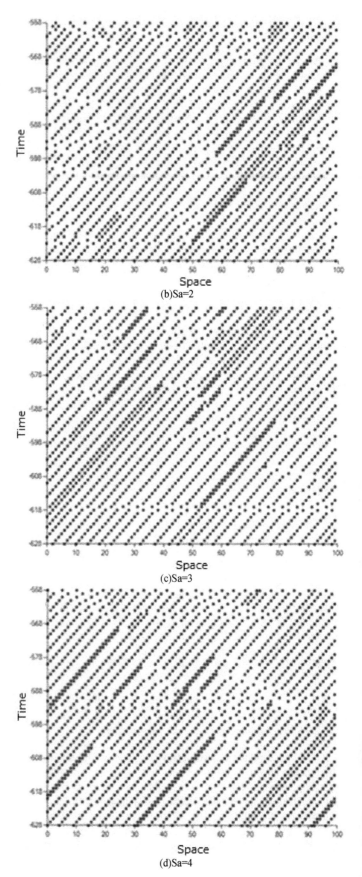

(b)Sa=2

(c)Sa=3

(d)Sa=4

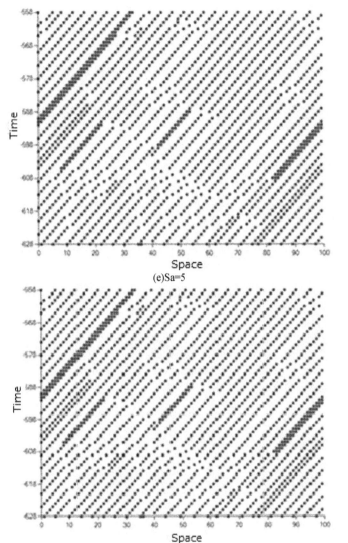

(e)Sa=5

(f)Sa=6

Fig. 4. Space-time diagram for light traffic condition (density ρ = 25%). Lane changing probability 100%. (a) for Scope awareness Sa=1 ; (b) for Scope awareness Sa=2 ; (c) for Scope awareness Sa=3 ; (d) for Scope awareness Sa=4 ; (e) for Scope awareness Sa=5 ; (f) for Scope awareness Sa=6

In the light traffic condition ρ=0.25, the increases of scope awareness distance affect the vehicles flow. Free flow phase showed in Sa=1 diagram (Fig.4 (a)), which are drawn as light area and have shallow negative inclinations. However, when the Sa value is increased then some solid area appears. The solid area with steep positive inclination reflects the traffic jam. Short-vehicle-life-lines frequently appear and disappear in this case. This implies that there are a great number of lane changing at this traffic density ρ=0.25. In accordance with increasing of the scope awareness, occurrence of the short-vehicle-life-lines becomes smaller than before (Fig.4 (f)). Meanwhile, in the heavy traffic condition, traffic situation is independent on scope awareness. In this traffic condition, lane changing is very rare. The result implies that in the heavy traffic condition, drivers' lane changing style have no influence to the traffic situation.

Table 1 (a) and (b) present the ratio of the spontaneous braking number over lane changing by using the scope awareness parameter (SA) and without using the scope awareness parameter, respectively. "0" means that the spontaneous braking did not happen, although the lane changing maneuver is still occurring in this certain traffic condition. Meanwhile, "N/A" means that the lane changing maneuver did not occur at all in this traffic condition.

TABLE I.   RATIO OF THE SPONTANEOUS BRAKING NUMBER OVER LANE CHANGING WITH THE SCOPE AWARENESS PARAMETER (SA) AND WITHOUT THE SCOPE AWARENESS PARAMETER

(A)WITH SA						
DENSITY(%)	SA-1(%)	SA-2(%)	SA-3(%)	SA-4(%)	SA-5(%)	SA-6(%)
10	56	40	23	24	20	0
20	82	74	63	38	18	3
30	81	60	12	0	0	0
40	82	35	3	0	0	0
50	92	11	0	0	0	0
60	83	8	0	0	0	0
70	85	9	0	0	0	0
80	80	0	0	0	0	0
90	88	0	0	0	0	0

(B)WITHOUT SA						
DENSITY(%)	SA-1(%)	SA-2(%)	SA-3(%)	SA-4(%)	SA-5(%)	SA-6(%)
10	53	45	36	48	13	0
20	82	83	80	63	17	6
30	71	64	55	9	0	0
40	74	63	6	0	0	N/A
50	85	33	0	0	N/A	N/A
60	89	14	0	N/A	N/A	N/A
70	91	20	0	N/A	N/A	N/A
80	100	0	0	N/A	N/A	N/A
90	100	N/A	N/A	N/A	N/A	N/A

It is found that the ratio is getting large in accordance with density while the ratio is getting small according to SA parameters. Also, the ratio of the case with SA is smaller than that without SA. These implies that the number of spontaneous braking can be suppressed by considering scope of awareness.

C. Vehicle Speed EstimationEerror

When the drivers make mistakes in speed estimation by visual perception, then the drivers fails their lane changing. Therefore speed estimation error is a key factor of lane changing. Fig.5 shows the effect of speed estimation error to the lane changing (a) and to the spontaneous braking action (b) in the light traffic ρ=0.25, moderate traffic ρ=0.5, and heavy traffic ρ=0.75, respectively. Fig.5 (c) shows the ratio between both of lane changing and spontaneous braking.

It is obvious that the number of lane changing and spontaneous braking are increased with increasing of the speed estimation error. Also, the number of lane changing and spontaneous braking in light traffic condition is larger than that in moderate and heavy traffic conditions. The reason for this is that it is getting hard for lane changing and spontaneous braking when the traffic condition is getting heavy. Also, it may say that it is getting hard for lane changing and spontaneous braking when the speed estimation error is increased.

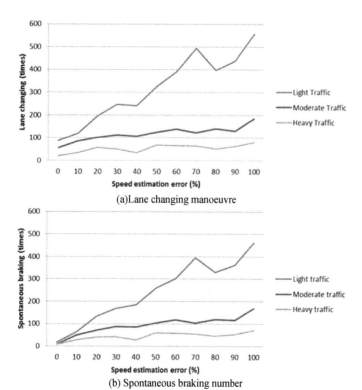

(a)Lane changing manoeuvre

(b) Spontaneous braking number

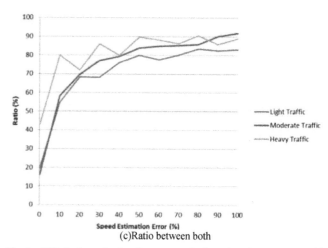

(c)Ratio between both

Fig. 5. Effect of speed estimation error to the lane changing manoeuvre (a) and to the spontaneous braking number (b) together with the ratio between both (c). These diagrams are simulated for the case $S_a=6$

Lane changing is much influencing to traffic situation than spontaneous braking, in general. Speed estimation error in lane changing is much influencing than that in spontaneous braking. Fig.6 shows Space-Time diagrams with consideration of speed estimation errors in the visual perception based driver's vehicle speed estimation raged from 0 to 1 for the light traffic condition (SA=6).

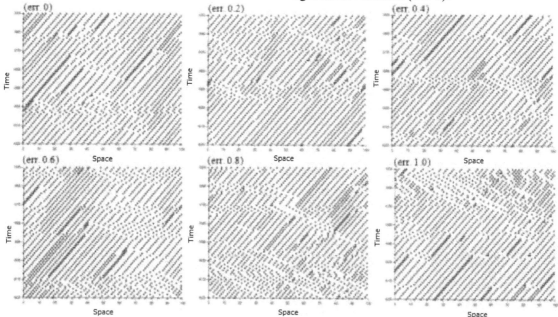

Fig. 6. Space-Time diagrams with consideration of speed estimation errors in the visual perception based driver's vehicle speed estimation raged from 0 to 1 for the light traffic condition (SA=6)

Through a comparison between Fig.4 and 6, it is found that traffic congestions seem to be independent to the speed estimation error and are happened occasionally. The phenomena of short-thin solid lines and wide solid lines also appear in these values of traffic density. The appearance of short-thin solid lines and the disappearance of wide solid line confirmed the conclusion that the short-thin solid line caused by the lane changing maneuver of another vehicle from adjacent lane so the subject vehicle has to make a spontaneous braking in order to avoid collision, and the wide solid line appeared as a result of deceleration into the minimum speed of the vehicle as the consequence of the reduced opportunities for lane changing maneuver.

V. SUMMARY AND CONCLUSION

The simulation model of the traffic cellular automata for representation of a driver behavior in a two lane highway is proposed. The term of scope awareness introduced to reflect the certain area of roadway that is considered by the driver to make a perception of road condition. This perception includes the estimation speed of vehicles that are located within the scope-awareness distance prior to make a lane changing maneuver. The relation between flow-density and space-time are investigated in order to examine the effect of scope awareness parameter in the traffic flow. The followings are concluded through this study,

1) This model describes the realistic traffic situation, in particular for capturing the situation when driver make a lane changing maneuver. Compared to the conventional approach, the usage of scope awareness model approach produces a better flow of vehicles.

2) The various parameters of the scope awareness may represent the characteristic and the experience level of the drivers. The increases of the scope awareness value means the driver become much aware to estimate the road condition in a lane changing maneuver.

3) The proposed model reveals the phenomena of the short-thin solid line jam and the wide solid line jam in the traffic flow. It is found that the short-thin solid lines are the result of the lane changing maneuver of another vehicle from adjacent lane which makes the subject vehicle which has to make a spontaneous braking in order to avoid collision. As the result of this spontaneous braking, it is causing the other following vehicles which have to adjust or decrease their speed with the vehicle ahead. This action introduces a short queue of vehicles. On the other hand, a wide solid line appears as a result of the deceleration vehicle speed to the minimum speed of the vehicles as the consequence of the reduced lane changing chances. The chance reduction introduces a transient bottleneck effect.

4) The simulation results show that lane changing maneuvers with taking into account another vehicle speed could reduce the level of traffic congestion. However, in the heavy traffic (high dense) situation, the chance of a lane changing is small which results in traffic congestions.

The simulation results show that when the driver become more aware during lane changing decision, then the traffic flow seems to behave a single lane traffic character. In the situation, the vehicles tend to decrease its speed rather than making a lane changing. Moreover, the results indicate that in the density level less than 75%, traffic congestion is reduced by managing of the driver behavior. Other than that, the management of driver behavior makes a significant impact. The simulation result can serve as a reference for transportation planning, evaluation, and control. Moreover, the result paves the way for accurate simulation of a more complex traffic system. Based on the result, the effect of road shape towards the vehicle deceleration is to be studied hereafter.

REFERENCES

[1] K. Arai and S. Sentinuwo, "Spontaneous-braking and lane-changing effect on traffic congestion using cellular automata model applied to the two lane traffic", (IJACSA) International Journal of Advanced Computer Science and Applications, Vol. 3 (8), 2012.

[2] Elvik, R., Vaa, T. (2004) The Handbook of Road Safety Measures. Elsevier Science, Oxford.

[3] Finch, D. J., Kompfner, P., Lockwood, C. R. & Maycock, G. (1994) Speed, speed limits and crashes. Project Record S211G/RB/Project Report PR 58. Transport Research Laboratory TRL, Crowthorne, Berkshire.

[4] Nilsson, G. (2004). Traffic safety dimensions and the power model to describe the effect of speed on safety. Lund Bulletin 221. Lund Institute of Technology, Lund.

[5] Salusjärvi, M., 1981. The speed limit experiments on public roads in Finland. Technical Research Centre of Finland. Publication 7/1981. Espoo, Finland.

[6] Lewin, I. (1982). Driver training a perceptual-motor skill approach. Ergonomics,25, 917–924.

[7] Elander, J., West, R., & French, D. (1993). Behavioral correlates of individual differences in road traffic crash risk: An examination of methods and findings. Psychological Bulletin, 113, 279–294.

[8] G. M. Fitch, S. E. Lee, S. Klauer, J. Hankey, J. Sudweeks, and T. Dingus. Analysis of Lane-Change Crashes and Near-Crashes. National Technical Information Service, Springfield, VA 22161, 2009

[9] Shinar, D., Psychology on the road: The human factor in traffic safety. Wiley New York, (1978)

[10] Sun, D. J., & Elefteriadou, L. (2011). Lane-changing behavior on urban streets: a focus group-based study. Applied ergonomics, 42(5), 682–91. doi:10.1016/j.apergo.2010.11.001.

[11] X. G. Li, B. Jia, Z. Y. Gao, and R. Jiang, "A realistic two-lane cellular automata traffic model considering aggressive lane- changing behavior of fast vehicle," PhysicaA, vol. 367, pp. 479– 486, 2006.

[12] K. Nagel and M. Schreckenberg, "A cellular automaton model for freeway traffic," Journal of Physics I France, vol. 2, no. 12, pp.2221-2229, 1992.

[13] K. Nagel, Wolf, Wagner, and Simon, "Two-lane traffic rules for cellular automata: A systematic approach," Physical Review E, vol.58, no.2, 1998.

[14] S. Maerivoet and B. D. Moor, "Transportation Planning and Traffic Flow Models," 05-155, Katholieke Universiteit Leuven, Department of Electrical Engineering ESAT-SCD (SISTA), July 2005.

[15] M. Rickert, K. Nagel, M. Schreckenberg, and A. Latour, "Two Lane Traffic Simulations using Cellular Automata," vol. 4367, no. 95, 1995.

[16] W. Knospe, L. Santen, A. Schadschneider, and M. Schrekenberg, "Disorder effects in cellular automata for two lane traffic," Physica A, vol. 265, no. 3-4, pp. 614–633, 1998.

[17] A. Awazu, "Dynamics of two equivalent lanes traffic flow model: selforganization of the slow lane and fast lane," Journal of Physical Society of Japan, vol. 64, no. 4, pp. 1071– 1074, 1998.

[18] E. G. Campri and G. Levi, "A cellular automata model for highway traffic," The European Physica Journal B, vol. 17, no. 1, pp. 159–166, 2000.

[19] L. Wang, B. H. Wang, and B. Hu, "Cellular automaton traffic flow model between the Fukui-Ishibashi and Nagel- Schreckenberg models," Physical Review E, vol. 63, no. 5, Article ID 056117, 5 pages, 2001.

[20] B. Jia, R. Jiang, Q. S. Wu, and M. B. Hu, "Honk effect in the two-lane cellular automaton model for traffic flow," Physica A, vol. 348, pp. 544– 552, 2005.

[21] D. Chowdhury, L. Santen, and A. Schadschneider, "Statistical physics of vehicular traffic and some related systems," Physics Report, vol. 329, no. 4-6, pp. 199–329, 2000.

[22] W. Knospe, L. Santen, A. Schadschneider, and M. Schreckenberg, "A realistic two-lane traffic model for highway traffic," Journal of Physics A, vol. 35, no. 15, pp. 3369–3388, 2002.

[23] W. Knospe, L. Santen, A. Schadschneider, and M. Schreckenberg, "Empirical test for cellular automaton models of traffic flow," Phys. Rev. E, vol. 70, 2004.

[24] Nagatani, T., "Self Organization and Phase Transition in the Traffic Flow Model of a Two-Lane Roadway," Journal of Physics A, Vol. 26, pp. 781-787, 1993.

[25] Shahar, A., Van Loon, E., Clarke, D., & Crundall, D. Attending overtaking cars and motorcycles through the mirrors before changing lanes. Accident; analysis and prevention, 44(1), 104–10, 2012.

[26] Rayner, K., Warren, T., Juhasz, B.J., Liversedge, S.P. The effect of plausibility on eye movements in reading. Journal of Experimental Psychology: Learning, Memory and Cognition 30, 1290–1301, 2004.

[27] Benjaafar, S., & Dooley, K. (1997). Cellular automata for traffic flow modeling. Minneapolis, MN, University of. Retrieved from http://ntl.bts.gov/lib/21000/21100/21189/PB99103996.pdf.

[28] Paz, A., & Peeta, S. (2009). Information-based network control strategies consistent with estimated driver behavior. Transportation Research Part B: Methodological, 43(1), 73–96. doi:10.1016/j.trb.2008.06.007.

Permissions

List of Contributors

Kohei Arai and Taka Eguchi
Graduate School of Science and Engineering Saga University Saga City, Japan

Dr. Md. Mijanur Rahman
Dept. of Computer Science & Engineering Jatiya Kabi Kazi Nazrul Islam University
Trishal, Mymensingh, Bangladesh

Fatema Khatun
Dept. of Electrical & Electronic Engineering Hamdard University Bangladesh Sonargoan, Narayanganj, Bangladesh

Dr. Md. Al-Amin Bhuiyan
Dept. of Computer Engineering King Faisal University Al Ahssa 31982, Saudi Arabia

Sylvia Encheva
Stord/Haugesund University College Bjørnsonsg. 45,5528 Haugesund, Norway

Christopher Cooper
College of Engineering North Carolina State University, Raleigh, NC, 27695 Kent Wise SGS Inc. The Woodlands, TX, 77381

John Cooper
Department of Chemistry and Biochemistry Old Dominion University, Norfolk, VA, 23529

Makarand Deo
Department of Engineering Norfolk State University, Norfolk, VA, 23504

Kohei Arai and Masanori Sakashita
Graduate School of Science and Engineering Saga University Saga City, Japan

Sylvia Encheva
Stord/Haugesund University College Bjørnsonsg. 45, 5528 Haugesund, Norway

Mohammed E. El-Telbany
Computers and Systems Department Electronics Research Institute Cairo, Egypt

Mahmoud Warda
Computers Department National Research Center Cairo, Egypt

Khaled M. Alhawiti
Computer Science Department, Faculty of Computers and Information technology Tabuk University, Tabuk, Saudi Arabia

Ronny Mardiyanto
Department of Electric and Electronics Institute Technology of Surabaya Surabaya, Indonesia

Yuzana Win and Tomonari Masada
Graduate School of Engineering Nagasaki University Nagasaki, Japan

Shariba Islam Tusiy
Ahsanullah University of Science & Technology (AUST), Dhaka, Bangladesh

Rasoul Kiani, Siamak Mahdavi and Amin Keshavarzi
Department of Computer Engineering, Fars Science and Research Branch, Islamic Azad University, Marvdasht, Iran

Karbhari V. Kale, Prapti D. Deshmukh , Yogesh S. Rode and Majharoddin M. Kazi
Dr. G. Y. Pathrikar College of Computer Science, MGM, Maharashtra, India - 431005

Shriniwas V. Chavan
Department of Computer Science and IT, Dr. B. A. M. University, Aurangabad, Maharashtra, India - 431004
Department of Computer Science, MSS's ACS College, Ambad, Jalna, Maharashtra, India – 431203

Atousa Zaeim and Samia Nefti-Meziani
School of CSE University of Salford Manchester United Kingdom

Adham Atyabi
School of CSE University of Salford Manchester United Kingdom School of CSEM
Flinders University of South Australia

Javier G. R´azuri, David Sundgren, Rahim Rahmani and Aron Larsson
Dept. of Computer and Systems Sciences (DSV) Stockholm University, Stockholm, Sweden

Antonio Moran Cardenasz
Pontifical Catholic University of Perú (PUCP) Lima, Peru

Isis Bonetx
Antioquia School of Engineering (EIA) Antioquia, Colombia

Nutan Farah Haq, Musharrat Rafni, Abdur Rahman Onik, Faisal Muhammad Shah, Md. Avishek Khan Hridoy and Dewan Md. Farid
Department of Computer Science and Engineering Ahsanullah University of Science and Technology Dhaka, Bangladesh

Yoshihiko Sasaki, Shihomi Kasuya and Hideto Matusura
Sasaki Green Tea Company, Kakegawa city – Japan

Md. Hasan Tareque
Department of Computer Science and Engineering IBAIS University Dhaka, Bangladesh

Ahmed Shoeb Al Hasan
Department of Computer Science and Engineering Bangladesh University of Business & Technology Dhaka, Bangladesh

Sovan Samanta
Department of Applied Mathematics with Oceanology and Computer Programming, Vidyasagar University, Midnapore - 721 102, India.

Madhumangal Pal
Department of Applied Mathematics with Oceanology and Computer Programming,
Vidyasagar University, Midnapore - 721 102, India.

Anita Pal
Department of Mathematics, National Institute of Technology Durgapur, Durgapur-713209, India.

Bi-Jun REN, Yan-Ling FU
Department of Information Engineering Henan College of Finance and Taxation Zhengzhou, Henan 451464, China

Ke-Yun QIN
College of Mathematics Southwest Jiaotong University Chengdu, Sichuan 610031, China

Manh Hung Nguyen
Posts and Telecommunications Institute of Technology (PTIT) Hanoi, Vietnam UMI UMMISCO 209 (IRD/UPMC), Hanoi, Vietnam

Dinh Que Tran
Posts and Telecommunications Institute of Technology (PTIT) Hanoi, Vietnam

Sajjawat Charoenrien and Saranya Maneeroj
Department of Mathematics Chulalongkorn University Bangkok, Thailand

Dewa Gede Hendra Divayana
Chair of Information Technology Department Indonesia Technology University Bali, Indonesia

I Putu Wisna Ariawan and I Made Sugiarta
Lecture of Mathematics Education Ganesha University of Education Bali, Indonesia

I Wayan Artanayasa
Chair of Sport & Health Education Department Ganesha University of Education Bali, Indonesia

Steven Ray Sentinuwo
Sam Ratulangi University, Kampus Unsrat, Manado, Indonesia

Nasif Shawkat, Md. Arman Ahmed, Biswajit Panday, Nazmus Sakib

Amiraj Dhawan, Shruti Bhave, Amrita Aurora and Vishwanathan Iyer

Index

A

Active Contour Models (ACM), 153
Aerosol Refractive Index, 35-36, 42
Affective Computing, 128, 134
Ambient Intelligence, 120, 127
Android, 1-5, 7, 10, 55
Artificial Bee Colony (ABC) Algorithm, 81-82, 86
Association Rules, 1-2, 6, 140, 144
Augmented Reality, 25-26, 33

B

Blocking Black Area, 15-20
Blood Pressure, 8-9, 87-96, 113, 118
Body Temperature, 9, 87-96, 113, 116, 118
Boundary Detection, 15, 17-18

C

Cirrus Cloud, 43, 47-49
Classification, 2, 7, 15, 20, 25, 33, 53-58, 63, 74-80, 97, 99-111, 120-121, 123-132, 134-137, 139-142, 144, 178, 200, 214
Computer Assisted Language Learning, 59-60, 64-65

D

Data Mining, 7, 54-55, 57-58, 97, 99, 102-103, 127, 135, 139-140, 145, 178, 207
Dataset, 2, 49, 55-57, 97-100, 105, 109-111, 126, 129, 131-132, 136, 140-144, 191, 200
Decision Making, 52-53, 64, 96, 98, 108, 118, 134, 140, 179, 200
Degree of Polarization, 35, 39, 146, 152
Devanagari Compound, 104, 106-107, 110
Diatom, 166, 168, 173
Digital Library, 199-206
Distributed Learning, 120, 126
Distribution Reduction, 174, 177-178
Driver Perception, 215

E

Expert System, 199-206

F

Feature Selection, 55, 58, 103, 128-129, 133, 136, 140, 143-144
Forward Chaining, 199-200, 202-206
Fuzzy Competition Graphs, 158, 164, 207
Fuzzy Competition Number, 158, 162, 164

Fuzzy Dual Graphs, 207
Fuzzy K-competition Graphs, 158, 214

G

Gaze, 66-68, 70, 72-73, 215
Generalized Indiscernibility Relation, 174, 178
Genetic Algorithm, 96-98, 100, 103, 119, 200

H

Handwritten Character, 104, 110-111
Human Computer Interaction, 7, 14, 66

I

Image Classification, 25
Image Registration, 25-28, 32-33
Improved Cuckoo Search (ICS) Algorithm, 81, 86
Information Retrieval, 65, 74, 78-80, 130, 135
Intrusion Detection, 136-142, 144-145
Isomorphism, 158, 164, 207-208, 212, 214

J

Jaccard Similarity, 74-76

L

Lane Changing Maneuver, 215-216, 219-221
Learning, 1-7, 14, 21-24, 54-56, 58-61, 64-65, 73, 80, 97, 109, 112, 120-121, 125-132, 134-137, 139-142, 144-145, 222
Liar, 179, 181, 183-185
Location Estimation, 8, 13
Lypunov Stability Theory, 153

M

Machine Learning, 1-7, 58, 120-121, 127-131, 135-137, 139, 144-145
Machine Learning In Embedded Systems, 1
Max-miner, 1-2, 6-7
Microwave Radiometer, 193-194, 197-198
Multiagent System, 179, 186

N

Natural Language Processing, 59-61, 64-65, 76, 80
Nature Inspired Algorithms, 81
Nitrigen Content, 146

O

Obstacle Avoidance, 66, 68-70, 73
Otsu's Algorithm, 15

P

Positive Region Reduction, 174-175, 177-178

Principal Component Analysis, 25-26, 28, 32-33

R

Random Forest, 54-58

Rapidminer, 97-99

Red Tide, 166, 168

Refinement Orders, 21

Relational Concept Analysis, 21, 23

Remote Sensing, 41-43, 49-50, 146, 152, 166, 173, 193, 197-198

Reputation, 179-187

Rescue System, 8-9, 13-14, 87, 96, 113, 118

S

Scope Awareness, 215-216, 218-221

Sea Surface Temperature, 49, 193-194, 196-198

Seonsor Network, 87, 113

Similarity Values, 188-190, 192

Simulated Annealing, 193, 195, 197-198

Size Distribution, 35-37, 39-40, 42, 44

Sky View Camera, 43-49

Soft Sets, 52-53

Speech Segmentation, 15-17, 19-20

Speed Estimation, 197, 215, 219-220

Syntactic Simplification Tools, 59

T

Topogramphic Representation of 3d Clouds, 43

Traffic Cellular Automata, 215-216, 220

Trust, 179-192

Trust Values, 188-192

Trust-based Recommender Systems, 188

U

Uncertainties, 36

W

Watershed Model, 153

Weighting, 77, 97, 102, 156

Word N-grams, 74-75

Z

Zernike, 104-112

Printed in the USA
CPSIA information can be obtained
at www.ICGtesting.com
JSHW051433221024
72173JS00006B/1457

9 781632 406460